Simulation Foundations, Methods and Applications

More information about this series at http://www.springer.com/series/10128

Stuart Berry · Val Lowndes
Marcello Trovati
Editors

Guide to Computational Modelling for Decision Processes

Theory, Algorithms, Techniques and Applications

Springer

Editors
Stuart Berry
Department of Computing and Mathematics
College of Engineering and Technology
University of Derby
Derby
UK

Marcello Trovati
University of Derby
Derby
UK

Val Lowndes
University of Derby
Derby
UK

ISSN 2195-2817 ISSN 2195-2825 (electronic)
Simulation Foundations, Methods and Applications
ISBN 978-3-319-55416-7 ISBN 978-3-319-55417-4 (eBook)
DOI 10.1007/978-3-319-55417-4

Library of Congress Control Number: 2017934055

Printed on acid-free paper

This Springer imprint is published by Springer Nature
The registered company is Springer International Publishing AG
The registered company address is: Gewerbestrasse 11, 6330 Cham, Switzerland

Preface

Outline of Content

This book is organised into three sections, Part I introduces modelling techniques and models used to represent application used later to evaluate solution processes and procedures.

Part I: Introduction to Modelling and Model Evaluation.

This part introduces modelling methodologies and models to be used as starting points to enable the derivation of efficient and effective solution techniques.

Part II: Case Studies.

This part presents a series of case studies to demonstrate how heuristic and analytical approaches may be used to solve large complex problems.

A series of case studies are presented where models are constructed and then analysed and evaluated to derive efficient and effective ways to produce good solutions.

Within Part I:

Chapter 1: Model Building. This chapter introduces the modelling methodologies:

Activity life cycles and problem analysis using activity life cycles.
Constructing models from "Big Data".
Blackboard modelling.

Chapter 2: Introduction to Cellular Automata in Simulation. This chapter aims to show how both these approaches can be used as modelling tools.

Introduction to cellular automata.
Cellular automata are introduced by way of Conway's Game of Life and applying agents.

Simulating the spread of an infection through a population.
Traffic modelling, investigating traffic flows.

Chapter 3: Introduction to Mathematical Programming. The aim in this chapter is to demonstrate how a mathematical programming model can be used to describe and explain the complexity of a planning problem and hence lead towards an efficient solution technique or methodology.

Whether (these) problems have easy solutions or because of the inherent complexity of the form of the required solution, they are better approached using heuristic techniques (generally accepting good rather than best solutions).

To achieve this objective, the following problems are presented:

Diet problems, the very obvious formulations (Stigler and Dantzig), lead to an undesirable solution (only one meal!), and a more heuristic approach is needed to add incorporate multi-objectives that typically minimise cost while maximising variety/taste.
Knapsack problems, showing how many problems are reducible to knapsack problems and are therefore appropriate for the use of heuristic solution techniques.
Network flow problems, again many planning problems can be modelled as network flow problems and hence can be solved easily.

Chapter 4: Heuristic Techniques in Optimisation. This part introduces approaches that can be used to obtain good solutions to hard or large problems comparing and contrasting the effectiveness and efficiency of heuristic approaches to problem-solving.

To achieve this objective, the following approaches are presented:

Genetic algorithms implementations illustrated through its application in producing solutions to knapsack problems, travelling salesman problems, scheduling problems, and quadratic assignment problems.
Tabu search implementations illustrated through its application in producing solutions to financial planning and travelling salesman problems.

Chapter 5: Introduction to the Use of Queueing Theory and Simulation. This chapter shows how queueing theory and simulation techniques can be applied to design efficient and effective service systems.

The major sections are concerned with the following:

Queueing theory leading to "quick" modelling, how queueing theory can be employed to carry out an evaluation of a manufacturing system.
Simulation modelling, introducing an alternative approach to modelling complex planning problems.

Part II: Case Studies

This part presents a series of case studies to demonstrate how heuristic and analytical approaches may be used to solve large complex problems.

A series of case studies are presented where models are constructed and then analysed and evaluated to derive efficient and effective ways to produce good solutions.

May be solved typical suggested applications, presenting alternative approaches to problem solving

Chapter 6 describes an investigation into the appropriateness of heuristic methodologies in the solution of

Travelling salesman problem.
Garbage collection problem, a multiple travelling salesman (type) problem.
Production planning and control problems, where a seemingly hard problem can be shown to be solvable using a simple heuristic approach reducing the need for cost data.

Chapter 7 describes how an efficient heuristic approach can be derived from an initial complex model using the following:

Flow shop scheduling, showing how a hard problem can be approached using heuristic methods.
Transport planning, deriving approaches to evaluate the benefits to be gained from the installation of an active traffic control system and the paradoxes resulting from changes to transport planning.

Chapter 8 describes an investigation into the production of an efficient and effective means of scheduling air traffic controllers, where the method used has to have the ability to respond (create a new schedule) rapidly to staff availabilities.

Chapter 9 describes how a multiple objective optimisation problem can be solved by the incorporation of techniques from genetic algorithms and fuzzy logic into a mathematical programming methodology.

This approach is illustrated by its application into the provision of solutions to a diet problem with the extended objectives:

Minimise cost.
Produce a healthy diet.
Produce a large variety of good diets.

Chapter 10 describes an investigation into the application of fuzzy logic showing how it can be used to derive a dynamic method of scheduling operations in a workshop.

This approach is applied to a workshop where there are multiple objectives:

Importance of customer and delivery due dates, using fuzzy logic to derive work schedules.

Chapter 11 describes how an approach based on tabu search methodologies can be employed to derive optimal control settings.

This approach is illustrated through its application to a Surround Sound 5 speaker system determining settings so that the system could produce "perfect" directional sound.

Chapter 12 describes how system dynamics modelling can be employed to describe the output from complex decision making processes.

Models are constructed to describe

the changes in the Dow Jones index, from growth to decline, and
the changes in the dominant mode of transport (with time) and the effect of these changes on the prior dominant modes.

Chapter 13 describes the use of queueing theory in the evaluation of traffic control systems (traffic lights) showing how the system could be improved through the use of "available forward road capacity" that is passing information between traffic lights.

Chapter 14 describes case study investigations into the use of cellular automata and agent-based simulations.

The case studies are based on message passing, by mobile devices, within a closed environment (a shopping centre for example) and
The spread of a fire and the improved positioning of the fire exits in a closed environment.

Chapter 15 discusses the use of "Big Data" to derive models.
Three case studies are provided.

Criminology,
Depression evaluation, and
University admissions.

Derby, UK

Stuart Berry
Val Lowndes
Marcello Trovati

Contents

Contributors

Ovidiu Bagdasar College of Engineering and Technology, University of Derby, Derby, UK

Andy Baker College of Engineering and Technology, University of Derby, Derby, UK

Stuart Berry College of Engineering and Technology, University of Derby, Derby, UK

Adrian Bird College of Engineering and Technology, University of Derby, Derby, UK

Richard Conniss College of Engineering and Technology, University of Derby, Derby, UK

James Hardy University of Derby, Derby, UK

Richard Hill College of Engineering and Technology, University of Derby, Derby, UK

Val Lowndes University of Derby, Derby, UK

Chris Parkes College of Engineering and Technology, University of Derby, Derby, UK

Mirko Paskota College of Engineering and Technology, University of Derby, Derby, UK

Nicolae Popovici Babes-Bolyai University, Cluj-Napoca, Romania

Kim Smith College of Engineering and Technology, University of Derby, Derby, UK

John Stubbs College of Engineering and Technology, University of Derby, Derby, UK

Marcello Trovati Department of Computer Science, Edge Hill University, Ormskirk, UK

Amanda Whitbrook College of Engineering and Technology, University of Derby, Derby, UK

Bruce Wiggins College of Engineering and Technology, University of Derby, Derby, UK

Part I
Introduction to Modelling and Model Evaluation

This section introduces modelling techniques and constructs models to represent and analyse planning problems in business, industry and the management of facilities.

These constructed models are evaluated; can they be solved in a reasonable time using standard analytical techniques or should the solution be approached using heuristic methods or heuristic methodologies?

Chapter 1
Model Building

Val Lowndes, Stuart Berry, Marcello Trovati and Amanda Whitbrook

Section 1.1 introduces the use of system dynamics in modelling and then uses this approach to construct models to describe real applications.
Section 1.2 introduces the concepts needed to construct models using available data, modelling using Big Data.
Section 1.3 introduces modelling using blackboard architecture; this provides a flexible, symbolic artificial intelligence (AI) method for the cooperative modelling and then solution of complex problems.

1.1 Introduction to System Modelling

The purpose of system dynamics modelling is to develop understanding and then the improvement of systems. The first stage in this process is the construction of a logical model (influence diagram) to describe a system.

V. Lowndes (Retired)
University of Derby, Kedleston Road, DE22 1GB Derby, UK
e-mail: v.p.lowndes@derby.ac.uk

S. Berry (✉) · A. Whitbrook
College of Engineering and Technology, University of Derby,
Kedleston Road, DE22 1GB Derby, UK
e-mail: s.berry@derby.ac.uk

A. Whitbrook
e-mail: a.whitbrook@derby.ac.uk

M. Trovati
Computer Science, Edge Hill University, St Helens Road, Ormskirk,
L39 4QP Lancashire, UK
e-mail: marcello.trovati@edgehill.ac.uk

S. Berry et al. (eds.), *Guide to Computational Modelling for Decision Processes*,
Simulation Foundations, Methods and Applications,
DOI 10.1007/978-3-319-55417-4_1

This model can then lead to sets of equations describing the operation of the system. These can be used to simulate the system to gain understanding of its dynamic behaviour and to be able to evaluate alternative policies, leading to improvements within the system.

A series of small examples are used to introduce this modelling process. Where information is available, the behaviour predicted by these models is compared with reality, i.e. what has happened in reality.

1.1.1 Introducing Influence Diagrams

Modelling using influence diagrams is introduced through the following illustrative examples:

- Stock control model: used to illustrate the basic modelling notation.
- Spending/saving model: used to illustrate the construction of an influence diagram and to introduce the concept of "positive" and "negative" feedback loops
- House building, financial models and population modelling: so that the predictions from the models (positive or negative loops) can be compared with reality.
- Transport modelling: extending the work to demonstrate the effect of government policy on transport provision (the "Beeching" cuts for example).

Example A: Stock Control Policies

A company holds stocks of finished goods to be able to satisfy demand; when stocks are low, more newly manufactured items are added to the finished goods stock; in this example, the available stock (for use) is "influenced" by production and demand (see Fig. 1.1).

The direction of the arrow from [despatched] to [production] indicating the production levels is influenced by the quantity of items dispatched, and the arc label (D) indicates the delays between each event.

Where

Demand	Influences	Number dispatched
Number dispatched	Influences	Production
Production	Influences	Available stock
Available stock	Influences	Number dispatched

The model constructed from this diagram will have the form:

Dispatched is described by the function f(Demand, Stock)

$Dispatched_i = Minimum(Demand_{i-3}, Stock_i)$

Assuming a delay of 3 between receipt of order and despatch.

Production is defined by f(Dispatched) or following through f(Dispatched, Demand, Stock) or by implication f(Dispatched, Demand, Stock, production))

$Production_i = Maximum(Dispatched_{i-2}, Demand_{i-5}, MinProduction)$

Assuming a delay of 2, dispatched and production request.

Stock is described by the function $f(stock_{i-1}, Production_{i-1}, Dispatched)$

$Stock_i = Stock_{i-1} + Production_{i-1} - Dispatched$

Assuming a delay of 1 between production and stock ready for use.

The next stage gives examples to introduce approaches to the production of "influence diagrams" and the notation used to analyse the resultant model.

1.1.1.1 Categorising Dependencies (Links) in a Model

An initial (causal) analysis is used to categorise an influence diagram and hence the underlying model, essentially the causal analysis asks: (Fig. 1.2).

If the input value increases, what is the effect on the output value? leading to the categorisation of the links as either positive (+) or negative (−) links.

Connecting between inflation rate and prices, as inflation rises, then so too do prices giving:

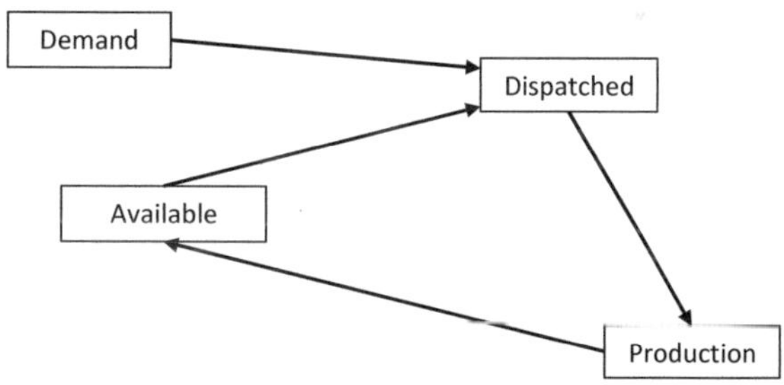

Fig. 1.1 Production

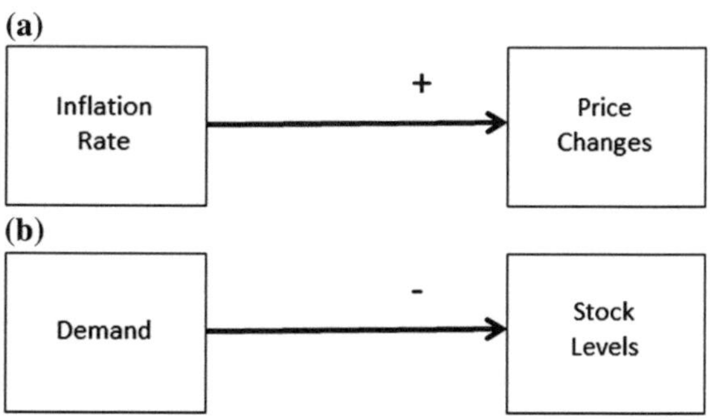

Fig. 1.2 **a** Positive arc: if "inflation rate" increases then "prices" will rise. **b** Negative arc: if "demand" increases then "stock levels" will fail

Connecting between demand and stock levels, as demand rises, it follows that stock levels will fall:

In carrying out this analysis always start with…"if the input rises…" starting with the opposite "…if the input falls…" can lead to double negative statements and some confusion in the following analysis.

1.1.1.2 Categorising a Model

Having categorised all the links, a loop in a model can fall into one of the two states: positive or negative feedback loops. In general, a negative loop indicates a "goal-seeking" model here there will be convergence, whereas a positive feedback loop indicates unrestricted growth or decay (Fig. 1.3).

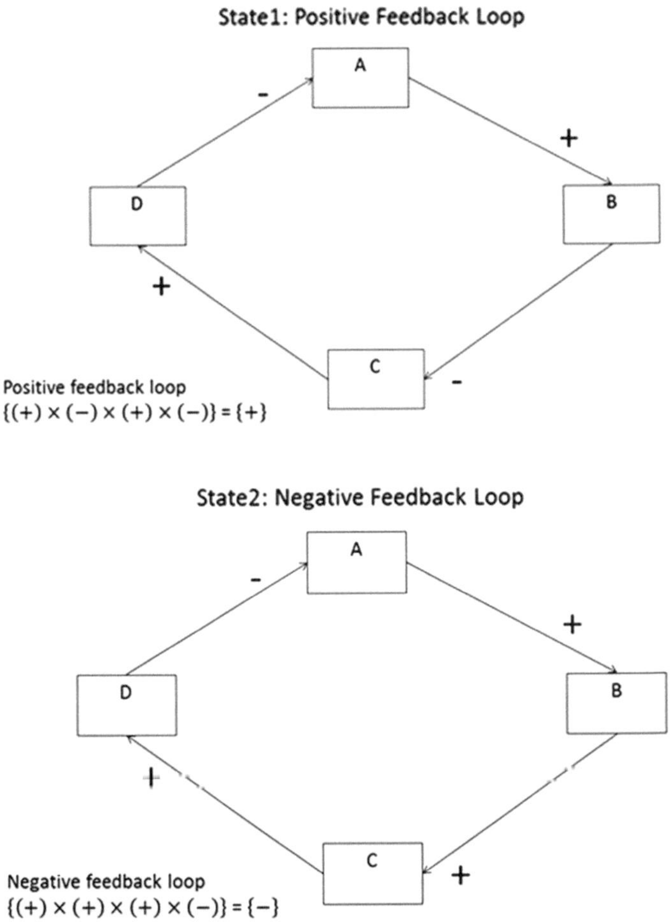

Fig. 1.3 Categorising feedback loops

- State 1, a positive feedback loop would lead to "unconstrained" growth or decline, while
- State 2, a negative feedback loop would lead to a steady-state solution (goal-seeking).

1.1.2 *Model Evaluation/Validation, Comparing the Model with Historic Data*

Here, a model is constructed and evaluated showing that its behaviour replicates the real situation.

Example B: House-Building Model

An initial model links house buying with house prices and housing stock giving a model with a negative feedback loop, ignoring the overall demand for housing (Fig. 1.4).

Analysing Influences in the diagram leads to the consequent effects:

House prices	Up	Then	House building	Down	Negative effect
House building	Up	Then	Housing stock	Down	Negative effect
Housing stock	Up	Then	House prices	Down	Negative effect

Thus giving the model cycle from and returning to a given starting position

House building	Up	Gives	Housing stock	Down
Housing stock	Down	Gives	House prices	Up
House prices	Up	Gives	House building	Down

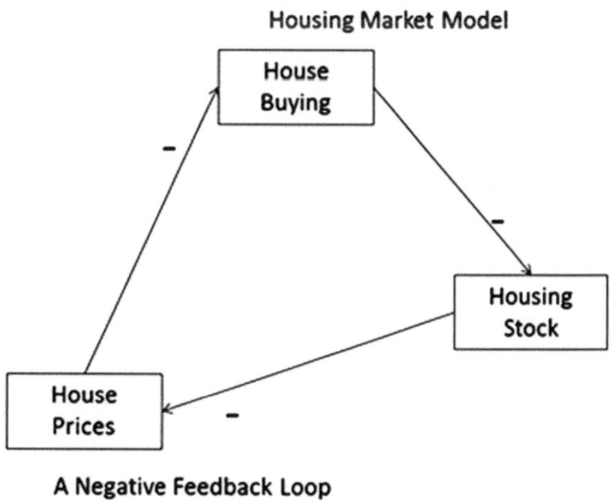

Fig. 1.4 First house price model

Continuing to complete the cycle of effects

House building	Down	Gives	Housing stock	Up
Housing stock	Up	Gives	House prices	Down
House prices	Down	Gives	House building	Up

Completing the cycle shows the goal-seeking behaviour of this model.

Extending the model through the addition of house construction gives a second loop (Fig. 1.5).

The additional loop adds the influences

House construction	Up	Then	Housing stock	Up
Housing stock	Up	Then	House prices	Down
House prices	Up	Then	House buying	Down
House buying	Down	Then	House construction	Down

Thus, adding to the existing model cycle starting from the same starting position

House buying	Up	Gives	House construction	Up
House construction	Up	Gives	Housing stock	Up
Housing stock	Up	Gives	House prices	Down
House prices	Down	Gives	House buying	Up

Finally, adding financial considerations into the influence diagram gives the model: (Fig. 1.6).

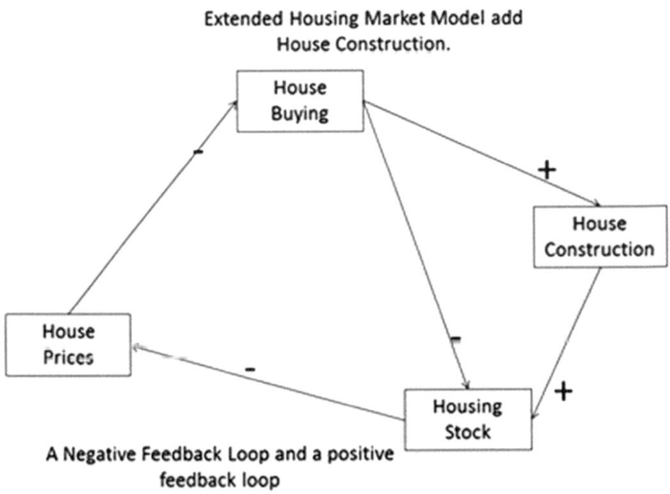

Fig. 1.5 Second house price model

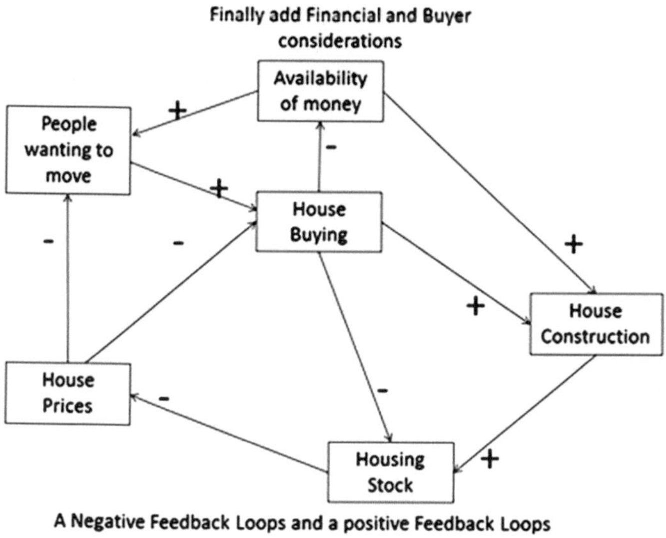

Fig. 1.6 House pricing model

All housing models contain a negative loop; however, a fuller model would include "people needing housing" and "construction of social housing" (see appendix for an extended model).

1.1.3 *Example C: Developing Financial Models*

These models investigate the movement of stock market share prices. The first model investigating the effect of investor confidence on share price movements and the second (extended) model adding the effect of short selling on the first model. Notice that these models demonstrate the effect of positive feedback loops and explain the movement of stock market prices (Fig. 1.7).

1.1.3.1 Effect of Investor Confidence on Financial Markets

This model can be constructed in stages.

Stage 1: share price and its effect on investor confidence, as the price increases, then confidence will also increase (Fig. 1.8)
Stage 2: now add the effect of investor confidence on the sale of shares, assuming that when confidence is rising, the investors will tend to hold onto the shares to give
Stage 3: as the sale of shares increase, then the share price will fall, giving the completed model (Fig. 1.9).

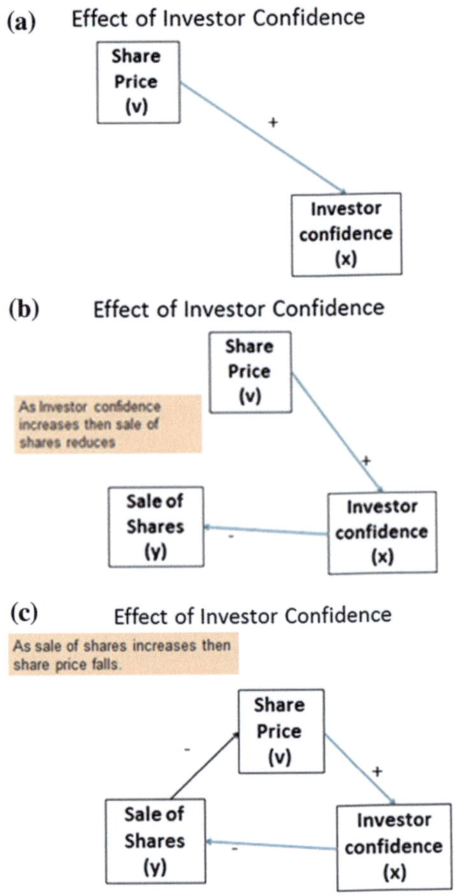

Fig. 1.7 **a** Start financial model **b** start financial model **c** first financial model

This model exhibits a positive feedback loop {+, −, −}, and hence, either an uncontrolled increase or decrease in share values will result from this model:

Increase in share prices gives				
Share price	Up	Leads to	Investor confidence	Up
Investor confidence	Up	Leads to	Sale of shares	Down
Sale of shares	Down	Leads to	Share price	Up
Decrease in share prices gives				
Share price	Down	Leads to	Investor confidence	Down
Investor confidence	Down	Leads to	Sale of shares	Up
Sale of shares	Up	Leads to	Share price	Down

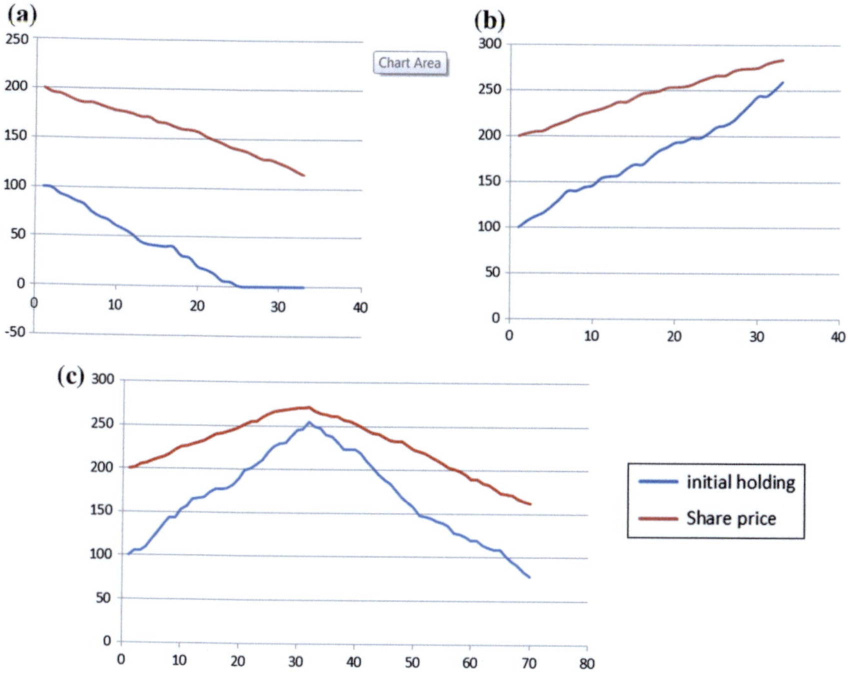

Fig. 1.8 **a** , **b** and **c**: Simulating stock holdings

Fig. 1.9 FTSE index

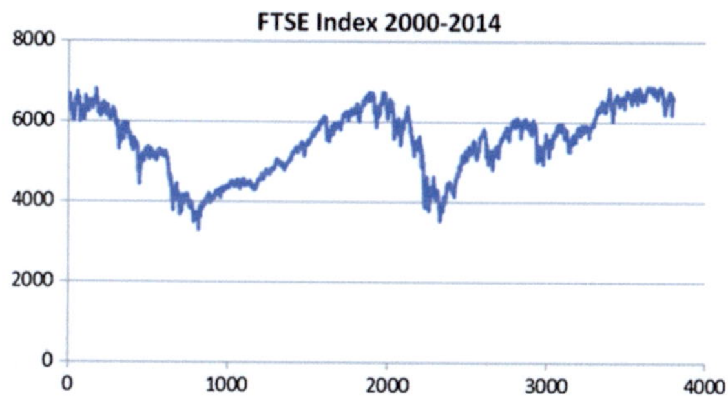

The direction (of movement of share prices) is changed only through the effect of external factors.

Using this model, simulation models were constructed first starting with low investor confidence; the results from this simulation produced the plots of "share price" and "share holding" showing how they both fall with time.

Conversely, if the investors start with high confidence, the plots show how both share price and share holding grow with time.

Finally, if the investor state starts in one mode (high confidence), then switching the next plot shows the effect on share prices and investor holding.

Validating the Model

This effect is demonstrated in both FTSE and Dow Jones historic data log plots; both the FTSE and the Dow Jones data can be shown to have several changes of direction moving from a period of continual growth to a period of continual decline; in addition, the Dow Jones plot also demonstrates that there can be periods where the plot is stationary (Fig. 1.10).

Turning points, changes from growth to fall or from fall to growth in the FTSE Index, occur at (about) 2003; 2007; 2009; 2010.

Similarly, plotting the Dow Jones, log(price) and movements for the period 1902–2012 gave (Fig. 1.11).

With Dow Jones turning points, marked above in Fig. 1.11, at (about) 1902; 1929; 1936; 1946; 1964; 1984; 2002;

Replotting the Dow Jones Index (log price) over a shorter time period, up to 1935, shows the dramatic effect of this model around the time of the Wall Street Crash (1929–1933); here, attempts to halt the fall failed until external events caused the changed direction.

Here, turning points seem to occur at 1924, 1929, 1932 and 1934.

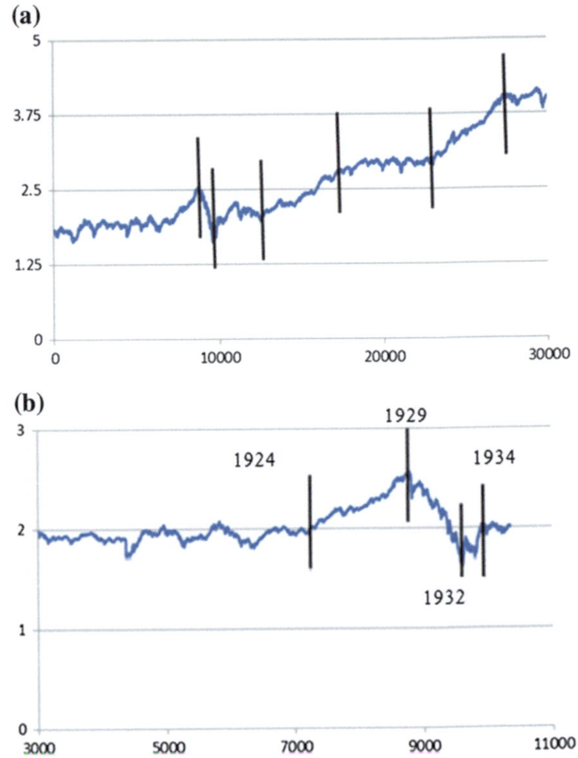

Fig. 1.10 **a** Dow Jones index **b** growth periods in Dow Jones index

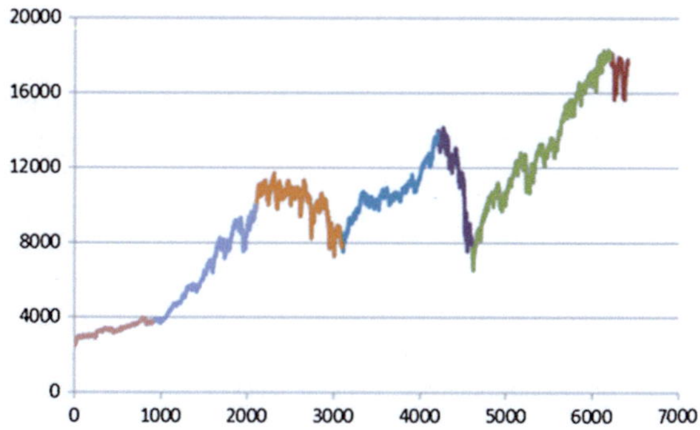

Fig. 1.11 Growth periods in Dow Jones index

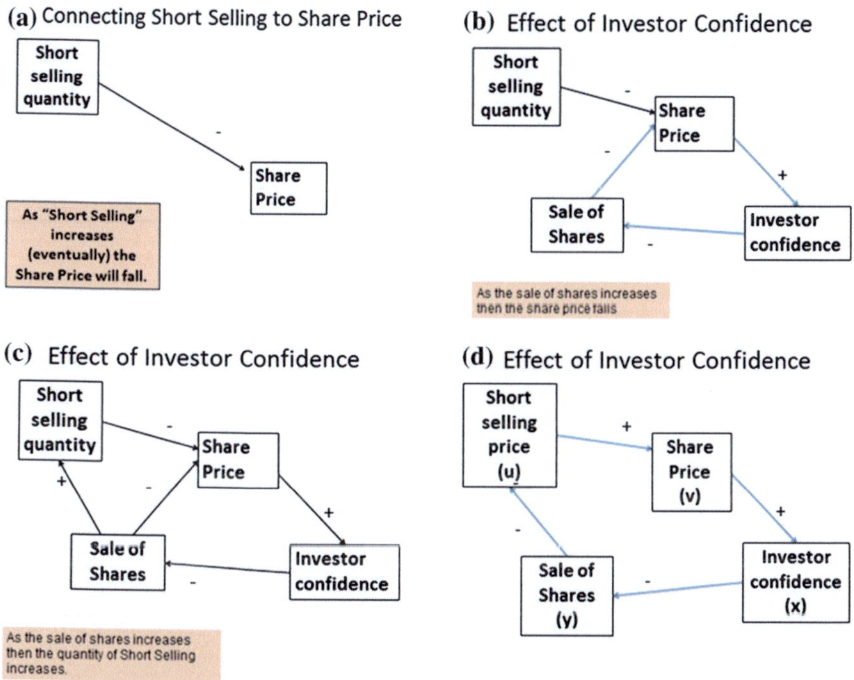

Fig. 1.12 **a** Modelling investor confidence and shares, **b** modelling investor confidence and shares and **c** modelling investor confidence and shares **d** alternative model for investor confidence and shares

Finally, plotting the Dow Jones Index over the period from 1991 not only shows the same effects during this period but also that the changes in direction tend to occur more frequently (Fig. 1.12).

Here turning points at 1991, 1994, 1999, 2003, 2007, 2009 and 2015 on average every 4 years.

1.1.3.2 Example D2: Effect of Investor Confidence and Short Selling

Definition: Short sale is the sale of securities or commodities not owned by the seller (who hopes to buy them later at lower price); hence, short-selling cause falls in asset prices.

Short selling and the results from short selling have been known for a long time:

- "He who sells what isn't his'n, must buy it back or go to pris'n."—Daniel Drew, 1797–1879, American financier.
- 1609—The Dutch East India Co protests to the Amsterdam Exchange after short sellers make enormous profits on its stock. Leading to the first ever regulations on shorting in the following year
- 1733—Britain bans naked short selling. "Investopedia" ref South Sea Bubble.
- 1932—US President Herbert Hoover condemns short selling for speculative profit on the New York Stock Exchange. www.reuters.com/article/us-sec-shortselling-history

Developing the influence model, from the base investor confidence model (Fig. 1.13).

Then, adding in investor confidence events gives:

Finally, as the sale of shares increases, then the short-selling quantity will increase.

This model exhibits only positive feedback loop {+, −, −} and {+, −, +, −}; hence, short selling will have the following effect on share prices:

Increase in short selling gives				
Short selling	Up	Leads to	Share price	Down
Share price	Down	Leads to	Investor confidence	Down
Investor confidence	Down	Leads to	Sale of shares	Up
Sale of shares	Up	Leads to	Short selling	Up

Notice that a similar model exists with respect to the short-selling price:

An analysis of this model gives:

Fall in short selling price gives				
Short selling price	Down	Leads to	Share price	Down
Share price	Down	Leads to	Investor confidence	Down
Investor confidence	Down	Leads to	Sale of shares	Up
Sale of shares	Up	Leads to	Short selling price	Down

Comparing this model with the stock market when "falls" occurred in the 1990s and 2000s and 2010s.

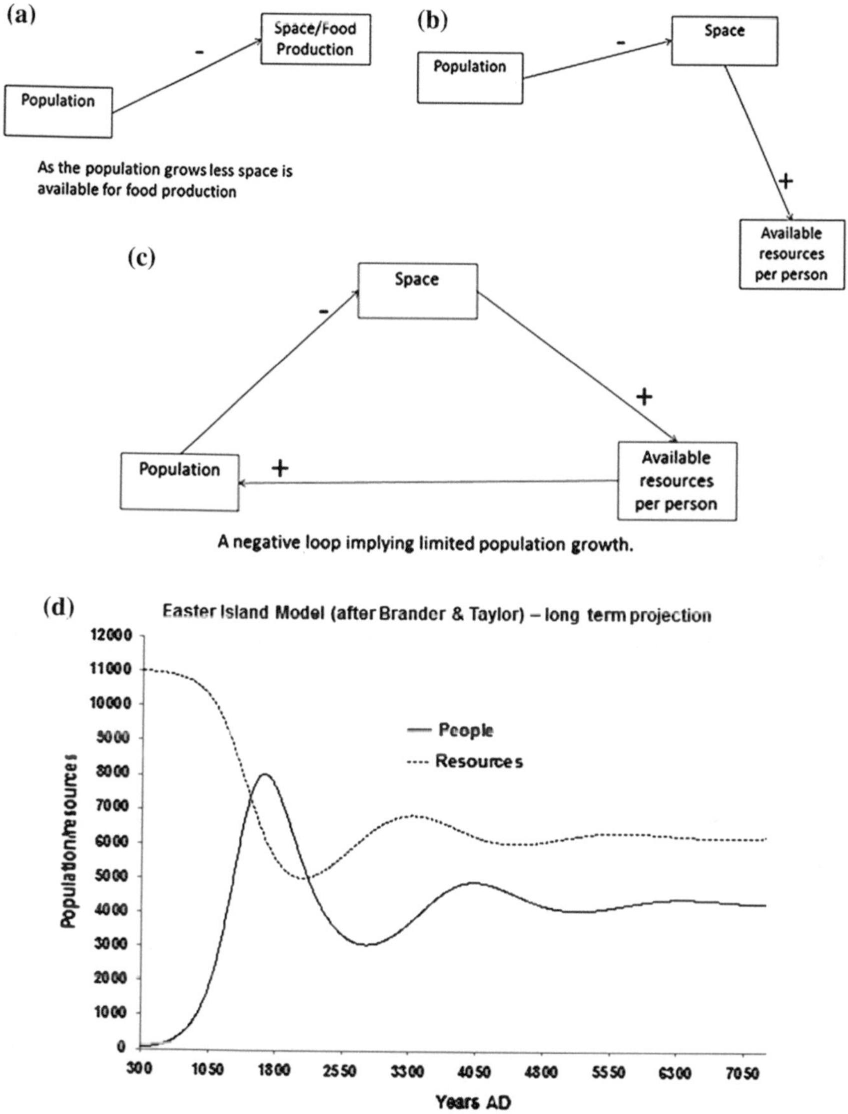

Fig. 1.13 **a–c** Developing the model, **d** results from a simulation based upon this model

1.1.4 Population Modelling

Again, two models are presented investigating population changes in a closed society, for example Easter Island an isolated Pacific Ocean island, in the first model no technological developments and in the second there are technological developments.

Basic "Easter Island" model

Stage 1: as the population increases, then the space available for food production falls.
Stage 2: as the space for food production increases, then the resources available per person also increases.
Stage 3: linking to give the final model, a negative goal-seeking loop (Fig. 1.14).

Notice that									
Population	Up	Leads to	Available space	Down					
Available space	Down	Leads to	Resources per person	Down					
Resources per person	Down	Leads to	Population	Down	Population	Down	Leads to	Available space	Up
Available space	Up	Leads to	Resources per person	Up					
Resources per person	Up	Leads to	Population		Up				

Applying this model using Easter Island, data gave the population plot

Where population and resources are tending towards a steady state, as expected with periods of population growth and population decline.

Note Easter Island was "discovered" at a time when the population was in the first "state of decline" shown in the population plot.

Allowing for technological development produces the model (Fig. 1.15):

Leading to an increasing population, compare this with the Earths continual development of food sources and the total population growth.

1.1.5 *Transport Modelling*

Travellers choose the most appropriate mode of transport to reach their chosen destination. They could travel by rail, or car, or bicycle or by other forms of public transport or even walk.

This model investigates the interaction between the use road (car) and rail transport and the subsequent expansions of the road and rail systems.

Stage 1: only road travel is considered giving (Fig. 1.16):

The model suggesting that road capacity is continually increasing, generally true, in the absence of any alternative mode of transport.

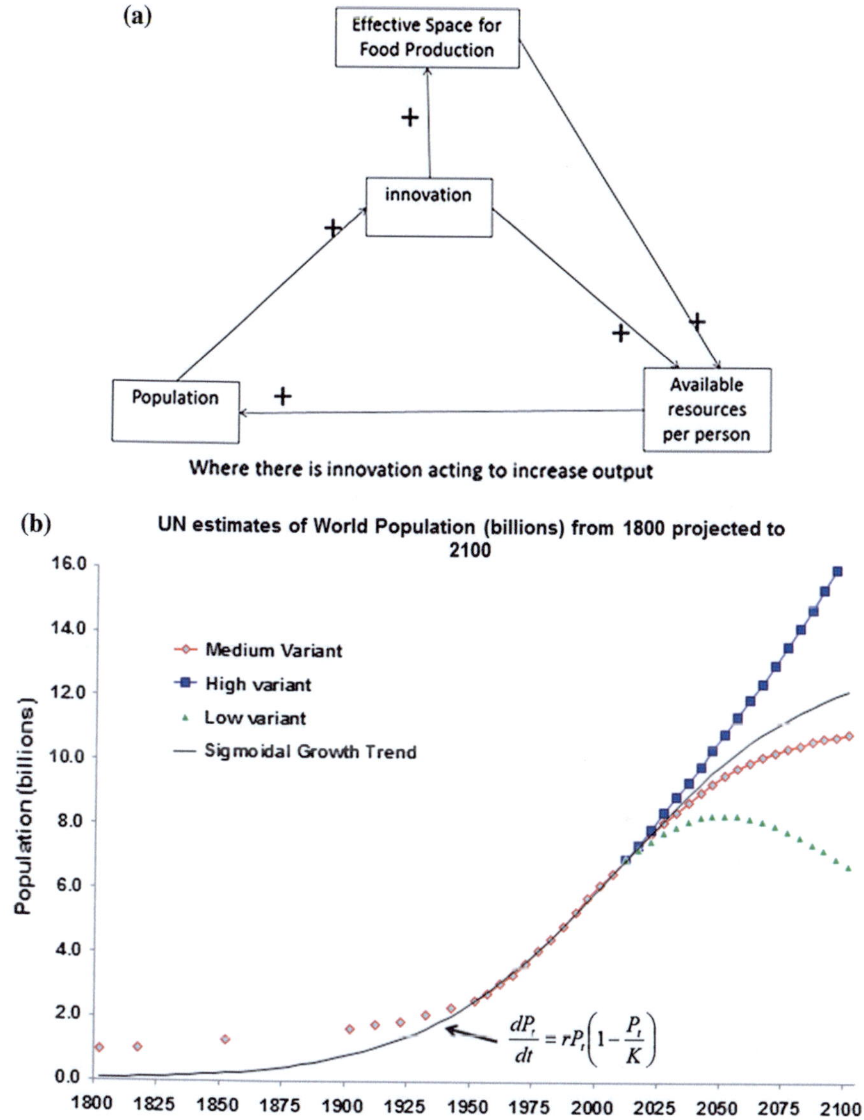

Fig. 1.14 **a** Second population model **b** world population estimates

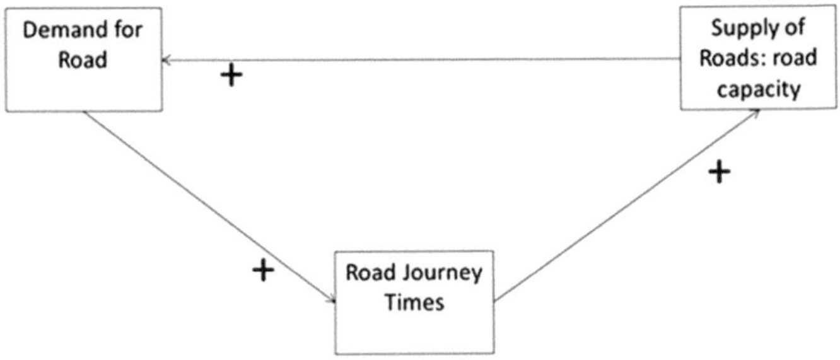

Fig. 1.15 Modelling road building

Journey times	Up	Leads to	Road Building	Up					
Road building	Up	Leads to	Demand for Roads	Up	Demand for Roads	Up	Leads to	Journey times	Up
Journey times	Up	Leads to…							

Similarly, extending the model by incorporating rail changes could produce another model containing only positive feedback loops, thus implying, as in the 1960s, that road provision is (continually) increasing and rail provision declining.

However, when increases in road capacities or the development of the road network are not, or no longer, a viable option, this model becomes:

Here there exists a negative feedback loop leading to an equilibrium state:

- {Demand for Road Travel; Road journey times; Demand for railways; Supply of Railways; Demand for road travel}
- { +, +, +, −}

The alternative modelling approach, using travel cost, also gives a model with a negative feedback loop

- {Supply of railways; Rail Journey costs; Demand for Railways}
- {+, −, +}

Here, paradoxically, the increased demand for rail travel can act to cause developments in the rail system and thus act to reduce demand!

Further development leads to the more comprehensive model given by:

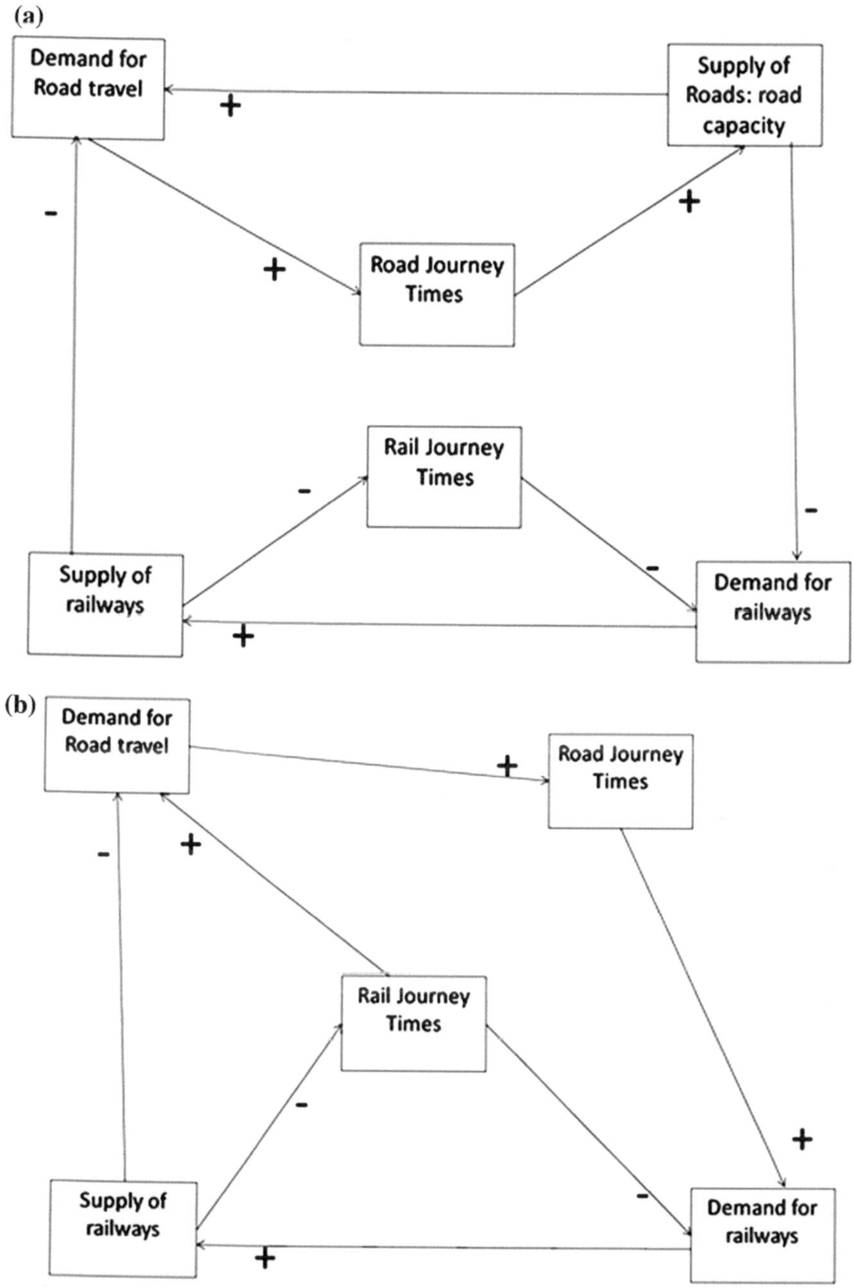

Fig. 1.16 **a** Modelling the effects of roads on rail demand **b–d** modelling the effects of roads on rail

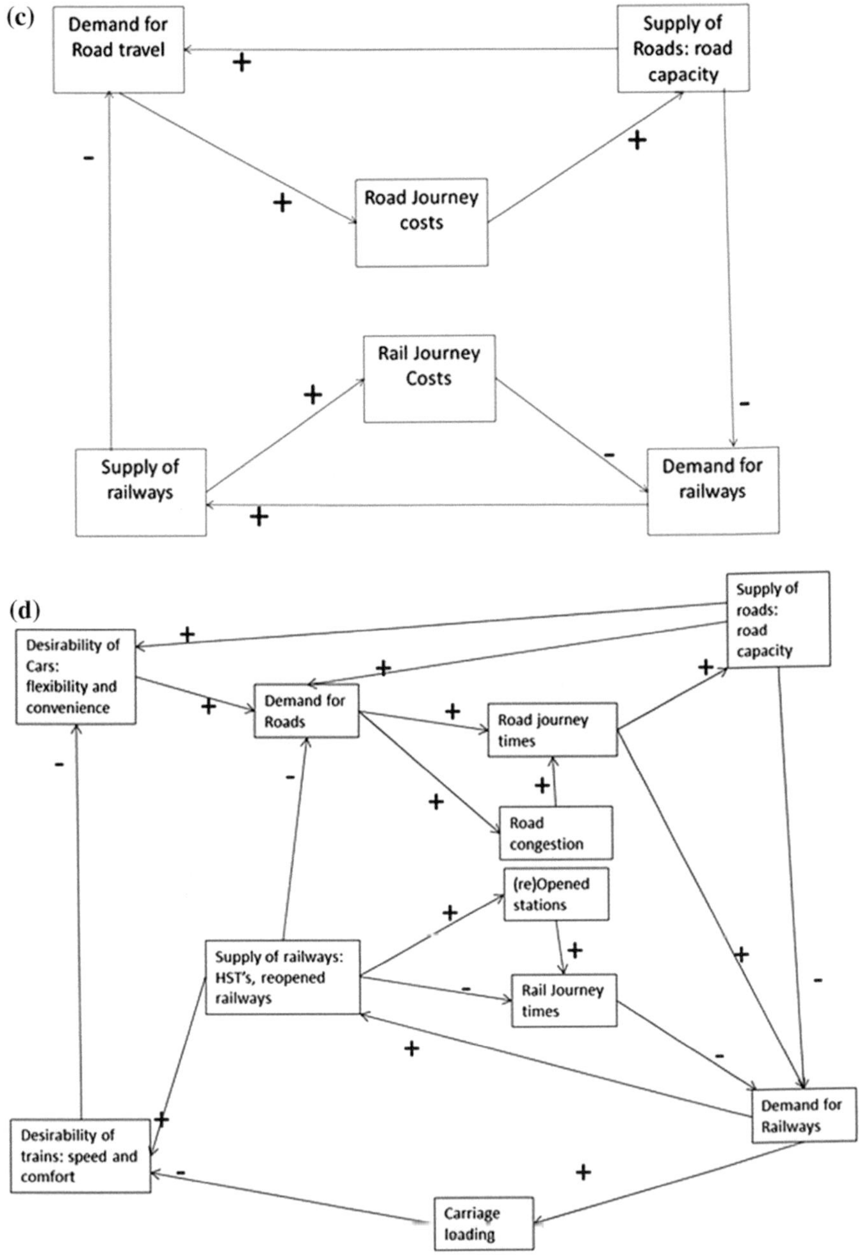

Fig. 1.16 (continued)

A negative feedback loop with respect to rail travel gives an overall goal-seeking model for both road and rail travel.

	{Desirability of cars;	
	Demand for roads;	
	Road journey time;	
	Supply of roads;	
	Demand for rail;	
	Carriage loading;	
	Desirability of trains;	
	Desirability of cars}	
Changes	{+, +, +, −, +, −, −}	A negative loop

1.2 Constructing Models from "Big Data"

1.2.1 Introduction

Data science is a relatively new research field consisting of well-established methods and approaches to address the need to identify actionable knowledge from the continuous creation of data. In particular, the use of the existing techniques has led the discovery of new research directions, with groundbreaking results.

There are a variety of mathematical and statistical tools to identify and discover knowledge, which can be used to facilitate the decision-making process. In particular, these include Bayesian and dependency networks, which provide modelling tools to determine how mutual relationships between concepts influence the knowledge captured by such networks. In the rest of this section, we will discuss their main theoretical properties, which be fully exploited in the two case studies discussed in part 2.

Bayesian and Dependency Networks

Bayesian networks (BNs) are graphical models, which capture independence relationships among random variables, based on a basic law of probability called *Bayes' rule* [1]. They offer an efficient modelling framework in risk and decision analysis with a variety of applications, such as safety assessment of nuclear power plants, risk evaluation of a supply chain and medical decision support tools [2]. More specifically, BNs are defined by nodes, representing objects based on a level of uncertainty, also called *random variables*, which are connected by edges indicating a dependence relationship between them. Furthermore, Bayesian networks also contain quantitative information, which represents a factorisation of the joint probability distribution of all the variables in the network.

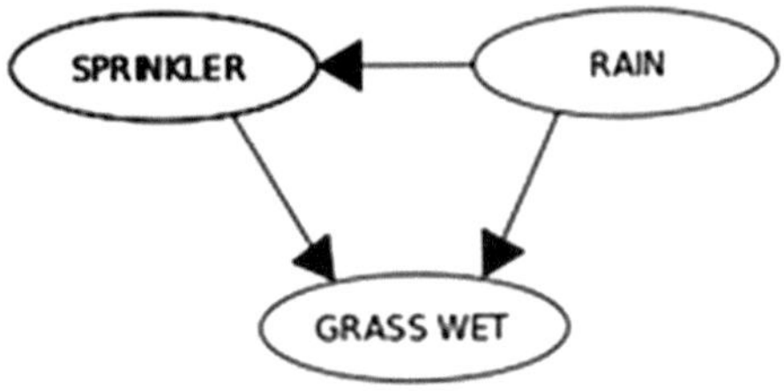

Fig. 1.17 Representing a Baysian network

Suppose, for example, we want to explore the chance of finding wet grass on any given day. In particular, assume the following

1. A cloudy sky is associated with a higher chance of rain,
2. A cloudy sky affects whether the sprinkler system is triggered and
3. Both the sprinkler system and rain have an effect on the chance of finding wet grass.

In this particular example, no probabilistic information is given. The resulting BN is depicted in the following Fig. 1.17.

It is clear that such graphical representation provides an intuitive way to depict the dependence relations between variables.

In the definition of BNs, the most complex statements do not refer to dependencies, but rather about independences (i.e. absence of edges in the graph), as it is always possible to determine dependence through the conditional probability tables when an edge is present, even though the reverse is not true.

The definition of a BN can be carried out either through explicit data analysis, or via literature review and expert elicitation. These are typically manually intensive tasks depending on the size and complexity of the data sets analysed, especially when they exhibit unstructured components. There is extensive research on the extraction of BNs from text corpora. For example, in [3], the authors suggest a domain-independent method for acquiring text causal knowledge to generate Bayesian networks. Their approach is based on a classification of lexico-syntactic patterns which refer to causation, where an automatic detection of causal patterns and semi-validation of their ambiguity is carried out. Similarly, in Kuipers [4], a supervised method for the detection and extraction of causal relations from open domain texts is presented. The authors provide an in-depth analysis of verbs, cue phrases that encode causality and, to a lesser extent, influence.

Dependency networks (DNs) have also been attracting increasing attention within several research fields. These types of networks are similar to BNs, which, however, allow cycles (that is a path starting and ending at the same node) enabling a more computationally efficient approach and making them more applicable especially when large and unstructured data sets are considered [5].

Text Mining

Text mining (TM) consists of computational techniques to achieve human language understanding via linguistic and semantic analysis. Such methods have been shown to be crucially important in the way we can represent knowledge described by the interactions between computers and human (natural) languages [6].

Language is based on grammatical and syntactic rules which can be captured by *patterns* or, in other words, templates that sentences with similar structure follow. Such language formats enable the construction of complex sentences, as well as framing of the complexity of language.

In order to understand the properties of human language, a variety of methods have been developed to address the complexity and challenges posed by it. These, broadly speaking, fall into three categories: *symbolic, statistical* and *connectionist.*

In the symbolic approach, linguistic properties are mapped onto precise and well-understood knowledge representation [7]. Once the linguistic rules have been defined, the hierarchical structure of the semantic concepts within the corresponding textual fragments is identified. Subsequently, the properties associated with the different textual components are investigated to provide an insight into their structure. Symbolic methods have been widely exploited in a variety of research contexts such as information extraction, text categorisation, ambiguity resolution, explanation-based learning, decision trees and conceptual clustering.

On the other hand, the statistical properties from observable data and the investigation of large documents can be used to develop generalised models based on smaller knowledge data sets and significant linguistic or world knowledge [8]. They have many applications such as parsing rule analysis, statistical grammar learning and statistical machine translation, to name but a few.

The connectionist approach integrates statistical learning with representation techniques to allow an integration of statistical tools with logic-based rule manipulation, generating a network of interconnected simple processing units (often associated with concepts) with edge weights representing knowledge. This typically creates a rich system with an interesting dynamical global behaviour induced by the semantic propagation rules. In Troussov et al. [9], a connectionist distributed model is investigated pointing towards a dynamical generalisation of syntactic parsing, limited domain translation tasks and associative retrieval.

General Architecture and Various Components of Text Mining

A grammar is a set of well-defined rules which govern how words and sentences are combined according to a specific syntax. A grammar does not describe the meaning of a set of words or sentences, as it only addresses the construction of sentences according to the syntactic structure of words. Semantics, on the other hand, refers to the meaning of a sentence [8]. In computational linguistics, semantic analysis is a much more complex task since its aim is the full understanding of the meaning of text.

Any text mining process consists of a number of steps to identify and classify sentences according to specific patterns, in order to analyse a textual source. Broadly speaking, in order to achieve this, we need to follow these general steps:

1. Textual data sources are divided into small components, usually words, which can be subsequently syntactically analysed.
2. These, in turn, create *tokenised* text fragments, which are analysed according to the rules of a formal grammar. The output is a parsing tree—in other words, an ordered tree representing the hierarchical syntactic structure of a sentence.
3. Once we have isolated the syntactic structure of a text fragment, we are in the position of extracting relevant information, such as specific relationships and sentiment analysis.

More specifically, the main components of text mining are as follows:

Lexical Analysis

Lexical analysis is the process which analyses the basic components of texts and groups into *tokens* [10]. In other words, lexical analysis techniques identify the syntactic role of individual words which are assigned to a single *part-of-speech* tag.

Lexical analysis may require a *lexicon* which is usually determined by the particular approach used in a suitably defined TM system, as well as the nature and extent of information inherent to the lexicon. Mainly, lexicons may vary in terms of their complexity as they can contain information on the semantic information related to a word. More research is currently being carried out to provide better tools in analysing words in semantic contexts [see 11 for an overview].

Part-of-Speech Tagging.

Part-of speech tagging (POS) allows to attach a specific syntactic definition (noun, verb, adjective, etc.) to the words which are part of a sentence. This task tends to be relatively accurate, as it relies on a set of rules which are usually unambiguously defined. Often, POS tasks are carried out via the statistical properties of the different syntactic roles of tokens [8]. Consider the word *book.* Depending on the sentence it belongs to, it might be a verb or a noun. Consider "a book on chair" and "I will book a table at the restaurant". The presence of specific keywords, such as "a" in the former, and "I will" in the latter, provides important clues as to the syntactic role that *book* has in the two sentences. One of the main reasons for the overall accuracy of POS tagging is that a semantic analysis is often not required, as it is based on the position of the corresponding token.

Parsing

Once the POS tagging of a sentence has identified the syntactic roles of each token, each sentence can be considered in its entirety. The main difference with POS tagging is the fact that parsing enables the identification of the hierarchical syntactic structure of a sentence. Consider, for example, Fig. 1.18b depicts the parsing tree structure of the sentence "This is a parsing tree". Note that each word is associated with a POS symbol which corresponds to its syntactic role [8].

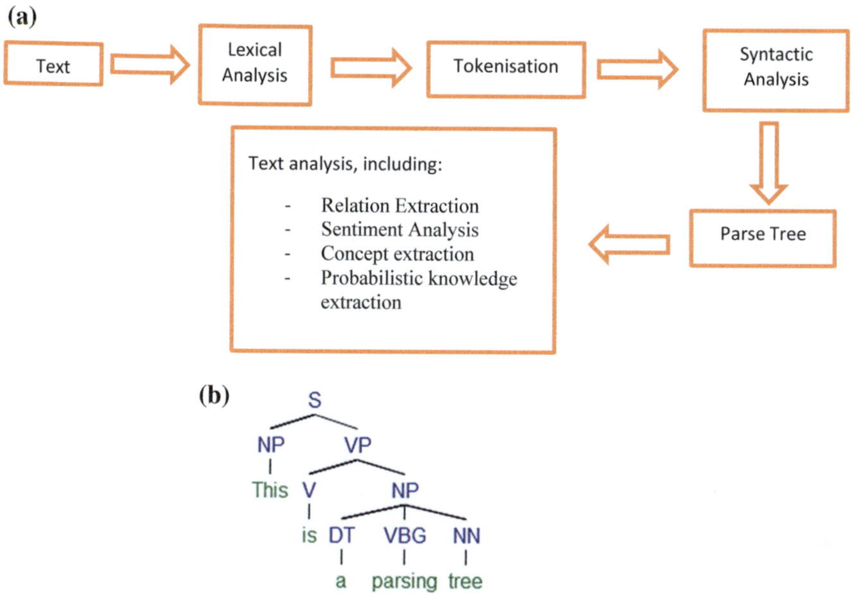

Fig. 1.18 **a** Text missing, **b** the parsing tree of the sentence "This is parsing tree"

Named Entity Recognition

An important aspect of text analysis is the ability to determine the type of the entities, which refer to words, or collections of them. For example, determining whether a noun refers to a person, an organisation or geographical location (to name but a few) substantially contributes to the extraction of accurate information and provides the tools for a deeper understanding. For example, the analysis of "dogs and cats are the most popular pets in the UK" would identify that dogs and cats are animals and the UK is a country. Clearly, there are many instances where this depends on the context. Think of "India lives in Manchester". Anyone reading such sentence would interpret, and rightly so, India as the name of a specific person. However, a computer might not be able to do so and determine that it is a country. We know that a country would not be able to "live" in a city. It is just common sense. Unfortunately, computers do not have the ability to discern what common sense is. They might be able to guess according to the structure of a sentence, or the presence of specific keywords. This is a very effective example of semantic understanding, which comes natural to humans, but a very complex task to computers.

Coreference Resolution

Coreference resolution is the process of determining which text components refer to the same objects. For example, *relation resolution* attempts to identify which individual entities or objects a relation refers to. Consider the following sentence,

"We are looking for a fault in the system". Here, we are not looking for *any* fault in the system, rather for a specific instance.

Relation Extraction

The identification of relations between different entities within a text provides useful information that can be used to determine quantitative and qualitative information linking such entities. For example, consider the sentence "smoking potentially causes lung cancer". Here, the act of smoking is linked to lung cancer by a causal relationship. This is clearly a crucial step in building BNs, even though such analysis requires a deep understanding of the associated textual information.

Concept Extraction

A crucial task in information extraction from textual sources is concept identification, which is typically defined as a one or more keywords, or textual definitions. The two main approaches in this task are supervised and unsupervised concept identification, depending on the level of human intervention in the system.

In particular, formal concept analysis (FCA) provides a tool to facilitate the identification of key concepts relevant to a specific topical area [12]. Broadly speaking, unstructured textual data sets are analysed to isolate clusters of terms and definitions referring to the same concepts, which can be grouped together. One of the main properties of FCA allows user interaction, so that user(s) can actively operate the system to determine the most appropriate parameters and starting points of such classification.

Sentiment Analysis

This instance of information extraction from text focuses on the identification of trends of moods or opinions associated with textual sources.

Broadly speaking, its aim is to determine the *polarity* of a given text which identifies whether the opinion expressed is positive, negative or neutral. This also includes emotional states, such as anger, sadness and happiness, as well as intent, such as planning and researching. Sentiment analysis can be an important tool is obtaining an insight into relationships among concepts in BNs, since it can support the process of relation discovery. In fact, not all the information contained in text is unambiguously described. Consider, for example, the sentence "I am very surprised by your irrational fear that spiders can kill you". Here, rather than drawing a definite conclusion that spiders are linked with death, a sceptical assessment of such relationship is noted.

Topic Recognition

This procedure attempts to identify the general topic of a text by grouping a set of keywords which appear frequently in the documents. These are then associated with one of more concepts to determine the general concept trend.

Semantic Analysis

Semantic analysis determines the possible meanings of a sentence by investigating the interactions among word-level meanings in the sentence. This approach can also incorporate the *semantic disambiguation of words* with multiple senses. Semantic disambiguation allows the selection of the sense of ambiguous words, so that they can be included in the appropriate semantic representation of the sentence [13]. This is particularly relevant in any information retrieval and processing system based on ambiguous and partially known knowledge. Disambiguation techniques usually require specific information on the frequency with which each sense occurs in a particular document, as well as on the analysis of the local context, and the use of pragmatic knowledge of the domain of the document. An interesting aspect of this research field is concerned with the purposeful use of language where the utilisation of a context within the text is exploited to explain how extra meaning is part of some documents without actually being constructed in them. Clearly, this is still being developed as it requires an incredibly wide knowledge dealing with intentions, plans and objectives [8]. Extremely useful applications in TM can be seen in inference techniques where extra information derived from a wider context successfully addresses statistical properties [4].

1.2.2 The Automatic Extraction of Bayesian Networks from Text

The mathematical constraints posed by Bayes' rule and general probability theory create a significant challenge. As a consequence, the identification of suitable Bayesian networks is often carried out manually usually by a modeller. However, this can be extremely time-consuming and based on only specific, often limited, sources depending on the modeller's expertise. As a consequence, the ability to automatically extract the relevant data would potentially add enormous value in terms of increased efficiency and scalability to the process of defining and populating BNs. However, extracting both explicit and implicit information, and making sense of partial or contradictory data, can be a complex challenge.

Dependence Relation Extraction from Text

As discussed above, nodes in BNs, which are connected by edges, indicate that the corresponding random variables are dependent. Such dependence relations must be therefore extracted from textual information, when present. The conditional dependencies in a Bayesian network are often based on known statistical and computational techniques, which are based on a combination of methods from graph theory, probability theory, computer science and statistics. Linguistically speaking, a dependence relation contains specific keywords which describe that two concepts are related to a certain degree. Consider the sentence "lung cancer is more common among smokers". There is little doubt that we would interpret this as clear relation linking lung cancer with smoking. However, there is not a precise linguistic definition to determine a relationship between two concepts from text, due to its content dependence. When a full automation of the process of textual information extraction is carried out, a clear and unambiguous set of rules ensures a reasonably good level of accuracy. As a consequence, it is usually advisable to consider *causal relationships,* which are a subgroup of dependence relationships [1]. In fact, they are likely to convey a much stronger statement, and they are more easily identified due to a more limited set of linguistic rules that characterise them. Going back to the above example, saying that smoking *causes* lung cancer assumes a direct link between them. We cannot arguably say the contrary, but there are other cases where there is a less marked cut-off. If we are only looking for causal relationships when populating a BN, we might miss out several dependence relations. However, accuracy is much more preferable. The integration of an automatic BN extraction with human intervention usually addresses such issue.

Variables' Identification

Mapping a representative to a specific variable is closely linked to the task of relation extraction. However, this is partially a modelling choice by the user based on the set of relevant concepts. Consider again the sentence "smoking causes lung cancer". If this was rephrased as "smokers are more likely to develop lung cancer", we would need to ensure that "smoking" and "smokers" are identified as a single variable associated with the act of smoking. In a variety of cases, this can be addressed by considering synonymy. However, such as in our example, it might also happen that they refer to the same concept, rather than being the *same* concept. Formal concept analysis (FCA) is one of the computational techniques that can be successfully applied in this particular context [8].

Probability Information Extraction.

An essential part in the extraction and subsequent creation of BNs involves the processing of the textual sources to determine probability of variables.

Sentences may capture some probabilistic relationships between concepts even though very few of them might provide conclusive and unambiguous information, which can be utilised to reason. In fact, the combination of qualitative and quantitative data creates a variety of challenges, which need to be addressed to produce relevant and accurate information.

The identification, assessment and ranking of specific keywords (and their combinations), when describing probability, can provide a useful insight into the structure of the corresponding relationships. However, Big Data research focuses on the interrelations of diverse and multidisciplinary topics, resulting in the intrinsic difficulty in finding a common ground in terms of the linguistic features that specific probabilistic description should have.

In [2], an automated method to assess the influence among concepts in unstructured sets is introduced. Despite not being directly related to BNs, it shows potential in the extraction of the mutual co-occurrence properties between nodes.

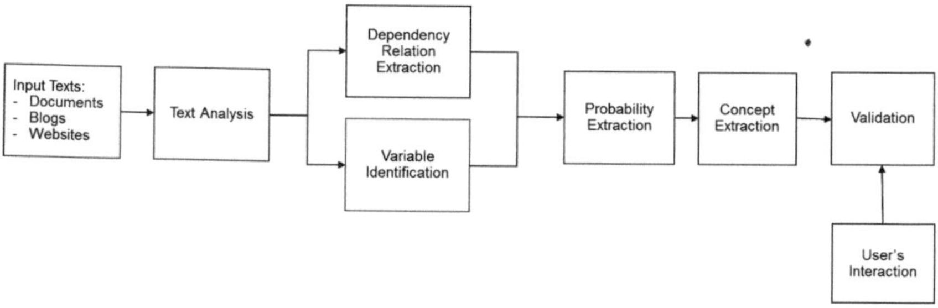

Fig. 1.19 Architecture of Bayesian network extraction from textual information

Aggregation of Structural and Probabilistic Data.

This step integrates the steps discussed above, to construct fragments of BNs via user's interaction.

Figure 1.19 depicts this process in a sequential set of steps. However, the repeated implementation of such steps in a series of loops might be required to obtain meaningful BN fragments.

General Architecture

The general architecture of the extraction of fragments of BNs from text corpora consists of the following components:

1. Existing and predefined information on specific topics would be incorporated into a database, or *Knowledge Database* (KDB) consisting of

   (a) Historical data from structured DBs,
   (b) Bayesian networks built on existing data and
   (c) Data entered by modeller(s) and manually validated.

   The KDB is an important component since it is based on information which is considered "reliable". In a variety of cases, the KDB is maintained by modelling experts to ensure that the data are regularly updated to prevent any inconsistency and ambiguity.

2. The user would interact with the system by specifying further textual sources and structured data sets.
3. The extraction and data aggregation stage consists of the identification of the appropriate textual data associated with such unstructured data sets, as well as the removal of any data duplication. An essential part of this process is to address any qualitative and quantitative inconsistency. As discussed above, BNs have strict mathematical constraints which make any fully unsupervised automatic extraction prone to inaccuracies and inconsistencies. As a consequence, human intervention is often advisable to minimise any such error.

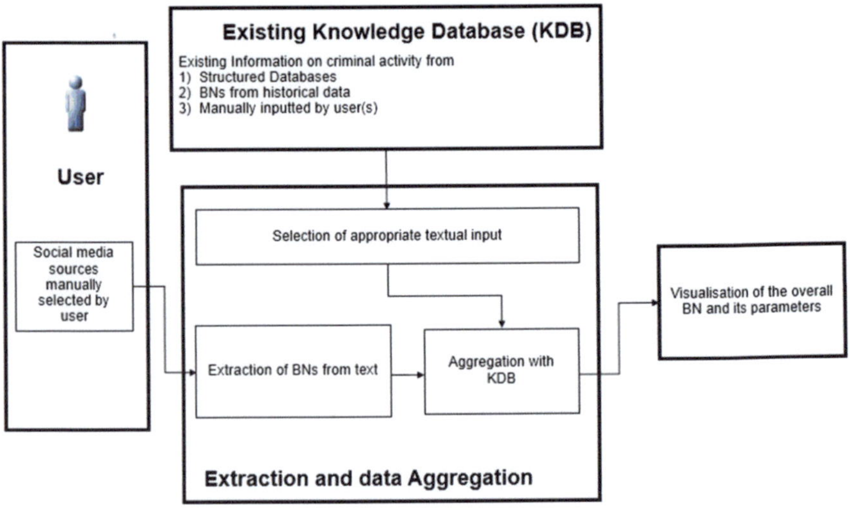

Fig. 1.20 General architecture of Bayesian networks for crime detection as discussed in Trovati [5]

4. Finally, the BN is visualised, providing
   (a) Relevant information on the structure of the BN,
   (b) Description of the different parameters and
   (c) Any required action in order to address any inconsistency which could not be resolved automatically. This is typically an interactive step, where the result can be updated by the user as well as focused on a specific part of the BN (Fig. 1.20).

1.3 The Blackboard Architecture

1.3.1 Introduction

The blackboard architecture represents a flexible, symbolic artificial intelligence (AI) method for the cooperative solution of complex problems. Systems that use this architecture have been in existence since the 1970s. In the beginning, they were used mainly for solving signal-processing problems, for example speech recognition with Hearsay-II (see [14]) and interpretation of sonar with HASP (see [15]). Following Hearsay-II and HASP, the blackboard architecture became very popular and was associated with many diverse application areas including mission control systems for satellites, military object tracking and detection, printed text recognition for scanners, fault diagnosis, assembly arrangement, and planning and scheduling, to name just a few. In the early nineties, the architecture endured a period of relative obscurity, which has mostly been attributed to a lack of formalism, but since the turn of the century it has enjoyed something of a renaissance, in some cases as a hybrid system coupled with a Bayesian Belief Network, with applications such as robotic mapping (see [16]) and logistics planning for the US military (see [17]). The blackboard architecture has also been adopted by the computer game AI community for solving decision-making and agent coordination problems (see [18]), although there is some controversy about whether game AI blackboard systems are actually "true" blackboard systems.

Blackboard systems (BBSs) have many properties that make them suitable for solving complex problems that require progression through different stages (with many paths to those stages) to reach the final solution [19]. These include problems that can be viewed as a search for the "best" solution given a set of constraints [20]. BBSs provide a flexible method for incremental reasoning about a problem and its solution, building up the solution in step-by-step manner by opportunistically examining different paths at different levels of abstraction. This means that the system's exploratory properties are high compared with systems or algorithms that used fixed, predetermined methods such as forward or backward chaining rule-based systems. BBSs can also deal with large quantities of diverse, uncertain, incomplete or inaccurate data [21] and can integrate different kinds of knowledge in order to solve a given problem. This makes them highly useful for solving problems with limited and imperfect input data, problems that require the combination of many separate diagnostic components, dynamic decision-making problems and problems involving systems of systems.

1.3.2 Architecture

The original blackboard architecture was developed in the early 1970s with the Hearsay-II (HSII) speech recognition project [14], but has evolved to some extent since then (literature review for further details). The aim of the HSII project was to build a system that could manage complex and ill-defined problems without the requirement for a formal model, so that enhancements could easily be made [19]. The deliberate flexibility of the design has led to a number of different interpretations of BBS terminology and definitions. This section aims to clarify its core definition and design and also to dispel some of the myths surrounding BBSs. First, the analogy to a set of experts gathered around a blackboard is explained, and then, the key components of the BBS are listed and described. The essential properties of the BBS are then reported, either as advantages or disadvantages of the architecture. This section also includes a review of the main problem types that are suited to solution with BBSs, provides some definitions and briefly discusses the history and availability of toolkits for frameworks for BBSs.

1.3.2.1 Analogy

The BBS architecture is based on the concept of a group of experts using a physical blackboard as a shared workspace for constructing a solution to a problem. The first step is to write a description of the problem and the initial data on the blackboard. Each expert waits until an opportunity arises to apply their particular knowledge to the problem, at which point they may contribute by writing information on the blackboard; this information may take many forms including a partial solution, an informed suggestion about which solution paths or avenues of exploration to select next, an alternative solution, a candidate complete solution, an agreement to a candidate solution or partial solution, or a disagreement about a candidate solution or partial solution. The aim of each contribution is to provide insights that will enable the experts to progress closer to a complete solution that is optimal in some way. Thus, the contributions continue until an agreed solution is reached.

An important consideration is managing the flow of the problem-solving, i.e. deciding who takes the chalk next in the event that more than one expert identifies an opportunity to contribute; they cannot interact with the blackboard simultaneously as this would cause confusion. The intuitive way to impose control is to nominate an independent arbiter, who decides which expert should contribute next. However, a consistent basis for making that decision is still needed. The arbiter needs to be able to assess the benefits of the potential contributions in some way. One method of proceeding is to ask all of the candidate experts to estimate the value of their contribution, selecting the one who makes the highest estimate.

The scenario described above has some important properties. First, the experts act independently and are self-contained. Their approach to solving the problem and the knowledge they possess about it (in terms of their prior experience, the level they have reached and the focus of their expertise) can differ vastly, which means that very diverse contributions can be made. Second, the problem is solved in an opportunistic fashion, i.e. the flow of expert contributions is not predetermined or prescriptive. They respond when they have something worthwhile to contribute based on their perception of previous contributions. This is also an event-based method of problem-solving. The event that triggers an expert to ask to contribute is the writing of a partial solution (or other information) that allows that expert to apply his or her knowledge to the problem.

1.3.2.2 Components

The blackboard AI architecture is analogous to the scenario described above. It consists of three main components:

- **The Knowledge Sources** (analogous to the experts in the human system)
- **The Control Component** (analogous to the arbiter in the human system)
- **The Blackboard** (analogous to the physical blackboard in the human system)

Each of these components is now described in detail.

Knowledge Sources

In the AI system, the contributions to the problem solution are made by a set of "artificial experts", i.e. a set of diverse problem-solving algorithms known as knowledge sources (KSs). As in the analogy, these programs are self-contained, operate completely independently from one another and may have different strengths and weaknesses to bring to the problem. Thus, within the system of systems, each KS may be viewed as a black box with its implementation details hidden from the others. A given KS may reason by any means, for example by top-down, bottom-up, goal-driven, data-driven or opportunistic methods [22]. However, each KS needs to be able to understand the information that is contributed by the others and also the current state of the problem-solving. Thus, the BBS requires a common information representation that is understood by all KSs. For example, a particular blackboard application may have one KS based on an artificial neural network (ANN) design, one that uses a forward-chaining production system, one that uses fuzzy logic reasoning and one that implements Dempster–Shafer models. The production system KS does not need to understand how the ANN works, but it needs to be able to interpret the contributions that are output by the ANN as the KSs are required to work together to solve the problem, as in the human analogy. If the KSs contribute different types of information, then each KS must be able to understand each of these different types.

The KSs contribute information when they are able (they have something to contribute) and when activated (they are selected for contribution). Each KS has a precondition attached to it to determine whether it should be activated and an action that describes what it will do when activated.

Control Component

In the AI system, the control component is analogous to the arbiter. It is independent of the KSs and is tasked with selecting an estimate of the "best" candidate KS to make a contribution to the problem solution. The mechanism for control in BBSs has evolved considerably over time but was fairly simple in early systems; each candidate KS was asked to determine an estimate of its cost (to system resources) and value (expected usefulness of the contribution). The control component merely combined these two metrics in some way to establish which KS should contribute next.

Blackboard

The component in the AI system that is analogous to the physical blackboard in the human system is also called a blackboard. However, in the AI system, this is a shared memory location that acts as a repository for all of the information about the problem including the input data, the problem statement and all of the KS-contributed information (solution path suggestions, partial solutions, alternative solutions, candidate complete solutions, final complete solutions, agreements and disagreements). The information deposited on the blackboard is often referred to as a *hypothesis*. The blackboard component can be thought of as a global database for KS hypotheses.

The blackboard serves as the main, in-direct communication mechanism between the KSs; their outputs are recorded on it, and these are read from it and understood by the other KSs. In addition, contributions made to the blackboard (events) may trigger a response from one or more KSs.

Some BBSs have blackboards that are subdivided into some way to categorise the hypotheses that they hold. The division may take the form of levels, sections, areas, a hierarchy of levels (as in HSII) or some combination of these, and the information in each part of the blackboard may also be sorted in some way, for example alphabetically or by the time it was contributed. In addition, some systems use a number of blackboards for categorisation. The reason for subdivision is to enable information to be located efficiently. Early BBSs tend to retain all of the information placed on them for the entire duration of the program life cycle. Information that appears to be irrelevant is kept because its use often becomes apparent later on during the program execution. However, some later systems permit deletion of information that is regarded as "outdated".

1.3.2.3 Similarities to Other Systems

There are some similarities between BBSs and symbolic rule-based decision-making systems, also known as production systems (see [23]), for example expert systems. Rule-based systems consist of a semantic reasoning subsystem and a set of rules with preconditions and actions. The semantic reasoning subsystem takes an action from the rule set based on its relationship to the input data, i.e. a rule is fired when conditions in the input data match its precondition. The rules in these systems are like the KSs in BBSs, which are triggered under certain conditions, but rule-based systems differ in that the rules are interdependent, which reduces the flexibility of these systems. As stated above, KSs operate independently of one another and are thus highly flexible. BBSs were, in fact, first proposed as a generalisation of rule-based systems with "rules" that could take any format and trigger [18].

1.3.2.4 Algorithm

As in the analogy, the BBS algorithm proceeds in an event-based manner. Any change to the blackboard is an event, and KSs can also cause events when they write their hypothesis to the blackboard. Any event can result in a match with a KS precondition. When its precondition is matched, a KS becomes a candidate for contributing to the blackboard, i.e. executing its action, and it competes with other KS candidates to execute that action on the blackboard.

The control component is responsible for selecting one of the candidate KSs, based upon the current state of the blackboard. In some systems, the control component is also charged with managing the events that trigger the KSs, but its chief function is to rank the candidates and select the most highly ranked. The methods for rating and ranking the KSs differ between BBSs. Some of the different approaches are discussed in the literature review.

1.3.2.5 Properties

This section is subdivided into the advantages and disadvantages of BBSs. The main contenders in each category are summarised as follows. Advantages: BBSs are flexible and general and solve problems incrementally in a step-by-step manner. They are opportunistic in the way that they tackle the search space leading to more sophisticated searches. They can also handle both a wide variety of data types and imperfect data. Disadvantages: the methodology suffers from a lack of formalism. In addition, the BBS architecture does not scale down to more simple problems and scales poorly for large search spaces. Further details about each of these properties are provided in the following subsections.

Advantages

BBSs have an important advantage over more traditional problem solvers that use rules, for example production systems. Production systems are prescriptive; there is a set of order for firing the rules, i.e. all reasoning is either forward or backward chaining. In BBSs, the problem-solving is opportunistic in that the most appropriate KS is selected, for example in speech recognition words that are understood well can be used to limit the search for interpretations of poorly sensed words. In addition, the solution is built incrementally in a step-by-step manner in response to events on the blackboard; there is no predetermined prescription for the triggering of KSs. KSs respond to events as and when they happen, and it is the interaction between the KSs (via the blackboard) that determines the solution paths explored and thus the final solution. The exploratory properties of the BBS search are therefore high in comparison with rule-based systems, which results in a much more sophisticated search of the solution space. In fact, BBSs show emergent properties in that their behaviour is governed by the interactions between a number of independent, interacting agents (the KSs) that follow relatively simple rules.

The BBS architecture is modular and is thus highly general and flexible when it comes to solving knowledge-based problems. As the KSs are independent, they can be adapted, added and removed from the system to improve its performance or to adapt it to solve other problems. This is in contrast to rule-based systems, where, for example, removal of a rule might affect the logic of the system leading to poor performance.

The control component and the structure of the blackboard can be designed in many different ways, permitting much freedom for the BBS application developer. BBS applications can also be built incrementally because of the system flexibility.

When designing, the application a developer may thus postpone decisions about, for example, which KSs to include or what control strategy to employ, until the system has reached an appropriate level of experimentation and testing. This is especially important for novice developers, who are learning to build BBS applications.

Another major advantage is the ability of BBSs to handle quite severe limitations in input data and also to handle a wide variety of data types. They are capable of dealing with ambiguity, incomplete data, uncertain and imprecise data and large quantities of data. They can also work with and can integrate different data types, which makes them particularly suited to the solution of problems that require the combination of many separate diagnostic components, dynamic decision-making problems, data fusion problems and problems involving systems of systems. They can also work with data that arrives asynchronously and sporadically [17].

BBSs are also capable of following multiple lines of reasoning concurrently. Once the relationships between hypotheses are established, each can be linked to others that justify or clarify it.

Disadvantages

Unfortunately, the chief advantage of BBSs, flexibility, is directly related to their main disadvantage in that it has led to a lack of formalism in defining them. This drawback is often cited as the major reason for their abandonment following the "AI Winter" (around 1990), where more formal frameworks were generally adopted for dynamic decision-making. The dismissal of less formal methods has had an impact on the perception of BBSs and their place in AI. Notably, BBSs receive only three sentences in Russel and Norvig [24], which is widely regarded as the leading textbook on AI, and in the latest edition [25], there is no mention of them. BBSs were recast as Bayesian Blackboard Systems (BBBSs) in response to this drawback (see [26]).

BBSs have traditionally been used for solving problems that are complex and ill-defined. Erman has suggested that the advantages of BBSs do not scale down to simpler problems [19]. Thus, unless a problem is sufficiently complex to warrant use of a BBS, then it is best to use a simpler architecture. A simpler architecture is also favourable when dynamic decisions do not need to be made or if an application contains only one system. Another well-documented drawback is the difficulty in estimating the value of potential KS actions. This has led to the design of more sophisticated control components for BBSs and also, later, their integration with Bayesian Belief Networks. These developments are discussed further in Sects. 3.3 and 3.4.1.

Although the early BBSs demonstrated that they worked well with complex, dynamic decision problems, they ran into difficulties when the search space became too large [20]. In fact, this was another reason for their decline in the 1990s, as the processors of the time were not powerful enough to deal with the extra computation burden.

1.3.2.6 Definitions

The flexibility that is a key feature of the BBS design has led to some confusion about its definition and some of the terminology surrounding it. This section aims to clarify some of the misunderstandings.

First, the game AI community and some other sources tend to define BBSs as merely a means for sharing common data among subsystems, i.e. the view is that they can be thought of simply as a global database or communication medium. In the seventies, eighties and nineties, this view was very much frowned upon; for example, Corkill states that a system with a global database also requires a set of KSs and a control component to be classed as a BBS [19]. However, as BBSs are now widely used in game development, this definition has become somewhat acceptable. Tuple spaces, a form of distributed shared memory, are also often confused with BBSs (see [27]). They provide a repository for tuples that can be accessed concurrently by, for example, different processors, but there is no requirement for a set of KSs to work together to solve a problem. Tuple spaces are said to use the blackboard metaphor (referred to in this chapter as the blackboard analogy), but the metaphor is, in fact, only partially implemented.

Second, there has been a lack of agreement between authors regarding some of the terminology. To counteract this, Engelmore has listed a set of definitions to distinguish between blackboard systems, models, frameworks, applications, architectures and shells [28]. This chapter adopts most of the definitions in Engelmore with some slight modifications. The terminology used in this chapter is summarised as follows: the term *architecture* is used to refer to the general core design and set of components of the blackboard AI method. The term *method* refers to the procedural details of the algorithm, which effectively represent a computational interpretation of the analogy [29]. The term *system* refers to a particular design implementation; for example, HSII is a blackboard system (BBS) with a specific design for the control component. *Application* is used to refer to the use of a given BBS to solve a particular problem; for example, HSII is an application when it is used to solve the speech recognition problem. *Framework* is used to refer to commercial or academic tools that enable a developer to build a BBS based on, for example, supplied program methods and subroutines. *Shells* are synonymous with *systems* in some sources (e.g. [30]), but the term is not used in this chapter.

1.3.2.7 Problem Types and Applications

This section examines the types of problems that are particularly suited to solution using a BBS. The discussion in the section "Analogy" highlighted the advantages of BBSs and linked the flexibility of allowed input data (e.g. incomplete and uncertain data), the modular design and the incremental solution-building approach with capacity for solving problems that require the combination of many separate diagnostic components. Thus, case-based reasoning and dynamic decision-making problems such as speech recognition, signal understanding, symbolic learning, robot vision, image understanding, structure identification, vehicle monitoring and tracking, robot map creation, fault and disease diagnosis, planning and scheduling, and information fusion problems are all suited to solution using a BBS. For example, a robot building a map of an area may rely upon infrared and sonar sensor data to attempt to trace an image of its environment. Both of these data types are unreliable, and the environment may be changing dynamically. However, a BBS is able to fuse the data from the two input types as it arrives, make sense of the resulting information and reason about it.

In general, any problem with a number of different, interacting components that requires the integration of diverse expertise is a candidate BBS problem. This includes complex problems with very large search spaces and problems that suffer from combinatorial explosion, as well as ill-defined and poorly structured problems, for example problems where the goals are not clear or where the solution path from the initial state to the goal state is highly irregular [21]. Cooperating agent systems and distributed and parallel AI problems are also good candidates for solution with a BBS. It is the integration of several different KSs that makes BBSs applicable to the solution of a wider variety of problem types. Most other symbolic AI methods, for example production systems, can only deal with knowledge about a single problem domain. This tends to limit them to the solution of diagnosis-type problems, which are much simpler than interpretation-type problems such as speech recognition. The reasons for this are discussed further in the section "Combination with Bayesian Belief Networks". Furthermore, many production systems are only capable of working offline; they cannot handle the arrival of new information on the fly, whereas BBSs have this capability.

It is also possible to solve problems with hard, real-time constraints (i.e. problems with deadlines for the solution) using the BBS model (see [31]), although more sophisticated and predictable control strategies are needed to allow accurate estimation of action durations, as the computation time and use of other resources need to be considered. The system also needs a mechanism for assessing the trade-offs between cost and effectiveness of actions. This requires it to be able to predict resource usage [20].

The literature review provides examples of BBSs solving some of the problem types listed in this subsection. It is presented in chronological order so that it also acts as a guide to the "History and Evolution" of BBSs as well.

1.3.2.8 BBS Frameworks and Toolkits

There was very little software to support the development of BBSs for a long time after their conception. Thus, initially, many researchers and application developers had to build them from scratch [19]. Hearsay-III (see [32]) was developed in response to the demand for a domain-independent framework upon which to build a BBS. The framework, which sat on top of a relational database called AP3, was used to develop many other systems including a BBS that was applied to a crisis management task (see Hayes-Roth et al. [33]). Another early development toolkit was Corkill's GBB framework (see [34], which was used to build a satellite control system for the Canadian Space Agency. It was also used for logistics planning in the US Army and for design engineering projects at Ford. Many toolkits are now available commercially, for example BEST (Blackboard-based Expert System Toolkit), designed by the Mihailo Pupin Institute. BEST, which is designed to run in the Windows environment, is implemented using C++ and Arity Prolog and uses the MEKON inference engine. Many open-source resources are also available to assist developers; for example, the source code, a toolkit and a software development kit for the computer game "No One Lives Forever 2" (NOLF2, developed by Monolith Productions), which implemented a BBS architecture, are available (see [35]).

1.3.3 Literature Review

BBSs began with the Hearsay-II (HSII) speech recognition system built at Carnegie Mellon University in the early 1970s, although the notion of an "artificial blackboard" being used to share ideas to solve a problem was first conceived by Allen Newell in 1962 (see [36]). The success of the Hearsay-II project inspired many other researches to adopt the blackboard architecture for applications such as signal interpretation, for example in the HASP/SIAP systems [15], and planning [37]. The model became very popular throughout the 1980s for solving complex AI problems, both with academic researchers and also in industry where it was employed by, for example, Cambridge Consultants Ltd. as the BLOBS system to implement reasoning about time-dependent data [38] and as the MUSE system to solve real-time problems [39]. Fujitsu in Japan also developed a BBS called ESHELL for problem diagnosis in cranes, as well other applications (see [28]).

There has been considerable development of the architecture since the early systems. Many of the initial enhancements concentrated upon modifying the control component, and there was a general move away from data-based control towards more goal-oriented control [20]. Other researchers focused on extending the architecture to distributed applications using, for example, networks of BBSs [40]. Parallel BBSs were also investigated in [41] and by Bell Laboratories (see [42]). BBSs became very scarce in the academic literature following the early 1990s. This has largely been attributed to their lack of formalism, but may also be related to the move of many of the early BBS proponents and researches from academia to industry. BBSs made a reappearance in the literature from around the year 2000 when several researchers attempted to merge the architecture with Bayesian Belief Networks (see, e.g. [16]) in an attempt to give it more rigour. Although academic papers about BBSs remain fairly scarce, the architecture continues to be very popular today in commerce and industry, both in the Bayesian hybrid format and in formats that are very close to the original architecture. For example, they are widely used by the computer gaming industry to control non-player character

(NPC) behaviour (see [35]), are used by Adobe to recognise text and have been used to manage satellite operation in Canada.

1.3.3.1 Prehistory

Allen Newell was the first person to use the term “blackboard” in AI literature [36]. In his 1962 paper about organisational problems in programs such as chess-playing programs, he refers to a set of workers looking at the same blackboard, each capable of reading what is on it, writing to it and judging when they have something worthwhile to contribute to it. He likens the situation to Selfridge’s Pandemonium [43], where a set of demons shriek with varying levels of loudness in response to what they see. He goes on to discuss ways to organise the synthesis of complex processes using hierarchically organised subroutines [44]. These early ideas eventually led to the development of production systems with preconditions and actions. Interestingly, the term “demons” is still used to describe the set of rules with satisfied preconditions in production systems.

The term “blackboard” was mentioned again by Herbert Simon in 1966 in an article that was later published in 1977 [45]. In this article, the information generated about problem-solving that was fixed in permanent memory was referred to as the “blackboard”. The article also talks about the creation of subgoals and the use of a hierarchy of goals and subgoals to achieve the original overarching goal. Simon suggested his ideas about blackboards to Raj Reddy and Lee Erman, when they were preparing for the Hearsay project, although many of the ideas that eventually emerged from Hearsay were centred around the needs of the application, i.e. speech understanding [44]. Thus, most of the core properties and components of blackboard systems, such as opportunistic problem-solving, different levels of abstraction and KS collaboration, were derived directly from those needs.

1.3.3.2 The First BBS—Hearsay-II

This section looks in detail at the system that is widely regarded as the first BBS, Hearsay-II (HSII). HSII was created for speech recognition, evolving directly from Hearsay-I (HSI) (see [46]). HSI was a prototype blackboard system, but is probably not cited as the first BBS because there was no dynamic control component and also because it did not work well. In HSI, information sharing among the KSs was carried out only at the word level, so it was difficult to add non-word KSs and determine the value of their hypotheses [47]. KSs were also activated using a hypothesise-and-test paradigm, rather than by a control component that dynamically selected the most appropriate KS.

The most popular reference for HSII is [14], which describes the final developed BBS. Earlier incarnations of HSII implemented a data-directed control approach, where KSs were activated in response to events on the blackboard if their preconditions were satisfied, and all KSs were checked for this. The data-driven approach tries to answer the question "what should the system do given the available data?"

The final HSII system (described in [14]) has an agenda-based control mechanism referred to as the *scheduler*. Rather than checking the preconditions of all the KSs, which can be time-consuming, this incarnation of HSII categorises events into types and the KSs provide lists of the event types they are interested in. The scheduler then needs only to check the preconditions of KSs with event types that match the given blackboard event. KSs with matching types and satisfied preconditions are instantiated, are referred to as Knowledge Source Instantiations (KSIs) and are placed on the agenda. Note that some sources refer to Knowledge Source Activation Records (KSARs) rather than KSIs, but the terms are interchangeable. Once instantiated, the scheduler has access to information about the action each KSI would execute and the likely changes this would make to the blackboard. The agenda thus consists of all the actions that the system could possibly take next. The scheduler chooses the best KSI action based upon ratings of its contribution and cost, i.e. there is an estimate of how much progress it is likely to make towards solving the problem and another estimate of computational cost; the winning KS is then removed from the agenda. The cycle repeats until a different event takes place or there are no more KSIs on the agenda. The scheduler's rating calculation is a weighted linear function of several variables and is known as the expected value of the KSI; the result of the calculation tells the system which particular line of reasoning it is best to pursue; i.e., the hypothesis with the maximum expected value is the one considered worthy of further investigation.

The blackboard in HSII consists of a hierarchy of levels; for example, there is phrase level, a word level and a syllable level. There are also classes for each hypothesis associated with each level, and the levels contain a set of dimensions so that information can easily be retrieved when needed.

The KSs consist of acoustic, lexical, syntactical and semantic reasoning systems [17]. These KSs examine the blackboard for information that they can work with, for example hypotheses about adjacent words. When activated, a KSI posts a new hypothesis onto the blackboard. For example, a KSI may use the rules of grammar to generate words likely to appear next in a phrase or sentence, and another may actually detect words directly from the source. When a KS posts a hypothesis, which could be a partial solution, it then attempts to verify it in a test stage, where it may be refined. The entire process represents a search for a hypothesis that explains the data at each level of abstraction [17].

ARPA funded the speech recognition umbrella research program of which Hearsay was a part, and at the end of the project, the various systems that emerged were analysed and compared [29]. HARPY (see [48]), which employed a Markov algorithm to perform its analysis, was deemed the best in terms of performance, although it was not as flexible as HSII. Control was generally considered a problem in HSII because the decisions were based upon the scheduler's rating of the local and immediate effects of KSI actions. Using such a limited expected value, function can lead to poor performance because actions are not independent of one another; indeed, they can have very complex interrelationships. A more accurate expected value would depend upon when the actions were executed and the next actions in the sequence. Thus, in HSII, the global effects of actions and the long-term state of the blackboard had to be abstracted by using models of the intermediate state to predict them. This was a way around the problem, but what was really needed was a method for linking potential actions and goals. HSII struggled because it was forced to deal with uncertainty about whether a given hypothesis was part of a solution and also with uncertainty about the expected values of actions. This led the HSII team and other researchers to enhance the control mechanism in HSII; effective control is especially important when uncertainty of the input data and problem-solving knowledge is high [20].

1.3.3.3 Development of Control Mechanisms

The BBSs that followed HSII tended to adapt and improve the control component in some way because of the flaws in the original design. In addition, HSII was only capable of handling a single input phrase and could not deal with strict scheduling deadlines. It thus became necessary to create more elaborate control mechanisms to address these problems since multiple inputs could potentially overload the blackboard and the agenda, and predictable control structures were needed for hard time constraints to allow accurate estimation of action durations.

In general, there was a move away from implicit representation of goals, as in HSII, towards more explicit representation of goals and their relationships to the overarching goal, so that more efficient and sophisticated goal-oriented control mechanisms could be implemented [20]. Thus, goal-directed reasoning, which can be defined as reducing a problem from a set of abstract high-level goals into more detailed low-level subgoals and planning, was introduced to BBSs. In contrast to the data-driven approach, the goal-directed approach tries to answer the question "what should the system do to solve the problem?" This is essentially achieved by identifying sequences of actions capable of satisfying goals and subgoals. The system terminates when all the goals have been achieved, and the resulting solution is deemed acceptable in some way. There may be other solutions that meet the constraints of the input data, but the system must be capable of selecting the "best" one. Thus, many different search strategies can be employed, for example a focused depth-first search that pursues a particular solution if it is preferred, or a more exhaustive breadth-first search that pursues all potential solutions until a particular one is favoured.

The control mechanisms that evolved from HSII began to use methods for making more accurate predictions about the long-term effects and global value of an action by representing goals in a more detailed and explicit way. They also began to limit the rerating of KSIs by restricting the number on the agenda to those more likely to be executed, increasing the efficiency of the algorithm. This subsection chronicles the development of the control component from its agenda-based approach in HSII through to the event-based approach seen in HASP/SIAP [15], the hierarchical approach used in CRYSALIS [49, 50] and the goal-directed architecture of DVMT [40], which also included incremental planning. The BB1 [22] and RESUN [51] and Carver and Lesser [52] control mechanisms are also discussed as further extensions to the goal-directed approach.

Event-based control in HASP/SIAP
The HASP/SIAP BBS (see [15, 53]) was built to interpret sonar signals and identify the ships and submarines that produced them. The control architecture extended that of HSII by specifying the KS preconditions as predefined event types, so that, in effect, preconditions and event types were merged. Thus, as soon as an event occurred, the activated KSs were immediately known making the control algorithm much more efficient. The system used blackboard events, clock events (a time and a set of events expected to occur at that time), expectation events (events expected to occur in the future) and problem events (e.g. missing information). HASP also used a limited hierarchical control structure to distinguish domain knowledge from knowledge about its application; this was a precursor to the hierarchical control structure used in CRYSALIS.

The main disadvantage of the modified control mechanism in HASP was that opportunism in the system became more limited because the predefined event types governed the KS sequence. In HSII, the KSs were triggered in an ad hoc fashion by general events.

Hierarchical control in CRYSALIS
CRYSALIS (see Engelmore and Terry 1979; [49]) was used for protein crystallography. Its control mechanism was built on a hierarchy of control knowledge sources (CKSs) that were tasked with selecting the domain knowledge source (DKS) to be executed. There were two levels of CKS, a single strategy CKS and a set of task CKSs corresponding to the system subgoals. The strategy CKS selected a sequence of task CKSs for execution, which in turn selected a sequence of DKSs for execution. This technique eliminated the need for KS preconditions. The strategy CKS looked at the current blackboard hypotheses and made a decision about where to focus the problem-solving, i.e. determined which subgoals had the maximum expected value and should be pursued next; it thus provided coarse focus for the problem-solving. Its selection led to the sequential execution of a set of task CKSs, which provided the fine focus. Each task CKS in the sequence examined the conditions on the event list and selected appropriate DKSs for sequential execution. As in the core BBS architecture, the DKSs examined the blackboard in order to add or change hypotheses.

The opportunism was reduced in CRYSALIS as in HASP because there was no method to change focus once a path forward was decided. In HSII, switching between paths was simple. Moreover, HSII could pursue multiple lines of reasoning concurrently without a need for backtracking to a previous one. In contrast, CRYSALIS could follow only one line of reasoning at a time and thus had to backtrack when required. However, this was not a major problem for CRYSALIS because it was not designed to solve real-time problems. When researchers began to require the solution of such problems using BBSs, a different approach to control was necessary.

Goal-directed control in DVMT

A goal-directed approach to control was first used for the application of distributed vehicle monitoring in the DVMT BBS (see [40]). This system inserted a goal blackboard and goal processor (for creating goals on the goal blackboard) into the core BBS architecture. The goal processor employed three mapping functions: hypothesis-to-goal, goal-to-subgoal and goal-to-KS. Data-directed goals were created on the goal blackboard following the addition or amendment of a hypothesis on the domain blackboard and use of the hypothesis-to-goal mapping. Goal-directed subgoals were created on the goal blackboard after the creation of others via the goal-to-subgoal mapping. KSs were selected to have their preconditions checked in response to the insertion of a goal on the goal blackboard and use of the goal-to-KS mapping. The resulting KSI ratings incorporated both a data-directed and goal-directed element by using information about the super-goal, the hypothesis that gave rise to the goal and the level of the blackboard. DVMT succeeded in representing explicit goals and the global effects of actions by linking local effects with higher-level goals.

Control in BB1

In 1986, Hayes-Roth published work on the application of a BBS to solving arrangement-assembly problems [22]. The task was to arrange a set of given objects such that a given set of constraints was satisfied. The BBS with domain-independent knowledge about arrangement assembly was named ACCORD, and several other BBSs were also created based on the same design but with focus in a particular domain; for example, the PROTEAN system was tailored to protein-structure analysis and the SIGHTPLAN system was used for designing construction site layouts. Both of these systems made use of ACCORD as a KS. Collectively, the set of assembly-arrangement BBSs were known BB1 systems.

BB1 modified the original HSII control architecture so that additional KSs were used to build the control plan for the system's behaviour. In BB1's control architecture, the domain problem and the control problem were both solved using the blackboard model. The architecture used CKSs as in CRYSALIS but also inserted a control blackboard. It implemented an agenda-based approach to solving the control problem as in HSII, but introduced a more complex control planning method where the CKSs incrementally developed control plans on the control blackboard. The overall architecture can thus be thought of as "blackboard within a blackboard". BB1 worked at three levels of abstraction: the strategy (long-term plans), the focus (goal) and the heuristic (rating function). The long-term plans were reduced to sequences of substrategies, which in turn were reduced to foci, a set of goals. Each focus had an associated set of heuristics that were used to rate potential KSI actions that matched the focus. These could be changed dynamically to suit different problem-solving stages. KSIs from both the DKSs and CKSs were placed on the same agenda and rated using the same function.

In addition to a more sophisticated control system, BB1 also introduced learning KSs to modify facts in the KS knowledge bases and provided an additional capability for explaining its actions. On each solution cycle, and in response to user requests, it was capable of providing information about how the selected actions matched with the control plan.

The control architecture of BB1 did not compromise the opportunism of the system as in CRYSALIS and HASP, i.e. actions and plans were both implemented opportunistically. However, although BB1 succeeded in implementing the determination of high-level, long-term goals, it did not carry out goal decomposition.

Control in RESUN

Planning represents a search for the best solution and a search for the best way to find it. The RESUN BBS (see [51, 52]) extended the HSII BBS by replacing the agenda strategy and engineered complex rating functions with an incremental control planner that simply carried out a number of less complex searches to make the best decisions. It was also able to preserve the core BBS opportunistic properties as it incorporated an additional refocusing mechanism that permitted a posteriori changes to the planner focusing decisions. This enabled postponement of decisions when there was insufficient information about a plan, or when two plans could not be rated against each other without sufficient refinement. Refocusing was allowed at both a data/event level and at a planning/hypothesis level so that the system could switch focus in response to developing plans, new hypotheses and input data.

RESUN enabled better resolution of uncertainty in the hypotheses. It worked with detailed information about the uncertainty of hypotheses and their alternatives and used explicit statements about the sources of the uncertainty called source of uncertainty statements (SOUs). The SOUs were attached to hypotheses so that goals could be generated to eliminate the uncertainty.

RESUN was the most goal-directed control approach of all the BBSs discussed in these subsections. Most of the others simply added goal-directed methods to the agenda method. RESUN started with the goal-directed approach and used focusing to allow data-directed control when necessary for opportunism.

1.3.3.4 More Recent Developments

As mentioned at the start of this section, following the "AI Winter" around 1990, documentation of BBSs began to decline in the academic literature, largely due to a lack of formal underpinnings for belief in and decisions about actions. This meant that the systems could only be assessed empirically. Although they remained popular in industry, academic researchers were beginning to prefer non-symbolic (modern AI) approaches to problem-solving or more rigorous statistics-based approaches. In addition, there was a perception that BBSs were large and unwieldy requiring vast quantities of code to manage them and their complex data structures [18]. Another problem was that the initial successes documented in the literature did not scale well to larger dimensioned problems because the computers in use at the time did not have the storage or processing power required. Most of the problems that had been solved with BBSs thus far were NP hard, which meant that solutions were not tractable in the large limit at the time. For example, in the speech recognition field, many researchers began to work with simple Hidden Markov models as they could outperform BBSs like HSII. BBSs were thus recast as Bayesian Blackboard Systems (BBBSs) in an attempt to define them in modern AI concepts [26].

During the "AI Winter" some researchers, no doubt dismayed by the sudden demise of BBSs, undertook work to begin to formalise BBSs. In 1991, Craig provided a relatively complete mathematical specification for a sample interpreter BBSs using the Z specification language [see 54]. Velthuijsen used a similar approach in 1992 but applied the specification to a number of concurrent BBSs using the CCS language [55]. Craig then extended his 1991 work to provide a formal account of BBSs that showed that control information could be derived and represented in temporal logic [56]. Following this, the interpretation problem was formalised by Whitehair (see [57]) and his paper also analysed BBS systems when applied to the problem. The paper demonstrated that it was possible to develop models for opportunistic AI architectures like BBSs.

Combination with Bayesian Belief Networks

A BBBS closely resembles a traditional BBS, with KSs, a blackboard and a control component. However, in a BBBS, the role of the KSs is to modify Bayesian Belief Networks (BBNs) on the blackboard rather than direct hypotheses (see [17]). The Bayesian component provides the hybrid architecture with a foundation in probability theory, validating the underlying reasoning, i.e. it allows probabilistic models (that can reason about uncertainty) to be built rather than symbolic ones. In addition, the integration not only formalises the BBS but extends traditional belief-based systems by allowing them to be built incrementally. However, Carver states that some of the flexibility and opportunism is lost in interpretation BBBSs that work in time slices as they have to compute exact probabilities for all interpretations of new data with no ability to search selectively [26].

A belief network is a directed graph where the set of nodes represents propositions with associated probability distributions (PDs), some of which may be conditional (dependent on the PDs of the nodes pointed to). Nodes with unconditional probabilities are termed evidence nodes. The conditional PDs are stored in tables known as conditional probability tables (CPTs).

One of the first attempts at integrating BBSs with BBNs was the AIID system (Architecture for the Interpretation of Intelligence Data), which was developed for information fusion in military scenarios (see [17]). The problem was to infer an enemy unit's strategy given military intelligence, and it required the system to be able to cope with heterogeneous data (of varying precision and reliability) arriving sporadically from many different sources. Thus, it was necessary to be able to understand the meaning of data when it arrived. This is not possible using a simple data-driven approach as the search space is too large. A method for reasoning about uncertainty was also needed. Traditional BBSs such as HSII dealt with uncertainty implicitly and heuristically via the scheduler rating function. However, in AIID, each hypothesis was directly associated with a probabilistic uncertainty.

In AIID, the data corresponds to evidence with conditional probability information, and the belief network is dynamically created and grown as data are received and processed. The network consists of nodes that have a type and set of arguments that identify it, with nodes being associated with hypotheses, previous observations or background knowledge. In addition to preconditions and actions, the KSs have a confidence property that embodies their usefulness. When KSIs are activated, they alter the network in some way; for example, they may post new nodes to the blackboard, remove existing nodes, add edges or change conditional probabilities. The KSs themselves can also exist as belief networks that represent small fragments of knowledge. When a node in the fragment matches a node on the blackboard, i.e. their types and arguments match, a KS becomes a candidate KS and may post its knowledge fragment on the blackboard. After being posted, a knowledge fragment's two conditional PDs (the one from the KS knowledge fragment and the one on the blackboard) are combined. The evidence combination methods vary between BBBSs. As in traditional BBSs, only one KS is run at a time to maintain tractability, but the difference is that control is maintained by computing the expected utility of an action, given the available evidence.

BBBSs arose from a need to improve existing methods for solving complex interpretation problems (such as vehicle tracking, robot map making and speech understanding) as well as a desire to formalise BBSs. These problems are inherently much harder to solve than diagnosis problems as an interpretation requires that instances of input data types are explained in terms of hypotheses about events that might have given rise to them [26]. Moreover, there are multiple possible interpretations for a given set of input data, as there are many different instances of each type and many uncertainties because of data noise. The solution space can become exponentially large for high-dimensioned problems. A system that is able to reason about the validity of evidence for alternative hypotheses can thus make better judgements regarding which solution path to follow. Diagnosis problems do not share the same complexity as there is a single, fixed set of hypotheses, for example possible faults in engine fault diagnosis. Furthermore, the relations between the data and the hypotheses are known, and complete, static probability models are easily constructed. Interpretation problems, on the other hand, lack a model for connecting data instances with hypotheses. This phenomenon is known as the data association problem (DAP). BBBNs that attempt to solve interpretation problems can thus only work with estimates of conditional probabilities, and when performing evidence propagation, the solutions can only approximate the optimal [26].

Other recent work

Culliton describes the hypothetical use of a BBS to coordinate intelligent units in a combat computer game [58], referring to BBSs as "the perfect system to use", although he states that, initially, the complexity of the architecture prohibited its use in game design because of limited time, resources and budgets. However, there is documentation of the use of a BBS for coordinated behaviour in the game NOLF2, developed by Monolith Productions in 2002 (see [35]). He states that a BBS was used to handle coordination with respect to the timing of behaviours, pathfinding and tactics. For example, the architecture solved problems such as preventing duplication and repetition of behaviour in agents. Contrary to the belief that BBSs are "code-heavy", Orkin stresses that use of a BBS allows a reduction in code volume, is simple to implement and is flexible and maintainable. Furthermore, he asserts that it simplifies the agent architecture, permits reuse and sharing and allows complex reasoning to take place. This conflict of opinion may have arisen because Orkin implemented a more simplified version of the classic BBS, whereas Culliton was discussing the original, more complex architecture. Indeed, Dill reports that the term "blackboard architecture" is used differently in game AI than in the academic literature [59], often being used merely to describe shared memory space that AI components can use to store knowledge. He cites line-of-site (LOS) checks, path-planning checks and the coordination of AI components as common uses for BBSs in game AI, emphasising that the latter may require the more complex, classic use of BBSs, i.e. independent KSs posting partial solutions to a problem on the blackboard rather than merely using it as shared memory space.

Millington and Funge on the subject of AI for games, reporting that this method has been used extensively by game programmers as a mechanism for coordinating the actions of several independent decision-makers [18]. Their description of BBSs implies that many systems developed for games follow a more simplified form of the more classical architecture as seen is HSII, rather than the modernised Bayesian hybrid architecture. This makes sense as there is generally a less rigorous requirement in game AI; the purpose behind it is to create believable behaviours rather than to solve complex problems with large search spaces. Millington and Funge cite a typical example use of BBSs in a computer game, ballistics planning, where three different AI systems work with a blackboard to fire at enemy tanks. There is a route planning subsystem, a target selection subsystem and a ballistics calculation subsystem. Running each system sequentially is not efficient as the game environment changes rapidly, and this approach does not allow information to flow back in the opposite direction. There is a need for all of the AI subsystems to communicate freely without having to set up individual communication channels. The solution to this problem is to use the blackboard architecture. Suggested actions are written to the blackboard by the subsystems where they are stored along with agreement flags. Actions are only executed if there is full agreement between relevant subsystems. For example, "fire at tank X" would have an agreement slot for the ballistics subsystem. The ballistics subsystem could agree, disagree or even remove the suggested action from the blackboard. It could also post a new suggested action, for example "move into firing position for tank X", leaving the original proposition still on the blackboard, but deferring agreement until the correct position had been reached. This example strongly suggests the game AI definition of BBSs, i.e. the shared data paradigm. Champandard also discusses the use of BBSs for computer game behaviour coordination [60], citing their main advantage in game AI as their modularity; as the various systems that need coordinating are independent, they only interact by exchanging information on the blackboard. Champandard assets that this reduces the coupling between them, making the coding structure much more straightforward. Another article that provides insight into the use of BBSs for game AI is David Mark's account of an interview with Damian Isla, a developer at the MIT Media Laboratory who worked on games such as Halo 2, released in 2004 (see [61]). Mark reports that Isla views BBSs as architectural constructs for decoupling information gathering and storage from the decision-making process; multiple decision-making subsystems can work with the same information simply by looking at the blackboard; for example, the threat level of an enemy character can be calculated once and saved for all the subsystems to work with. Isla also likened the blackboard to a unified interface for all game data, storing the data in a contextually significant way so that relevant subsystems know what it means. This is an allusion to the definition of BBSs in the game AI sense of a shared data source for different decision-making components.

Aside from game AI, there have been other recent uses of BBSs. For example, Khosravi and Kabir used a classic BBS integrated with offline ANN training for optical character recognition (OCR) of typed text in the Farsi language [62]. In this BBS, some of the KSs that were used were generated offline a priori as a training exercise. These took the form of multilayer perceptrons (MLPs), a type of ANN. Other KSs, mostly classifiers boosted by the MLP training, ran online. In addition, some KSs were static and some were changed dynamically during run-time. The control component used confidence values that were usually taken directly from the classifier outputs to rate the value of potential actions on a scale 0 through 100. As an example, one KS was a segmentation and recognition module with the task of breaking words down into individual characters and then recognising them. Another KS was a vocabulary "expert" containing 55,000 words. Its task was to match recognised words with words in its database. If a recognised word did not exist (e.g. because of misclassification of characters), then it would find the word most similar to the recognised one. The system also used tools to help with the text recognition, for example a spellchecker and a line detector. The system was capable of recognising 10 popular Farsi fonts and was tested on 20 real-life documents producing a recognition rate of about 97% at the word level and about 99% at the character level.

Fox provided an interesting extension to BBBSs in 2012 by using a hierarchical Bayesian Blackboard integrated with a Metropolis–Hastings algorithm and a BBN for map building with a whiskered robot known as CrunchBot [16]. The system was required to process very large quantities of sparse sensory information in order that CrunchBot could recognise table-like objects (such as tables, chairs, desks) in its environment, using only four whisker-like tactile sensors. The solution architecture was composed of a hierarchical BBS that implemented hypothesis priming and pruning heuristics integrated with a Metropolis–Hastings algorithm and a Monte Carlo Markov chain (MCMC) sampling Bayesian network. The observations for the MCMC were the position and orientation data (with respect to the contact surfaces) from the whisker sensors. This information was fused with information about the pose of the robot and hierarchical models relating to furniture objects (e.g. the recognition of a table leg object infers the presence of a complete table object) in order to make inferences about the objects encountered and thus build the map. Each time step was treated as independent inference problem.

The problem tackled was difficult as many incorrect hypotheses can arise when sparse data are used, and there was often not enough information about the furniture objects to resolve ambiguities. The authors previously tried to solve the same mapping problem using particle filtering techniques, but this did not produce the same level of success as the BBBN system.

1.3.4 Summary

BBSs have been used extensively for solving a wide variety of complex, uncertain, real-time, dynamic and ill-defined problems, and over the many years since their conception, they have produced some excellent results. Their problem-solving power lies in their flexibility, modularity, incremental solution-building approach, and their ability to handle imperfect data, large quantities of data and data arriving sporadically and/or asynchronously. They have proved to be a very valuable tool in domains that require complex, multidimensional searches, for example complex scheduling and interpretation problems. Moreover, the optimal solution of such problems remains an open research question, which means that there is ample scope and opportunity for further research into BBSs, with many different potential design choices for developers. Although the blackboard architecture is not as popular with the academic community as it was in its early days, it remains an excellent vehicle for interesting research projects with a lot to offer for academics, industry and the game AI community. Symbolic methods may have declined, but the BBS, even in its core incarnation, is a hybrid architecture capable of incorporating elements of both modern and classic AI approaches. A recurrent theme throughout the BBS literature is that the control of the system is at least as important as the domain knowledge if the system is to be useful, effective and efficient.

References

1. Pearl J (1998) Probabilistic reasoning in intelligent systems: networks of plausible inference. Morgan Kaufmann Publishers Inc., San Francisco
2. Trovati M, Bessis N (2015) An influence assessment method based on co-occurrence for topologically reduced sets. Soft Comput 20(5):2021–2030
3. Sanchez-Graillot O, Poesio M (2004) Acquiring from text. LREC
4. Kuipers BJ (1984) Causal reasoning in medicine: analysis of a protocol. Cogn Sci 8:363–385
5. Trovati M (2016) An overview of some theoretical topological aspects of big data, Big-Data analytics and cloud computing, theory, algorithms and applications, computer communications and networks, Springer
6. Liddy ED (2001) A robust risk minimization based named entity recognition system. In: Encyclopedia of library and information science. Marcel Decker, Inc., New York

7. Laporte E (2005) Symbolic natural language processing. In: Lothaire (ed) Applied combinatorics on words, pp 164–209
8. Manning CD, Schutze H (1999) Foundations of statistical natural language processing. MIT Press, Cambridge
9. Troussov A, Levner E, Bogdan C, Judge J, Botvich D (2010) Spreading activation methods. In: Dynamic and advanced data mining for progressing technological development: innovations and systemic approaches, pp 136–167
10. Dale R, Moisl H, Somers HL (2000) Handbook of natural language processing. Marcel Dekker, Inc., New York
11. Korhonen AYK (2006) A large subcategorisation lexicon for natural language processing applications. In: Proceedings of LREC
12. Stumme G (1998) Efficient data mining based on formal concept analysis. Lecture Notes in Computer. Springer, New York
13. Wilks Y, Stevenson M (1998) The grammar of sense: using part-of-speech tags as a first step in semantic disambiguation. Nat Lang Eng 4:135–143
14. Erman LD, Hayes-Roth F, Lesser VR, Reddy DR (1980) The Hearsay-II speech-understanding system: integrating knowledge to resolve uncertainty. ACM Comput Surv 12(2):213–253
15. Nii HP, Feigenbaum EA, Anton JJ, Rockmore AJ (1982) Signal-to-symbol transformation: HASP/SIAP case study. AI Mag 3:23–35
16. Fox CW, Evans MH, Pearson MJ, Prescott TJ (2012) Towards hierarchical blackboard mapping on a whiskered robot. Robot Auton Syst 60(11):1356–1366
17. Sutton C, Morrison C, Cohen PR, Moody J, Adibi J (2004) A Bayesian blackboard for information fusion. In: Svensson P, Schubert J (eds) Proceedings of the seventh international conference on information fusion, pp 1111–1116
18. Millington I, Funge J (2009) Artificial intelligence for games, 2nd edn. CRC Press, Boca Raton, pp 459–466
19. Corkill DD (1991) Blackboard systems. AI Expert 6(9):40–47
20. Carver N, Lesser V (1994) Evolution of blackboard control architectures. Expert Syst Appl 7:1–30
21. Pang GK-H (2009) Blackboard architecture for intelligent control. In: Unbehauen H (ed) Control systems, robotics and automation: and intelligent control systems, vol 17. EOLSS, Oxford, pp 303–316
22. Hayes-Roth B, Johnson V, Garvey A, Hewett M (1986) Application of the BB1 blackboard control architecture to arrangement-assembly tasks. Artif Intell 1(2):85–94
23. Newell A, Simon HA (1972) Human problem solving. Prentice-Hall, Englewood Cliffs
24. Russell S, Norvig P (1995) Artificial intelligence: a modern approach. Prentice-Hall, New Jersey
25. Russell S, Norvig P (2016) Artificial intelligence: a modern approach, 3rd edn. Pearson, Harlow
26. Carver N (1997) A revisionist view of blackboard systems, In: Proceedings of the 1997 midwest artificial intelligence and cognitive science society conference. The AAAI Press, Dayton, Ohio
27. Gelenter D (1983) Generative communication in Linda. ACM Trans Program Lang Syst 7(1): 80–112
28. Engelmore RS, Morgan AJ, Nii HP (1988a) Introduction. In: Engelmore R, Morgan T (eds) Blackboard systems. Addison-Wesley, Boston, pp 1–22
29. Craig ID (1995) Blackboard systems. Ablex Publishing Corporation, Norwood, NJ
30. Jones J, Millington M, Ross P (1986) A blackboard shell in PROLOG. In: Proceedings ECAI-86, pp 428–436
31. Dodhiawala RT, Sridharam N, Pickering C (1989) A real-time blackboard architecture. In: Jagannathan V, Dodhiawala R, Baum L (eds) Blackboard architectures and applications. Academic Press Inc., San Diego

32. Balzar R, Erman L, London P, Williams C (1980) HEARSAY-III: a domain-independent framework for expert systems. In: Proceedings of the first annual conference on artificial intelligence, pp 108–110
33. Hayes-Roth F, Waterman DA, Lenat DB (1983) Building expert systems. Addison-Wesley, Reading
34. Corkill DD, Gallagher KQ, Murray KE (1986) GBB: a generic blackboard development system. In: Proceedings of AAAI-86, pp 1008–1014
35. Orkin J (2003) Applying blackboard systems to first person shooters. [online] slidepayer.com. Available at: web.media.mit.edu/~jorkin/utgameAI03-Orkin.ppt and http://slideplayer.com/slide/6102412/. Accessed 16 Apr 2016
36. Newell A (1962) Some problems of the basic organization in problem-solving programs. In: Yovits M, Jacobi G, Goldstein G (eds) Proceedings of the second conference of self-organising systems, pp 393–423
37. Hayes-Roth B, Hayes-Roth F, Rosenschein S, Cammarata S (1979) Modelling planning as an incremental, oppotunistic process. In: Proceedings IJCAI-79, pp 375–383
38. Zanconato (1988) BLOBS—an object-oriented blackboard system framework for reasoning in time. In: Engelmore R, Morgan T (eds) Blackboard systems. Addison-Wesley. Reading, pp 335–345
39. Reynolds D (1988) MUSE: A toolkit for embedded, real-time AI. In: Engelmore R, Morgan T (eds) Blackboard systems. Addison-Wesley, Reading, pp 519–532
40. Lesser VR, Corkill DD (1983) The distributed vehicle monitoring testbed: a tool for investigating distributed problem solving networks. AI Mag 4(3):15–33
41. Nii HP (1986b) CAGE and POLIGON: two frameworks for blackboard-based concurrent problem solving. Technical Report KSL-86-41. Stanford University, Stanford
42. Ensor JR, Gabbe JD (1986) Transactional blackboards. Int J Artif Intell Eng 1(2):80–84
43. Selfridge O (1959) Pandemonium: a paradigm for learning. In: Proceedings of symposium on the mechanisation of thought processes, pp 511–529
44. Nii HP (1986) Blackboard systems: the blackboard model of problem solving and the evolution of blackboard architectures. AI Mag 7(2):38
45. Simon HA (1977) Scientific discovery and the psychology of problem solving. In: Models of discovery, Reidel, Boston
46. Reddy DR, Erman LD, Neely RB (1973) A model and a system for machine recognition of speech. IEEE Trans Audio Electro Acoust AU-21:229–238
47. Engelmore RS, Morgan AJ, Nii HP (1988) Hearsay-II. In: Engelmore R, Morgan T (eds) Blackboard systems. Addison-Wesley, Reading, pp 25–29
48. Lowerre BT, Reddy R (1980) The HARPY speech understanding system. In: Lea W (ed) Trends in speech recognition. Prentice-Hall, Englewood Cliffs
49. Terry A (1988) Using explicit strategic knowledge to control expert systems. In: Engelmore R, Morgan T (eds) Blackboard systems. Addison-Wesley, Reading, pp 159–188
50. Englemore RS, Terry A (1979) Structure and function of the CRYSALIS system. In: Proceedings of IJCAI-79, pp 250–256
51. Carver N (1990) Sophisticated control for interpretation: planning to resolve sources of uncertainty. Ph.D. Thesis. University of Massachusetts, Computer and Information Science Department, Amherst
52. Carver N, Lesser V (1990) Control for interpretation: planning to resolve sources of uncertainty. Technical Report No. 90-53. University of Massachusetts, Computer and Information Science Department, Amherst, MA
53. Feigenbaum EA, Nii HP (1978) Rule-based understanding of signals. In: Waterman D, Hayes-Roth F (eds) Pattern-directed inference systems. Academic Press, New York
54. Craig ID (1991) Formal specification of advanced AI architectures. Ellis Horwood, Chichester
55. Velthuijsen H (1992) The nature and applicability of the blackboard architecture. Ph. D. Thesis. Faculty of General Science, Limburg University, Maastricht
56. Craig ID (1993) Formal techniques in the development of blackboard systems. Int J Pattern Recogn Artif Intell 7(2):197–219

57. Whitehair R (1996) A framework for the analysis of sophisticated control. Ph. D. Thesis. University of Massachusetts. Computer Science Department
58. Culliton P (2003) Implementing a blackboard-like system for squad-level combat AI Part I. [online] GameDev.net. Available at: http://www.gamedev.net/page/resources/_/technical/artificial-intelligence/implementing-a-blackboard-like-system-for-squad-r1931. Accessed 15 Sept 2016
59. Dill K (2014) Structural architecture—common tricks of the trade. In: Rabin S (ed) Game AI PRO: collected wisdom of game AI professionals. CRC Press, Boca Raton
60. Champandard AJ (2007) Using a static blackboard to store world knowledge. [online] aigamedev.com. Available at: http://aigamedev.com/open/article/static-blackboard/. Accessed 16 Apr 2016
61. Mark D (2010) Damián Isla Interview on Blackboard Arch. [online] intrinsicalgorithm.com. Available at: http://intrinsicalgorithm.com/IAonAI/2010/02/damian-isla-interview-on-blackboard-arch/. Accessed 19 Sept 2016
62. Khosravi H, Kabir E (2009) A blackboard approach towards an integrated Farsi OCR system. IJDAR 12(1):21–32

Chapter 2
Introduction to Cellular Automata in Simulation

Val Lowndes, Adrian Bird and Stuart Berry

Cellular Automata and Agents can be used to construct simulations where the movement/change of status in an individual is to be observed. This section introduces both Cellular Automata and Agent-based simulations.

Cellular automata and Agent-based simulations have been used to study:

- Traffic flow investigating the formation of traffic jams, the cellular automata representing vehicles (initially a one-dimensional model). N-S
- Conway's Game of Life. (often two-dimensional models) G-M
- The evolution of epidemics, the cellular automata representing people (often one dimensional models)
- Bird/fish flocking, the cellular automata representing a single bird/fish
- Evacuating buildings, the cellular automata representing people.

Section 2.2 introduces Conway's Game of Life.

Section 2.3 investigates population changes.

Section 2.4 introduces models for Epidemics and fire evacuations and traffic simulations based around the Nagel and Schreckenberg model for traffic flow.

V. Lowndes
University of Derby, Kedleston Road, Derby DE22 1GB, UK
e-mail: V.P.Lowndes@derby.ac.uk

A. Bird · S. Berry (✉)
College of Engineering and Technology, University of Derby, Kedleston Road, Derby DE22 1GB, UK
e-mail: s.berry@derby.ac.uk

S. Berry et al. (eds.), *Guide to Computational Modelling for Decision Processes*, Simulation Foundations, Methods and Applications,
DOI 10.1007/978-3-319-55417-4_2

2.1 Defining the Operation of a Cellular Automata-Based Simulation

A cellular automaton (CA) is a collection of cells arranged in a grid, such that each cell changes state as a function of time according to a defined set of rules that includes the states of neighbouring cells.

For example, applying rules to the system status at time zero produces the system status at time one as shown in Fig. 2.1.

The objective in these simulations is the determination of the behaviour of the system over a period in time. In particular, what is the final or steady state of the system, for example does it achieve a steady state.

The simplest cellular automata are one-dimensional, with cells on a straight line, where each cell can have only two possible states (such as high/low or black/white). But in theory, a CA can have any number of dimensions, and each cell can have any number of possible states. The state of each cell changes in discrete steps at regular time intervals.

The state of a cell at any given time depends on two things:

1. its own state in the previous time step, and
2. the states of its immediate neighbours in the previous time step.

2.2 Conway's Game of Life

This section uses "Conway's Game of Life" to introduce the use of a two-dimensional grid. This game became widely known in 1970 when it was published in the popular journal *Scientific American*. See Gotts, in Griffeath and Moore [1] for background information.

Conway devised the rules for the Game of Life (GOL), aiming to meet the following three criteria:

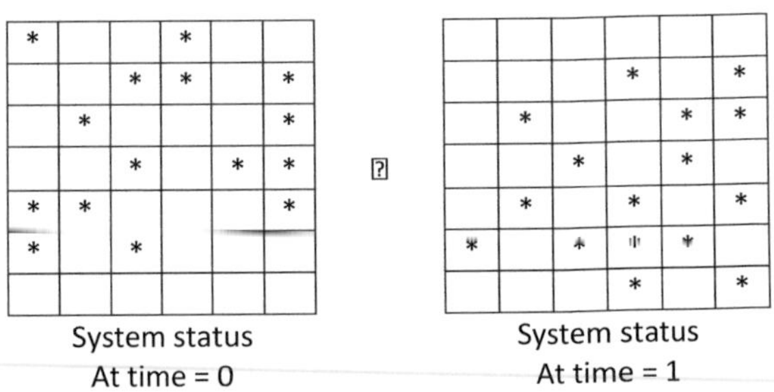

Fig. 2.1 Sample system transition

Criteria for the Game of Life:

1. There should be no initial pattern for which there is a simple proof that the population can grow without limit.
2. There should be initial patterns that *apparently* do grow without limit.
3. There should be simple initial patterns that grow and change for a considerable period of time before coming to an end in three possible ways:
   - Fading away completely (from overcrowding or becoming too sparse)
   - Settling into a stable configuration that remains unchanged thereafter
   - Entering an oscillating phase in which they repeat an endless cycle of two or more periods.

These criteria aimed to ensure that the rules cause the behaviour of the population to be unpredictable. It was, however, later shown by Charles Corderman that Conway's rules did not guarantee that the criteria were satisfied [1, p. 10].

The game itself consists of an infinitely large 2-dimensional grid of squares, in which each represent a cell. The status of each cell at time $t + 1$ depends upon the status of the system of cells at time t. In Fig. 2.2, the status of cell D5 is derived from its own status and that of its 8 surrounding cells at time t:

A cell will be denoted as ***alive*** if it contains a1 (or a*)

The GOL uses rules to generate the system status at time $t + 1$ based on the information (the system status) at time t.

***Rules for Conway's Game of Life*:**

When a cell is alive at time t:

If **two or three** of its eight **neighbours are alive**, **it remains alive**.
If **more than three** of its eight **neighbours are alive**, **it dies** from overcrowding.
If **less than two** of its eight **neighbours are alive**, **it dies** from loneliness.

When a cell is dead:

If **exactly three** of its eight **neighbours are alive**, **it comes to life**.
In **all other cases**, the **cell remains dead**.

Example 1 The rules are applied to the data given in Fig. 2.3 (status at time 0), to generate the state of the system at time 1 (see Fig. 2.4).

Applying the rules to the data shown in Fig. 2.3 gives:

Fig. 2.2 Cells relevant to status of cell D5

Fig. 2.3 Status at time 0

	1	2	3	4	5	6	7	8
A								
B		*	*					
C			*					
D								
E			*					
F			*					
G								
H								

Fig. 2.4 Status at time 1

	1	2	3	4	5	6	7	8
A								
B		*	*					
C		*	*					
D								
E								
F								
G								
H								

Applied to *live cells*:

Cells B2, B3 and C3 surrounded by 2 "*live*" cells, this stays *alive*
Cell E3 and F3 surrounded by only one "*live*" cell, this *dies* (now empty)

Applied to *not alive cells*

Cell C2 surrounded by 3 "*live*" cells, this becomes *alive*

Combining these results gives Fig. 2.4: the status at time 1.

Applying the rules again to the data displayed in Fig. 2.4 give the status at time 2 as shown in Fig. 2.5.

Example 2 Consider the starting pattern shown in Fig. 2.6. At each iteration, count the number of non empty cells. A plot of the number of nonzero (live) points at each iteration is given in Fig. 2.7.

Plotting the number of nonzero (live) points at each iteration gives the graph.

Figure 2.7 shows that a steady state is developing after 36 iterations: approximately 15 live cells.

When a GOL simulation is performed, the rules are applied until the population has become 'stagnant', which means that it will have adopted one of the following states:

Fig. 2.5 Status at time 2

	1	2	3	4	5	6	7	8
A								
B		*	*					
C		*	*					
D								
E								
F								
G								
H								

Fig. 2.6 Two parallel lines

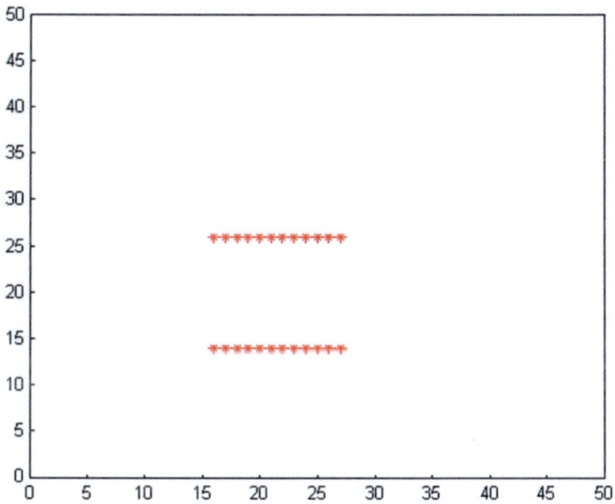

Fig. 2.7 Plot of number of live cells

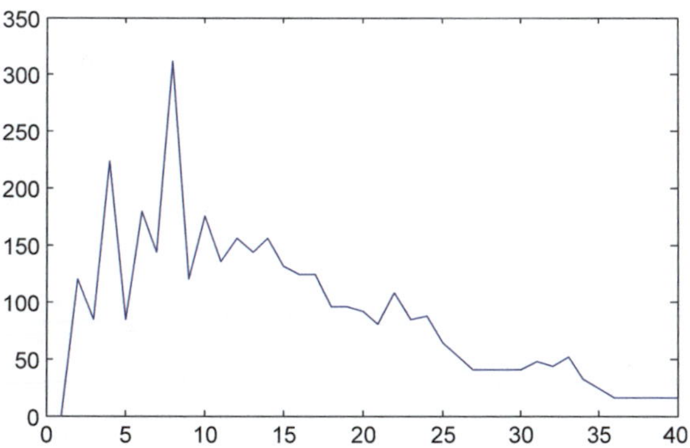

- There are no more living cells—complete annihilation.
- Cells form a grouping which exhibits no change from one generation to the next —still life.
- Cells form a grouping which mutate through a given number of generations resulting in a repetitive cycle.
- Cells traverse the 'universe' in a repetitive but translated (in position) manner and will never collide with any other cells—Gliders and spaceships.

2.3 Investigating Population Growth and Decay

The initial pattern was randomly generated with the number of live cells selected representing an *average* 15% loading.

Figure 2.8 shows one initial pattern generated with random 15% loading. By carrying out several simulations for a particular loading, the most likely state after a given number of iterations and the terminal state can be identified.

Figure 2.9 shows the number of live cells at each iteration, for the simulation generated by the initial pattern shown in Fig. 2.8.

This simulation was run for 500 cycles and Fig. 2.9 shows that for this particular starting pattern, the system converged to a steady state at approximately 470 cycles.

The pattern of live cells (system after 500 iterations) is as shown in Fig. 2.10.

This consists of both static and cyclical shapes. Remember, orientation is not fixed, so shape X is the same as shape Y.

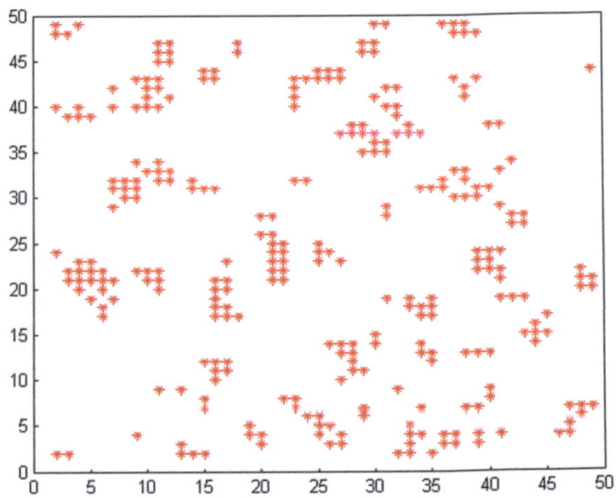

Fig. 2.8 Initial state 15% loading

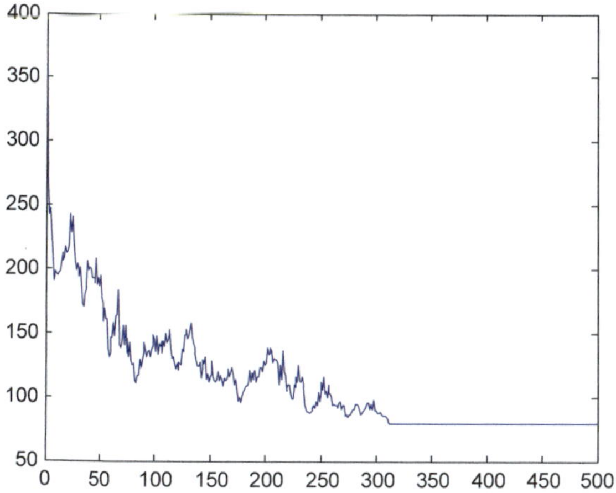

Fig. 2.9 Plot of number of live cells

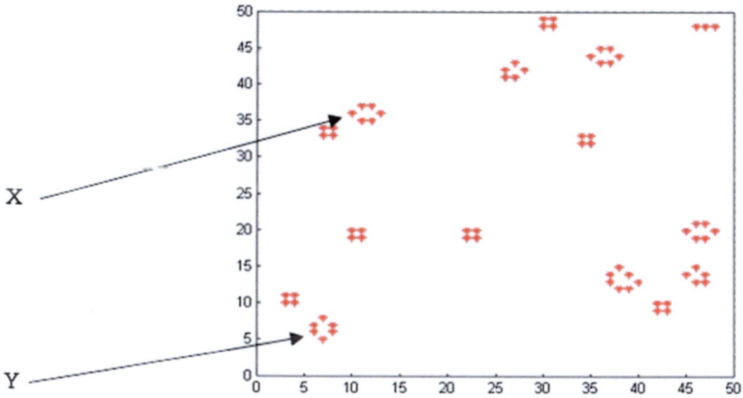

Fig. 2.10 Finishing state, stable shapes

2.4 Applying Cellular Automata and Agents in Modelling

2.4.1 Agent-Based Modelling: Modelling the Spread of Infections

The population is represented on an $n \times n$ where a healthy person is indicted by a cell value of 0 and an infected person by a cell value of 1. Each individual (agent) has an attached record of current duration of their infection.

For example, in Fig. 2.11

Fig. 2.11 Tables A and S showing infected cells and current duration of infection

0	0	0	0	0	0
0	1	0	0	0	0
0	0	0	0	1	0
0	0	1	0	0	0
0	0	0	0	0	0

0	0	0	0	0	0
0	3	0	0	0	0
0	0	0	0	2	0
0	0	5	0	0	0
0	0	0	0	0	0

A (4, 3) = 1 indicating the cell occupant is infected
S (4, 3) = 5 indicating that the infection has lasted 5 time periods

Probability of infection
This is dependent upon the status of the surrounding cells and the variables

$$\text{cv}(i,j) = a(i,j+1) + a(i,j-1) + a(i+1,j) + a(i-1,j)$$
$$\text{Tr} = \text{transmission rate}$$
$$P(i,j) = f_I(\text{cv}, \text{Tr})$$

Probability of recovery
This is dependent on the status of the surrounding cells and the duration of the illness

$$\text{cv}(i,j) = a(i,j+1) + a(i,j-1) + a(i+1,j) + a(i-1,j)$$
$$\text{Rr} = \text{Recovery rate}$$
$$P(i,j) = f_R(\text{cv}, \text{Rr})$$

Figure 2.12 shows the normalised spread, number contracting the illness and the duration of the illness, with reducing levels of infection, for a population of 6400.

The worst case, no treatment and highest rate of infection lead to:

Peak number of patients 2460 (normalised to have value 1)
Percentage of population 38.4%
Maximum duration of epidemic 615 time units

The effect of improved levels of treatment for each level of infection is shown in Fig. 2.13 where the treatment has improved from that used to create Fig. 2.12.

Peak number of patients 2052 (normalised value 2460 as above)
Percentage of population 32.0%
Maximum duration of epidemic 555 time units

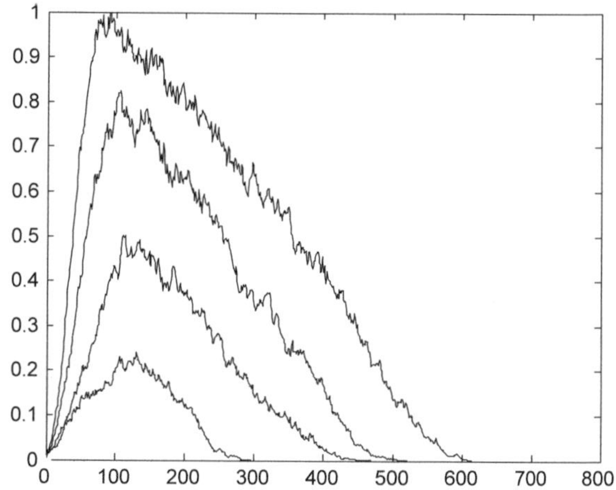

Fig. 2.12 Modelling the effect treatment on infection spread

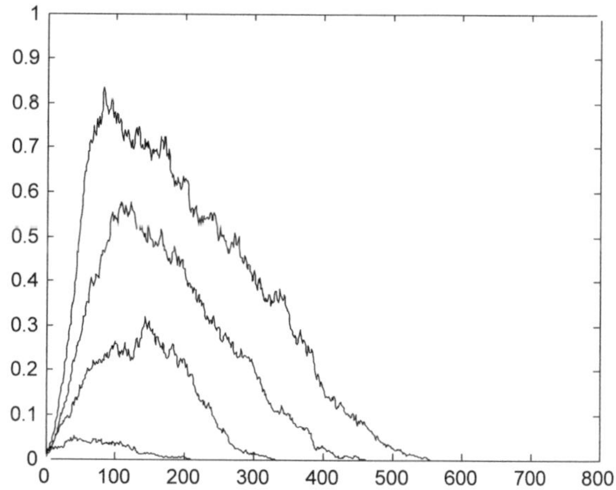

Fig. 2.13 Modelling the effect improvement in treatment

In each case both the duration of the epidemic and the maximum number infected has been reduced.

This agent-based model demonstrates the effect of both preventative and remedial treatment regimes in restricting the spread of an infectious disease.

2.4.2 *Traffic Flow*

This application is used to introduce the use of cellular automata because it employs a simple one-dimensional model, representing a single carriageway road (Note: this is assuming a model where initially cars are not allowed to overtake).

This section shows that although this is the simplest cellular automata-based model, it is a model where the results can have an immediate consequence, to the understanding of traffic flow and hence to the derivation of traffic management policies. The models presented here use the same principles as those in Nagel and Schreckenberg but the analysis of the results is by way of the average velocity.

2.4.2.1 Notation

v_{ij} the velocity of a car j at time i
$v_{i+1,j}$ the velocity of a car j at time $i + 1$
$d_{i,j}$ the distance to the next car $j + 1$, bumper to bumper, at time i
$v_{\max}$ the maximum attainable velocity (the speed limit)
x_{ij} the position of car j at time i.

2.4.2.2 Basic Model

In this model, each cell represents a section of road and each cell will either contain a car or be empty. The system is updated at regular time intervals, using a set of rules, the effect of the update being to allow the cars to move along the road.

In Fig. 2.14, the coloured markers show the progress of the cars with time. Notice that

- the "blue" car travels a total distance of 4 units in two time steps (average velocity 2)
- the "red" car travels a total distance of 2 units (average velocity of 1).

so a traffic jam is forming at time 2.

The rules to enable this simulation (passage of time) are defined in Nagel and Schreckenberg and are given by the four steps **N-S Rules**:

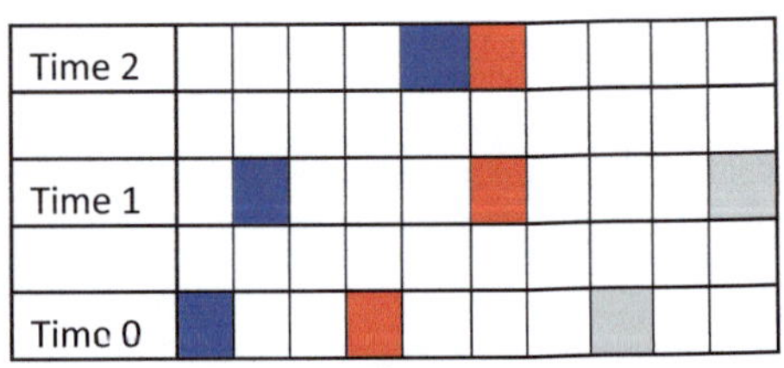

Fig. 2.14 Sample car movements

Step 1: acceleration
All cars that have not already reached the maximal velocity v_{max} accelerate by one unit:
$v_{i+1} \Rightarrow v_i + 1$

Step 2: safety distance
If a car has d empty cells in front of it and if its velocity v_{i+1}(after step 1) is larger than d, then the velocity reduces to d:
$v_{i+1} \Rightarrow \min\{d, v_{i+1}\}$

Step 3: randomisation
With probability p, the velocity is reduced by one unit (if v_{i+1} after step 2):
$v_{i+1} \Rightarrow v_{i+1} - 1$

Step 4: driving
After steps 1–3, the new velocity v_{i+1} for each car j has been determined.
Car j moves forward by v_{i+1} cells:
$x_{i+1,j} \Rightarrow x_{i,j} + v_{i+1}$.

Example 3: Illustrating the operation of the N-S Rules The section of road shown in Fig. 2.15a contains 4 cars with the given velocities, at Time = 0. Work through each of the steps (work through the four steps for each of the 4 cars, checking your velocity calculations against each of the Fig. 2.15b and e. Figure 2.15c gives the final state at Time = 1 after completion of the N-S Rules. Notice that the road is in effect a loop, with car 4 in Fig. 2.15a having one empty cell between itself and car 1.

Step 1: For car 1: $v_1 := v_0 + 1 = 3$; $v_{\max} = 5$
Step 2: For car 1: $v_1 := \min(d_0, v_1) = \min(2, 3) = 2$
Step 3: Random reduction: if rand $< p$ then $v_1 := v_{1-1}$
Step 4: *Drive forward*: For car 1: $x_{1,1} \Rightarrow x_{0,1} + v_{1,1} = 1 + 2 = 3$

2.4.2.3 Simulation of Traffic Flow Using MATLAB

The following analysis was completed using a set of MATLAB programs. Each MATLAB program required the following information to be defined:

- the road length
- traffic density, as a percentage,
- the probability of a car slowing.

At each time step, the simulation records:

- the status of each cell in array a,
  - $a(i) = 1 \Rightarrow$ presence of a car,
  - $a(i) = 0 \Rightarrow$ empty cell, no car

(a)

Road position Time t=0	Car 1			Car 2						Car 3		Car 4	
$v_{max} = 5, p = 0.5$													
v_0	2			4						3		1	
d_0	2			5						1		1	

(b) ***Step 1:*** For car 1: $v_1 := v_0 + 1 = 3$; $v_{max} = 5$

Road position at time t = 0	Car 1			Car 2						Car 3		Car 4	
v_1	3			5						4		2	
d_0	2			5						1		1	

(c) ***Step 2:*** For car 1: $v_1 := \min(d_0, v_1) = \min(2, 3) = 2$

Road position at time t = 0	Car 1			Car 2						Car 3		Car 4	
Velocity V_1	2			5						1		1	
d_0	2			5						1		1	

(d) ***Step 3:*** Random reduction: if rand<p then $v_1 := v_1 - 1$

Road position at time t = 0	Car 1			Car 2						Car 3		Car 4	
P=0.5: rand	0.83			0.62						0.31		0 .54	
Velocity v_1	2			5						0		1	
d_0	2			5						1		1	

(e) ***Step 4:*** *Drive forward*: For car 1: $x_{1,1} \Rightarrow x_{0,1} + v_{1,1} = 1 + 2 = 3$

Road position at time t= 1			Car 1						Car 2	Car 3			Car 4
Velocity at time 1 v_1			2						5	0			1
d_1			5						0	2			2

Fig. 2.15 **a** Sample data at time = 0. **b** Sample new velocities at time = 0, after step 1. **c** Sample new velocities at time = 0, after step 2. **d** Sample new velocities at time = 0, after step 3. **e** Final velocities and positions for sample at time $t = 1$

- the status of each car in array *vl*
  - $v(i)$ gives the current velocity of the car in cell i; $v_i \in (0, 5)$.

Figures 2.16 and 2.17 are typical graphical outputs from this program. For cellular automata, conventionally the vertical axis represents time units and the horizontal axis distance units. The results displayed in these graphs were generated using $p = 0.5$ (probability of a car randomly slowing) with 10% cell loading (traffic density) (Fig. 2.16) and then with 20% cell loadings (Fig. 2.17).

Notice

(a) In this and the following figures that the vertical axis measures time and the horizontal axis distance.

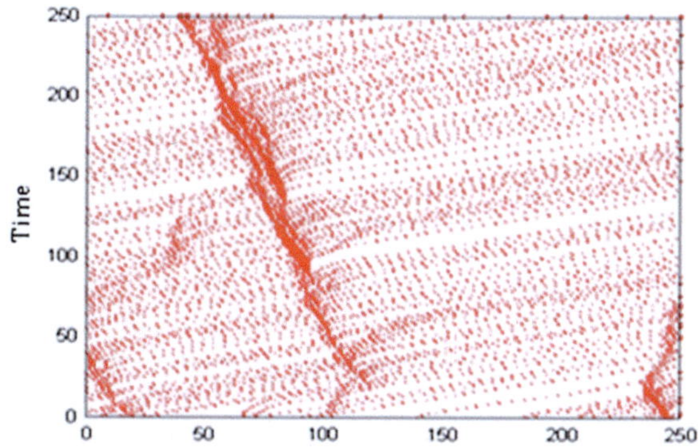

Fig. 2.16 10% cell loading, free-forming traffic jam

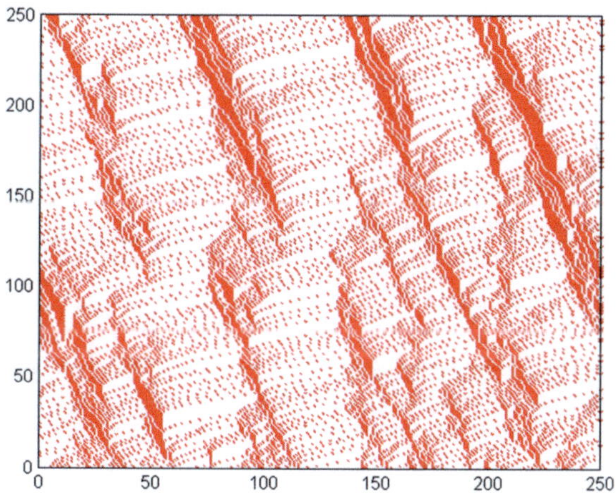

Fig. 2.17 20% cell loading. Higher density implies more traffic jams

(b) The deeper shading indicates the presence of a jam.
(c) A free-forming jam "moves" backwards in time.

Notice that the jams, deeper shaded areas" are parallel.

Extending the "road" length in the second simulation replaces Fig. 2.17 with Fig. 2.18.

The plot shown in Fig. 2.18 suggests the existence of many regular "self-forming" queues of traffic in this model and a congested traffic flow.

The program uses two parameters to define the simulation (CA model) to generate this plot

Density of traffic = 0.20	20% of cells filled	Traffic demand
Random slowing = 0.50	50% of cars slow down	Driver behaviour

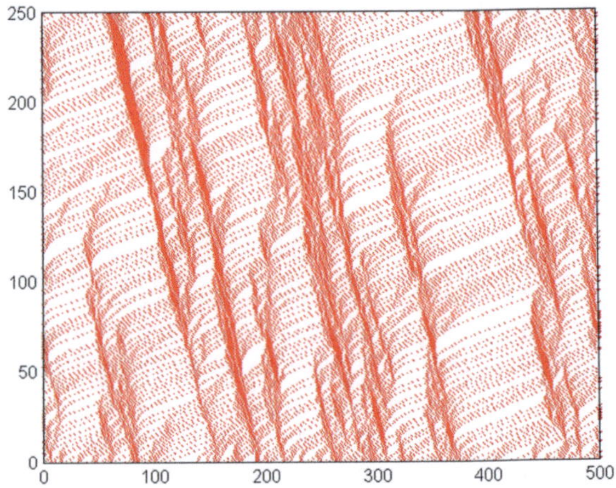

Fig. 2.18 20% cell loading extended road length

2.4.2.4 Analysis of Simulation Results

The traffic simulation program was run several times varying the traffic density, but keeping p (the probability of a random reduction in speed), constant (at 10%). The results generated have been used to produce the plot of average car velocity v traffic density in Fig. 2.19a.

Within the results shown in Fig. 2.19a, there seems to be two phases, traffic density (road loading x) up to 12% and over 12% giving the models for travel time T as:

Road loading less than 12%	travel time	$T = 4.9027 - 0.0044x$	$r^2 = 0.94$
Road loading more than 12%	travel time	$T = 8.2325e^{-0.046x}$	$r^2 = 0.99$

for the results shown in Fig. 2.19b, these models become:

Road loading (x) less than 5%	travel time	$T = 4.4048 - 0.0119x$	$r^2 = 0.92$
Road loading more than 5%	travel time	$T = 68.208x^{-1.427}$	$r^2 = 0.97$

and for the results shown in Fig. 2.19c, these models become:

Road loading (x) less than 15%	travel time	$T = 4.9979 - 0.0008x$	$r^2 = 0.84$
Road loading more than 15%	travel time	$T = 11.379e^{-0.05x}$	$r^2 = 0.98$

Figure 2.19d displays all three models for comparison showing that at normal loadings, up to 40%, random slowing has a great effect on a roads carrying capacity. Thus, indicating that at times of high traffic density, the "random" behaviour of drivers becomes less important, with respect to travel time.

Note 1: when there is no random slowing and all cars are equally spaced, the implied road loading will be 16.7%.

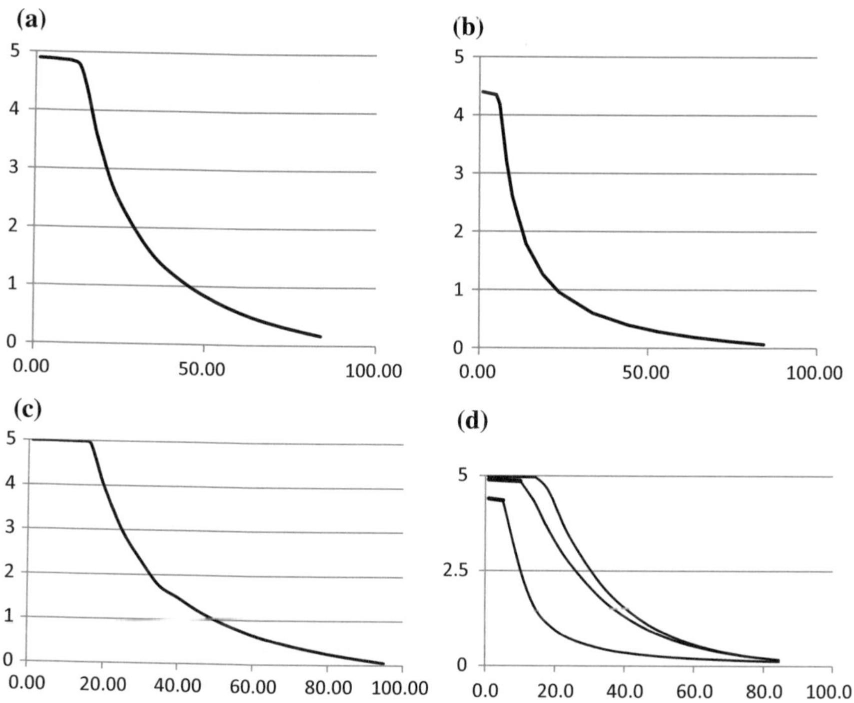

Fig. 2.19 Average speed against traffic density. **a** 10% random slowing. **b** 60% random slowing. **c** Average speed against traffic density, no random slowing. **d** Average speeds against traffic density, all cases

Note 2: the stopping distance s (minimum separation in feet) in terms of the vehicles velocity v (in miles per hour) is given by

$$s = 0.05v^2 + 2v + 15,$$

and as the average car length is 13 feet, thus the loading per mile for free-flowing traffic, model 7c, is given by

$$\text{MF} = \frac{5280}{((0.05v^2 + 2v + 15) + 13)}$$

The loading per hour is therefore given by

$$\text{MFH} = \frac{5280v}{((0.05v^2 + 2v + 15) + 13)}$$

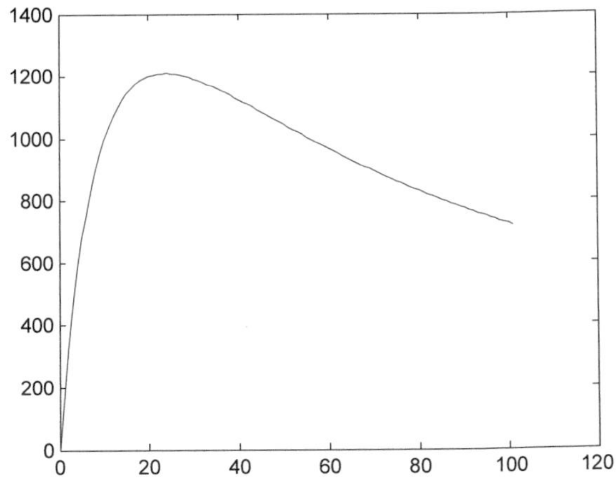

Fig. 2.20 Road capacity per hour against traffic speed

At 70 mph	MF = 16.1 per mile	MFH = 1127
At 30 mph	MF = 39.7 per mile	MFH = 1191
At 24 mph	MF = 50.4 per mile	MFH = 1209

Note the optimal velocity is 24 mph; a plot of "flow per hour" against velocity is shown in Fig. 2.20. This demonstrates that low speeds, less than 15 mph are very inefficient, low capacity per hour, as are speeds greater than 100 mph.

Note 3: loadings within 10% of the optimal occur at speeds between 12.5 and 44.5 mph.

2.4.2.5 Analysing Traffic Flows

An area of interest is the frequency of occurrence of self-forming jams. Autocorrelating the historic data can be used to investigate the frequency of jams within free-flowing traffic. Figure 2.21a shows a plot obtained from correlating the position data at time t with the position data at time $(t–k)$ over a time span of T time units. When the data has a perfect pattern, jams occur at regular intervals have the same length and traffic flows freely between jams the plot of the aurocorrelations will have this form the jam frequency being indicated by the spacing of the peak values in the plot.

Figure 2.21b shows the plot obtained for a more normal case when the jams are not (quite) evenly spaced and not (quite) the same length, but although the peaks do not have the same magnitude, the plot does give an indication of the jam frequency; here, the data can be analysed using a Hilbert–Huang transform, see Bird [2] for details.

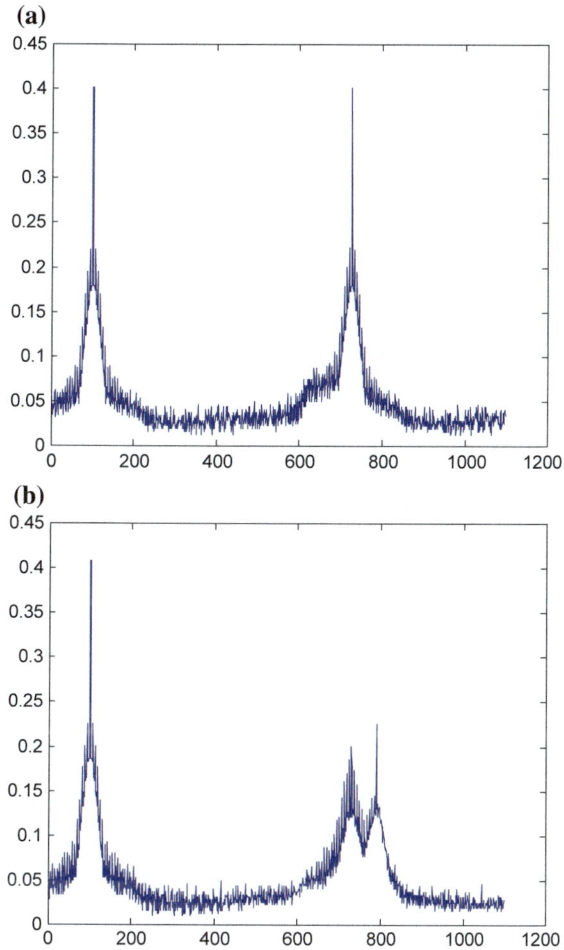

Fig. 2.21 **a** Autocorrelation perfect flow pattern. **b** Autocorrelation not (quite) perfect flow

Discussion Points:
Average Velocity and Traffic Flow

- Why would the maximum average velocity, in this model, be 4.9?
- What would be the highest traffic density that would enable traffic to flow freely without any need to reduce velocity.

Extending the Model
Obvious extensions are:

- Allow overtaking on a single lane road
- Allow for more lanes
- Include different type of vehicles.

One lane overtaking model N-S Rules (overtaking)
To enable overtaking, on a single carriageway, the model is amended so that each vehicle carries two possible velocities

Step 1: acceleration
All cars that have not already reached the maximal velocity v_{max} accelerate by one unit:
$v_{i+1} \Rightarrow v_i + 1$

Step 2: safety distance
If a car has d empty cells in front of it and if its velocity v_{i+1}(after step 1) is larger than d, then the velocity reduces to d:
$v^1{}_{i+1} \Rightarrow \min\{d,\ v_{i+1}\}$
$v^2{}_{i+1} \Rightarrow v_{i+1}$

Step 3: randomisation
With probability p, the velocity is reduced by one unit (if v_{i+1} after step 2):
$v^1{}_{i+1} \Rightarrow v^1_{i+1} - 1$
$v^2{}_{i+1} \Rightarrow v^2_{i+1}$

Step 4: driving
After steps 1–3, the new velocity v_{i+1} for each car j has been determined.
Car j moves forward by $v^k{}_{i+1}$ cells:
If the space is free move with velocity v^2 to
$x_{i+1,j} \Rightarrow x_{i,j} + v^2_{i+1}$.
otherwise
$x_{i+1,j} \Rightarrow x_{i,j} + v^1_{i+1}$.

Multi Lane Models
There are two cases:

Two Lane Model, here there are additional rules to

Allow a vehicle in the inside lane to move out
Allow a vehicle in the outside lane to move in
Prohibit "undertaking".

Three, or more, lane model, here there are the road rule sets

Inside lane, cars can move out
Middle lane, cars can move in or out
Outside lane, cars can move in
No undertaking

Note: having developed a three lane model, a multilane model follows using the same rule sets.

References

1. Griffeaths D, Moore C (2003) New constructions in cellular automata. Oxford
2. Bird A (2011) Early detection of aero engine damage from acoustic emission signatures, PhD thesis, Derby

Chapter 3
Introduction to Mathematical Programming

Val Lowndes and Stuart Berry

Mathematical programming can be used to determine the optimal solution to planning problems. Typically, mathematical programming modelling has been applied to problems in the areas:

Production Processes
Production Planning Models
Diet Problems
Manufacturing Processes
Staffing Problems
Transportation Models
Production Planning Models
Travelling Salesman Problems

Many of these applications lead to large complex models, typically Travelling Salesman Problems, Production Planning Problems and Nurse Scheduling, when efficient solution methodologies may need to be developed.

Here, models are presented and standard mathematical programming approaches to solution are evaluated. In general, an investigation follows the following steps.

V. Lowndes
University of Derby, Kedleston Road, Derby DE22 1GB, UK
e-mail: V.P.Lowndes@derby.ac.uk

S. Berry (✉)
College of Engineering and Technology, University of Derby,
Kedleston Road, Derby DE22 1GB, UK
e-mail: s.berry@derby.ac.uk

S. Berry et al. (eds.), *Guide to Computational Modelling for Decision Processes*,
Simulation Foundations, Methods and Applications,
DOI 10.1007/978-3-319-55417-4_3

Stages in an Investigation:

(i) Modelling

(a) Derive Constraints
Is the problem either
Linear, or
Nonlinear
(b) What is the Objective
Single objective, or
Multiple objectives
(c) Variables
Are they
Real, or
Integer, or
Binary/Logical

(ii) Deriving a Solution, does this use

(a) Software, Xpressive, or
(b) Heuristics

(iii) Evaluation (of solution), ref the Modelling Cycle (POPS stage 1).

(a) Have we solved the problem
(b) Is the solution acceptable

This section uses a series of scenarios to demonstrate the stages in a modelling exercise using mathematical programming indicating its benefits and drawbacks.

These examples are as follows:

Diet Problems: historically, these were an early application of Linear Programming. They are still valid because their solution highlights the problem of usefulness of an optimal solution and the need to evaluate the generated solution.
Knapsack Problems: another early application of mathematical programming but an approach with many varied applications.
Production Planning: a basic application chosen to demonstrate the suitability of the solution generated by the standard approach.

In each example, the size of problem and any structure in the model (its complexity) will be used as a measure of suitability of a LP approach to problem-solving.

3.1 Applications Model Furniture Manufacture

This example is used to illustrate the stages in the production and evaluation of a solution from a problem specification.

N&F producces both tables and chairs. Each item passing through 3 stages, in the order 1 then 2 then 3, with the times at each stage is given in Table 3.1:

Note only one item can be processed at a stage at any time.

For example, if the firm has the production schedule:

Table, Chair, Table, Chair, …

The daily working could be represented by a Gantt chart, for example (Fig. 3.1):

Normally, it is assumed that the company is (already) in business and that all production time is available leading to the model:

$$\begin{gathered}\boldsymbol{Maximise}(\mathbf{5x+7y})\\ \mathbf{x+3y\le 60}\\ \mathbf{5x+2y\le 100}\\ \mathbf{4x+4y\le 100}\end{gathered}$$

Table 3.1 Production requirements and revenues

Stages	Time in stage per unit of product		Available time
	Table	Chair	
1	1	3	60
2	5	2	100
3	4	4	100
Profit per unit	£5	£7	

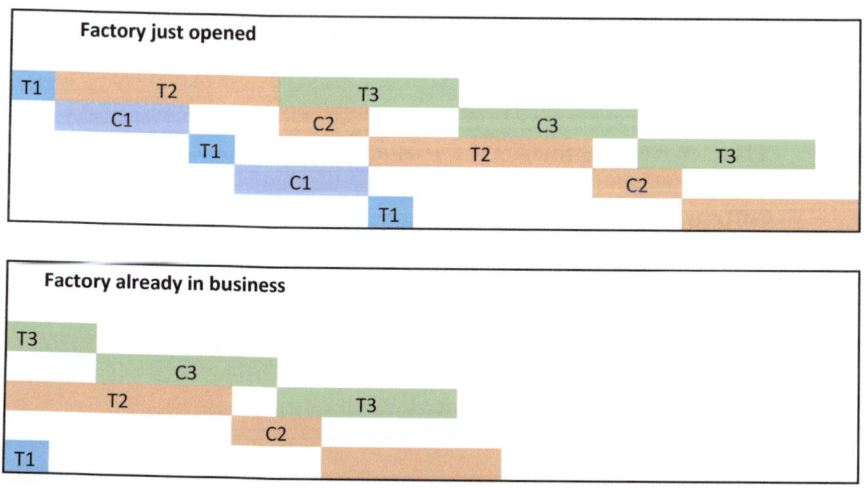

Fig. 3.1 Production plans for both new and existing factories

with the optimal solution given by:

$$\textbf{Profit} = \textbf{160}: \quad \textbf{Product(1)} = \textbf{7.5}: \quad \textbf{Product(2)} = \textbf{17.5}$$

Evaluation: A question, can the firm produce half a chair?
If it can be assumed that this will be finished on the following day, YES
if NO and the variables will have to be declared as integers!
Giving the new solution

$$\textbf{Profit} = \textbf{159}: \quad \textbf{Product(1)} = \textbf{8}: \quad \textbf{Product(2)} = \textbf{17}$$

Summary of the investigation:

(i) Modelling

(a) Constraints
Linear

$$\begin{aligned} \mathbf{1x + 3y < 60} \\ \mathbf{5x + 2y < 100} \\ \mathbf{4x + 4y < 100} \end{aligned}$$

Nonlinear or linear
(b) Objective or multiple
Single
Profit = 5 + 7*y* to be maximised
(c) Variables; real, integer, binary, logical
Real
Initially, assume that the variables are real
Integer
Then, move onto integer if the solution above is not acceptable
(d) Linear constraints and objective function therefore solution obtained using
Software, for example XpressIVE
Heuristics

(ii) Evaluation (of solution)
Have we solved the problem, YES
Is the solution acceptable, YES but say which is the solution.

These questions are to be asked for each model.

3.2 Applications of Mathematical Programming-Based Modelling

These examples are presented to illustrate the development of a fully representative model to describe an application and how this model is refined through the evaluation of the generated solutions. The first model to consider is the "Diet Problem" because this is a longstanding problem and was an initial problem modelled and solved using computer-based Linear Programming; this problem demonstrates both the advantages and disadvantages of this LP approach. Later, it is shown how the problems from this, LP, approach are overcome.

3.2.1 *Modelling Diet Problems*

A second model construction and evaluation is derived from the "diet construction problem"; this shows how a model can be successfully developed but the results (from this basic model) are in practice unacceptable, providing the same diet every day is not acceptable, to the recipients of the diet, although this diet satisfies dietary conditions and is inexpensive.

Stigler [1] and Dantzig [2]

Assuming that the aim is to minimise the cost of the diet, the notations are used:

x_j	the quantity of food type j in the diet
a_{ij}	the quantity of nutrient i in one unit of food type j
$\min Q_i$	the minimum allowable quantity of nutrient i in a healthy diet
$\max Q_i$	the maximum allowable quantity of nutrient i in a healthy diet
c_j	the cost of one unit of food type j

The model becomes:

$$\sum_j a_{ij}x_j \leq \max Q_i \quad \text{all nutrients } j$$

$$\sum_j a_{ij}x_j \geq \min Q_i \quad \text{all nutrients } j$$

$$\text{Minimise}\left(\sum_j c_j x_j\right)$$

Note: nutrients can be vitamins, minerals, fats, carbohydrates and proteins, for example, and there may be restrictions on the total volume/weight of the diet.

Thus, if m_j is the mass of food item j and M is the maximum allowed mass, then add the constraints:

$$\sum_j m_j x_j \leq M$$

However, this model is extended to incorporate the fact that the percentage of 'type A' fat is less than 40% of the total fat content of the diet. Assuming that there are two types of fat A and B, the constraint becomes:

$$\sum_{\text{fat A}} a_{ij} x_j \leq 0.4 \left(\sum_{\text{fat A}} a_{ij} x_j + \sum_{\text{fat B}} a_{ij} x_j \right)$$

Or if the selection of a unit of food 1 implies that k units of food 2 must be selected for inclusion in the diet, the model is extended through the addition of the constraints:

$$x_2 \geq k x_1$$

Or if only one food from a group of food types may be chosen, for example food 1 or food 2 or food 3, the model is extended through the addition of the constraints:

$$\begin{aligned} x_1 &\leq M\delta_1 \\ x_2 &\leq M\delta_2 \\ x_3 &\leq M\delta_3 \\ \delta_1 + \delta_2 + \delta_3 &\leq 1 \end{aligned}$$

Notice that if one of these foods must be included, then add the constraints:

$$\begin{aligned} x_1 &\geq \min Q_1 \delta_1 \\ x_2 &\geq \min Q_2 \delta_2 \\ x_3 &\geq \min Q_3 \delta_3 \end{aligned}$$

here $\min Q_2$ for example is the minimum allowable portion of food 2 if food 2 is chosen to be included in the diet.

Or if the selection of food 1 implies that food 2 must be chosen, then the model is extended to include the constraints,

$$\begin{aligned} x_1 &\leq M\delta_1 \\ x_2 &\geq \min Q_2 \delta_1 \end{aligned}$$

Summary of the models:

(i) Modelling

(e) Constraints
Linear

Nonlinear or linear

(f) Objective or multiple
Single
Cost to be minimised

(g) Variables; real, integer, binary, logical
Real
Initially, assume that the variables are real
Integer
Then, move onto integer if/when the solution above is not acceptable

(h) Linear constraints and objective function therefore solution obtained using
Software, for example XpressIVE

Heuristics

(ii) Evaluation (of solution)

Have we solved the problem?	YES
Is the solution acceptable?	probably NO, lack of variety and/or taste in the derived diets.

Summary: The solution of Diet Problems is concerned with the determination of the best selection of items from a given set of available items, but not the palatability of the diet.

3.2.2 *Blending 1*

A mining company will be operating in an area for the next 5 years. It has just signed contracts with the local power generation and heavy industrial companies for this period. Each mine has an annual production capacity, these are estimated as being

Mine	Capacity (million tonnes)	Cost per Million tonne
1	2	60
2	2.5	50
3	1.3	80
4	3	30

The quality of coal from each mine is given by:

Mine	Quality
1	1.0
2	0.7
3	1.5
4	0.5

x_i Production quantity at each mine is
q_i Quality at each mine
R Required quality
c_i Cost per unit at each mine
D Demand per year

In each year, quantities of these ores are blended to produce an acceptable quality of fuel, initially assume that the required quality is 0.9. If the blended coal will sell for £10 a tonne, determine the optimal purchasing policy for the company.

$$\sum_i x_i \geq D$$

$$\sum_i q_i x_i \geq R \sum_i x_i$$

$$\text{Profit} = 10\,D - \sum_i c_i x_i$$

Maximise Profit or if it quantity is fixed Minimise total costs.

3.2.3 *Blending 2: Animal Feed*

There are several available foodstuffs with differing nutrient content. These are to be mixed to produce a nutritionally acceptable animal feed.

Notation a_{ij} quantity of nutrient in one unit of food j
$\min_i$ minimum allowed percentage content of nutrient i
$\max_i$ maximum allowed percentage content of nutrient i
c_j cost per unit of food j
R required quantity to be produced

Table 3.2 Ingredient contents

	Food sources and nutrient content			
	Calcium	Protein	Carbohydrates	Cost per unit
Ingredient 1	0.045	0.25	0.16	10.00
Ingredient 2	0.005	0.09	0.22	30.00
Ingredient 3	0.003	0.50	0.08	15.00

$$\text{Model} \quad \begin{aligned} &\sum_j a_{ij}x_j \geq \min_i \sum_j a_{ij}x_j \\ &\sum_j a_{ij}x_j \leq \max_i \sum_j a_{ij}x_j \\ &\sum_j x_j \geq R \\ &\text{Cost} = \sum_j c_j x_j \end{aligned}$$

Example An animal feed has to have a weight between W_1 and W_2
Required ingredients and restrictions

Calcium	between 0.8 and 1.2% per weight of the final mixture
Protein	at least 22%
Carbohydrates	at most 15%

Possible resources are given in Table 3.2:

$$\begin{aligned} 0.045x_1 + 0.005x_2 + 0.003x_3 &\leq 0.012(x_1 + x_2 + x_3) \\ 0.045x_1 + 0.005x_2 + 0.003x_3 &\geq 0.008(x_1 + x_2 + x_3) \\ 0.25x_1 + 0.09x_2 + 0.50x_3 &\geq 0.22(x_1 + x_2 + x_3) \\ 0.16x_1 + 0.22x_2 + 0.08x_3 &\geq 0.15(x_1 + x_2 + x_3) \\ x_1 + x_2 + x_3 &\leq W_2 \\ x_1 + x_2 + x_3 &\geq W_1 \\ \text{Minimise Cost} &= 10x_1 + 30x_2 + 15x_3 \end{aligned}$$

But will the livestock be prepared to eat the resultant feedstuff?

3.3 Problems Reducible to Diet Problems

3.3.1 Financial Planning Modelling

There exists a sum of money to be allocated between several investment opportunities; this section shows how a basic Linear Programming model can be used to

Table 3.3 Cash flows

Cash flows				
Time	Project *W*	Project *X*	Project *Y*	Project *Z*
Cost	10	20	30	50
Return	12	26	35	62

select the best investments. As an example, a firm has £90 to invest into one or more of the projects listed in Table 3.3 costs, and expected returns are as given.

Let x_j be the sum of money invested in Project J and assuming that each project is available more than once (that is x_j can have any value), the investment decision can be modelled as a knapsack problem where the available money corresponds to the size of the container, here

$$\text{Profit}\, P = \text{Return} - \text{Cost}$$

$$\text{Maximise} \quad \left\{\sum_j P_j x_j\right\}$$

$$\sum_j C_j x_j \leq 90$$

The optimal investment policy is as follows: $X_x = 4.5$ Profit $= £27$

<u>Questions:</u>

1) Can a project be partly chosen? Here, 0.5 of a project
2) Can a project be chosen more than once? Here $X_X > 1$

Example 3.1a Here, restrict the solutions to integer values. This gives the solution.

Solution:

$$\text{Profit} = 26 \quad X = \{1,\ 4,\ 0,\ 0\}$$

To extend the investigation, consider the effect of not allowing a project to be chosen more than once.

Here, there are only two possible investments in each project:

- the full amount, or
- nothing.

Therefore, each variable can only have the value 1 or the value 0, binary variables.

Solution, notice now the reduction in profit and the "left over money".

$$\text{Profit} = 20 \qquad X = \{1, 1, 0, 1\}$$
$$\text{Unused Money} = XX$$

As a final extension to this model if any unallocated monies are invested at 2% the solution becomes

$$\begin{array}{ll} \text{Profit} = 20 & X = \{1, 1, 0, 1\} \\ \text{Invested money} = 10 & \end{array}$$

3.3.2 Modelling Investment Planning 2

This example introduces the use of logical variables to enable more realistic solutions, typically what is the minimum investment allowed in a project.

That is if the firm chooses to invest into Project *X*, then it must invest at least the quoted minimum quantity, see Table 3.4.

For Project *W*, either invest:

$$\begin{array}{ll} \text{Between 5 and 10,} & 0.5 \leq x_1 \leq 1.0, \\ \text{or Nothing} & x_1 = 0. \end{array}$$

logical (binary) variables (only have the value 0 or 1) are needed, let

$\delta_W = 1$ indicate that Project *W* has been chosen for investment.

The Model now becomes

$$\text{Maximise}\{\text{Profit} = \text{Return} - \text{Investment}\}\text{giving}$$

$$\text{Maximise} \quad \left\{ \sum_j P_j x_j \right\}$$

$$\sum_j C_j x_j \leq 90$$

$$x_W \leq M\delta_W$$

$$x_W \geq R_W \delta_W \quad R_W - C_{W\min}/C_W$$

$$x_j \qquad \text{binary}$$

Thus, if an investment is made into Project *W*, then δ_W is forced to have the value of 1; otherwise, it has the value 0, and if δ_W is 1, then x_j is forced to have the value of at least 0.5, otherwise it has the value 0.

Table 3.4 Cash flows

Cash flows				
Maximum	Project *W*	Project *X*	Project *Y*	Project *Z*
Maximum Investment (*C*)	10	20	30	50
Minimum Investment (if chosen) ($C_{\min}$)	5	8	18	16
Return at maximum investment	12	26	35	62

3.3.3 *Restricting the Investment by Adding Extra Conditions*

The firm has to choose

at least one project from group 1 projects	$\{W, Y\}$
but no more than one from group 2 projects	$\{X, Z\}$.

for example projects in group 2 are high risk and should be balanced by investments in the lower risk projects, group 1.

Using the logical variables in example 3 adding the constraints

$$\delta_1 + \delta_3 \geq 1$$
$$\delta_2 + \delta_4 \leq 1$$

produces the required model.

Solution

Profit	= 19	Unused money = 0	Plus Interest = 0
Product(1)	= 1	dx(1) = 1	{Profit = 12 – 10 = 2}
Product(2)	= 0	dx(2) = 0	
Product(3)	= 1	dx(3) = 1	{Profit = 35–30 = 5}
Product(4)	= 1	dx(4) = 1	{Profit = 62–50 = 12}

Alternatively to balance high- and low-risk projects, there is the additional requirement that if Project Z is chosen, then Project X must be chosen.

This can be achieved through the addition of the constraint

$$\delta_4 \leq \delta_2$$

giving the final formulation:

$$\text{Maximise}\left\{\sum_j P_j x_j\right\}$$
$$\sum_j C_j x_j \leq 90$$
$$x_j \leq M\delta_j$$
$$x_j \leq R_j\delta_j$$
$$R_j = C_{\min}/C \text{ all } j$$
$$\delta_1 + \delta_2 + \delta_3 \geq 1$$
$$\delta_1 + \delta_2 + \delta_4 \leq 2$$
$$\delta_4 \leq \delta_2$$

Solution

Profit = 18 Unused money = 20 Plus interest = 0.4
$X = \{0, 1, 0, 1\}$, $dx = \{0, 1, 0, 1\}$

Summary

This model indicates the possible types of variables used to model investment decision-making.

Real	1.7 units is allowed
Integer	0 or 1 or 2 or 3 units (only) allowed
Real or integer limited	$x_1 < 2$ for example
Binary	invest or don't $x_1 = 0$ or 1
Real with lower limit	at least 0.5 of the project

The effect of adding more constraints is to reduce the profit, compare 5a with 5, when in 5a the variables *x*(*i*) are not restricted to integers.

Profits from each model:

Model 1	Profit = 27
Model 1a	Profit = 26
Model 2	Profit = 21.67
Model 2a	Profit = 20
Model 2b	Profit = 22
Model 3	Profit = 20 + 2
Model 4	Profit = 19 + 0
Model 5	Profit = 18 + 0.4

Example Using the data given in Table 3.5, produce a model for their decision-making process given that they have 1200 units available for investment and they wish to invest in at least 5 of the projects.

3.3.4 Modelling Investment Planning 2 (Allocating Money Between Projects)

There exists a sum of money to be allocated between several investment opportunities with investments occurring over two time periods, any unused money is forfeit in the first examples and then invested in the later examples.

The aim being to maximise the discounted value of the total returns in the first examples and then to maximise the terminal value in the later examples.

This section shows how a Linear Programming model can be used to select the best investments using the 6 time period and 6 investment opportunities example that can be modelled as a multiple knapsack problem.

Table 3.5 Investment data and returns

Time	Projects						
	A	*B*	*C*	*D*	*E*	*F*	Available
0	−10	–	−6	−12	–	–	40
1	−4	−3	−4	−2	−5	−5	20
2	8	−6	4	6	−7	−7	
3	14	8	8	14	8	8	
4	12	5	6	8	10	10	
5	6	2	3	5	14	18	
Return discounted	22	3	7	13	14	20	

Investment opportunities

Investment projects

	A	*B*	*C*	*D*	*E*	*F*	*G*	*H*	*I*	*J*	*K*
Invest	100	200	50	250	410	260	300	420	150	80	160
Return	200	320	120	320	560	410	420	490	220	130	320

Example 1

The decision process can be modelled by the constraints:

$$\begin{aligned} 10x_1 + 0x_2 + 6x_3 + 12x_4 + 0x_5 + 0x_6 &\leq 40 \\ 4x_1 + 3x_2 + 4x_3 + 2x_4 + 5x_5 + 5x_6 &\leq 20 \\ 8x_1 - 6x_2 + 4x_3 + 6x_4 - 7x_5 - 7x_6 &\geq 0 \end{aligned}$$

Likewise for periods 3, 4 and 5.
With the aim of maximising the discounted return

$$22x_1 + 3x_2 + 7x_3 + 13x_4 + 14x_5 + 20x_6$$

General model has the form

$$\sum_j C_{ij}x_j \geq -A_j$$

$$\text{Maximise}\left(\sum_j R_j x_j\right)$$

Solution

$$\text{Profit} = 104 \quad X = \{4, 0, 0, 0, 0, 0.8\}$$

Invest 4 units in Project *A* and 0.8 units in Project *F*.
(are these values acceptable? and allowed?)

3.4 Knapsack Problcms

A Second Example of Mathematical Programming Modelling concerns the knapsack problem. The knapsack problem or rucksack problem is a problem in combinatorial optimisation: Given a set of items, each with a weight and a value, determine the number of each item to include in a collection so that the total weight is less than or equal to a given limit and the total value is as large as possible. It derives its name from the problem faced by someone who is constrained by a fixed-size knapsack and must fill it with the most valuable items.

The problem often arises in resource allocation where there are financial constraints and is studied in fields such as combinatorics, computer science, complexity theory, cryptography and applied mathematics.

3.4.1 Example: Herring Plc

Formulating and developing knapsack models, these models using the data where 7 items are available for packing

Item	1	2	3	4	5	6	7
Weight	10	20	50	20	5	2	1
Value	20	26	40	22	10	8	6
Total weights	108						
Container capacity	82						

Base model, constraint and objective function

$$10x_1 + 20x_2 + 50x_3 + 20x_4 + 5x_5 + 2x_6 + 1x_7 \leq 82$$
$$\text{Value} = 20x_1 + 26x_2 + 40x_3 + 22x_4 + 10x_5 + 8x_6 + 6x_7$$
$$\text{Maximise (Value)}$$

Modelling questions
Variables are they:

Real	is 3.74 allowed?	NO
Integer	is 7 allowed, that is multiple items	YES
Binary	single items only, pack or don't pack.	NO

Solution and Evaluations (multiple copies of each item can be packed).

$$\text{Profit} = 492$$
$$x_i = 0 \quad i = 1 \text{ to } 6$$
$$x_7 = 82$$

Comments…is it feasible to pack, just, one type of item?
Summary of the investigation:

(i) Modelling

(i) Constraints
Linear

Nonlinear or linear

(j) Objective or multiple
Single
Return to be Maximised
(k) Variables; real, integer, binary, logical
Real: Initially assume that the variables are real
Integer: Move onto integer if the solution above is not acceptable

(a) Linear constraints and objective function therefore Solution obtained using XPRESSIVE

Model 2: **Development of model**: To restrict the solution to one item of each type, define the variables to be binary.
Variables are they:

Real	is 3.74 allowed?	NO
Integer	is 7 allowed, that is multiple items	NO
Binary	single items only, pack or don't pack.	YES

Solution

$$\text{Profit} = 94 X = \{1, 1, 1, 0, 0, 1, 0\}$$

Summary of the investigation:

(i) Modelling

(b) Constraints
Linear

Nonlinear or linear

(c) Objective or multiple
Single
Profit to be Maximised

(d) Variables; real, integer, binary, logical
Binary, only one item of each type.

(e) Linear constraints and objective function therefore Solution obtained using
Software, for example XpressIVE

3.4.2 *Generic Models for the Basic Problem*

A container is to be packed with several items so that the value of its contents is optimised.

Using the notation

Value of each item is	v_i
Weight of each item is	w_i
Quantity of item *i* packed	x_i, integer valued
Capacity of the container is	C

This can be is modelled by:

$$\max\left[\text{Value} = \sum v_i x_i\right]$$

$$\text{Subject to } \sum_i w_i x_i \le C$$

Here, the aim is to maximise the value of the goods packed into the knapsack.

Notice that there are 8 variables and 1 constraint; using LP there will be only one nonzero value in the optimal solution.

Here, the solution will be given by:

$$x_1 = C/2, \quad \text{see note 1.}$$

and if C is an odd number, then this solution is impossible, 7.5 items is meaningless and averaging is impossible (assuming more than one container).

Therefore, the answer has to have an integer form thus implying a harder problem to solve.

Expanded problem (normal case)

At most, only one of each type of item may be chosen to be packed giving the full formulation

$$\max\left[\text{Value} = \sum v_i x_i\right]$$

$$\text{Subject to} \sum_i w_i x_i \leq C \quad x_i \text{ binary}$$

(note that within packages the variables can be defined to be binary/logical).
These problems can be solved using

Integer Programming,
Dynamic Programming,
Heuristics,
and Others

Notice that in an n variable problem, there are 2^n possible solutions where each variable can be [0, 1].

Notice that $N = 12$ is four times larger than the $N = 3$ problem but would take 500 times as long to solve given a full search.

Thus, there may be a need for an alternative approach to solving these problems, heuristic techniques?

3.4.3 Multiple Knapsack Problem

A more general formulation, of knapsack problems, leads to a model with many applications, not all applications seem at first to be related to knapsack problems.

Here, there exists a set of items are to be transported, as before, but here using a set of containers where container j has a capacity of C_j.

Here, the objective may be to

- Maximise the value of the packed items given a fixed number of containers, or
- Minimise the required number of containers to pack all the items, or
- Minimise the cost of the containers needed to pack all the item.

A model can be defined using the binary variables

$X_{ij} = 1$ indicate that item i is packed in container j
$X_{ij} = 0$ indicate that item i is not packed in container j

and
It follows that

$\Sigma X_{ij} \leq 1$,	to ensure that item i is packed in one of the containers
$\Sigma W_i X_{ij} \leq C_j$	to ensure that container j is not over-packed
$\text{Value} = \Sigma\Sigma V_i X_{ij}$	Value of items chosen to be shipped, choose the best items to maximise value

Question: "How many containers will be needed to pack all these items?"
To Minimise the number of containers needed to pack all the items
If all are to be packed using a minimum number of containers, then

$\Sigma X_{ij} = 1$ Each item has to be packed
$\Sigma X_{ij} \leq \delta_j M$ How does this work?
$\Sigma W_i X_{ij} \leq \delta_j C_j$ If used do not exceed capacity

Note: $\delta_j = 1$ implies container j is used.
Number of containers = $\Sigma \delta_j$
Minimise {number of containers}
Or to Minimise the cost of containers needed to pack all the items
If all are to be packed at a minimum cost (containers)

$\Sigma X_{ij} = 1$ Each item has to be packed
$\Sigma W_i X_{ij} \leq \delta_j C_j$ If used do not exceed capacity

Note $\delta_j = 1$ implies container j is used.
Cost of containers = $\Sigma C_j \delta_j$
Minimise {cost of containers}
Or to minimise a function of both number and cost of containers, in this case, the problem may need to be solved using a heuristic approach.

3.4.4 Logical Constraints

There are situations where logical constraints are needed to ensure that the container is packed safely.

For example, considering only one item of each type:

If item 1 is packed into container j, then item 2 must be packed into container j, this requires the additional constraints,

$$x_{2j} \geq x_{1j} \quad \text{all } j$$

or if items 1 and 2 cannot be packed into the same container, this requires the additional constraints

$$x_{2j} + x_{1j} \leq 1 \quad \text{all } j$$

Notice that if multiple items could be packed, additional "logical" variables would be needed, when items 1 and 2 cannot be packed into the same container giving:

$$x_{2j} \leq M\delta_{2j} \quad \text{all } j$$
$$\delta_{2j} + \delta_{1j} \leq 1 \quad \text{all } j$$

3.5 Problems Reducible to Knapsack Problems

3.5.1 Allocating Workers to Tasks

There exists a set of n jobs and a pool of (n) workers to be allocated to these tasks, each worker could carry out more than one task and the tasks could be completed in any order.

Models to describe this task can be based around those for the multiple knapsack problem but also considering:

Work loads evened out,
Staff capabilities, and
Staff changeovers, Manning machines implications.

What's the objective, how to describe a good solution?
Why solve it using LP methods? Is this an appropriate method/approach?
Basic Model

Let	$x_{ij} = 1$	worker i is working at time j
	$\delta_i = 1$	worker i is working at (any) time
	C_i	cost per session for worker i
	m_j	number of workers needed at time j
	n	number of available workers
	w	number of time periods

Then $\sum_i x_{ij} = m_j$ all i, adequate staffing level at all times

$\sum_j x_{ij} \leq M\delta_i$ all j, recording workers i s status, working or not working

With the objective

$$\text{Minimise}\left(\sum_i \delta_i\right) \text{number of workers or}$$

$$\text{Minimise}\left(\sum_j \sum_i C_i x_{ij}\right) \text{cost of workers}$$

Extended model

To share out the work and to prevent any worker being overloaded, an alternative model has the form

$$\sum_i x_{ij} = m_j$$

$$\sum_j x_{ij} + T \leq w \quad \text{all } i, \text{ adequate staffing level at all times}$$

$$\text{Maximise}(T) \quad \text{as } T \text{ is common to all constraints; its value will be defined by that worker assigned the greatest workload}$$

Limiting shift length and allowing breaks

Each worker is limited to working for two time periods from any three, effectively allowing regular breaks from work. This is implemented through the addition of the constraints

$$x_{ij} + x_{i,j+1} + x_{i,j+2} \leq 2 \quad \text{all } i, j = 1, \ldots, w-2$$

Currency of worker skills

Within some applications, it is essential that the workers retain "currency" for the task in hand, for example, air traffic controllers who have carry out a particular role every K days; otherwise, they have to be retrained. However, if they carry out the task at the start of their working day, their currency value needs to be reset to correctly represent their training needs.

The following example is used to show the need for updating the currencies.

A model with 5 work stations and 11 workers where each worker cannot work for more than 2 successive sessions was formulated, and the objective was to maintain the workers currency.

The solution, see Table 3.6, illustrates the deficiencies associated with this model; that is:

- The solution could be deduced; the computer model does not add to a "simple" solution.
- Although workers 5, 11, 10, 9 and 4 have re-validated their currency at the end of session 1, the model still allocates them to the same task later in the day, in preference to using another worker and increasing their currency.
- The solution does not employ all workers.

Table 3.6 Allocation of workers to tasks and workers currency

		Time							
		1	2	3	4	5	6	7	8
Work station	1	5	5	8	5	5	8	5	5
	2	11	11	1	11	11	1	11	11
	3	10	10	7	10	10	7	10	10
	4	9	9	3	9	9	3	9	9
	5	4	4	6	4	4	6	4	4

This requirement, changing the cost function value, can be achieved through the addition of the additional constraints and the extended notation:

x_{ijk}	representing the state of worker i with respect to task j at time k
$\sum_i x_{ijk} = 1$	all j, k assign only one worker to each job at each time
$\sum_j x_{ijk} \leq 1$	all i, k assign at most one task to each worker at each time.

In total $jk + ik$ constraints, for a typical problem of 15 workers, 8 time slots and 12 tasks, give 216 constraints.

$\sum_1^{k-1} x_{ijk} \leq M\gamma_{ijk}$	all i all n, number of controllers,
	all k in T the number of daily work slots.
	all j in S the number of different tasks
$\gamma_{ik} = 1$	controller i has completed task j at time k.

An additional njk constraint, for a typical problem of 15 workers, 8 time slots and 12 tasks, gives 1440 additional constraints.

The currency for worker i at task j in time slot k now becomes

$$c_i(1 - \gamma_{ijk}) + K\gamma_{ijk} \quad \text{where } K \text{ is the maximum currency}$$

Then, if the controller has not worked at task j the current currency is retained throughout the day. However, if the controller has worked at this task earlier in the day the currency is reset to its maximum value. Consequently, controllers are primarily assigned to tasks so that their currencies (all tasks) are maximised.

But this will generate many constraints and increase the solution time.

3.5.2 *Allocation of Workers to Teams*

Teams, of m players, are to be selected from a pool of n players for a competition lasting over d days, so that the skill rating for each team is levelled out. Each player must be selected at least once but no player may be selected on more than 2 days in any three days. All players have a skills rating from 1, highest skill, to 10, lowest skill.

Let	$x_{ij} = 1$	indicate that player i is to be a team member on day j
	s_i	skill rating for player i
	T	number of players in a team

Basic formulation	$\sum_j x_{ij} \geq 1$	all i, each to be selected at least once
	$\sum_i x_{ij} = T$	all j, select a full team
	$\sum_i s_i x_{ij} = V_j$	all j, calculate the total team skill rating
	$V_j \leq VB$	all j, find the lowest skilled team

Minimise(VB) even out the teams skills.
Extended formulation, add the 2 days from 3 conditions

$$x_{ik} + x_{i,k+1} + x_{i,k+2} \leq 2 \quad \text{for} \quad k \epsilon \{1, \ldots, d-2\}$$

Notice that if there were more than one type of player, attacker and defender for example, the constraints would become

$\sum_j x_{ij} \geq 1$	all i, each to be selected at least once
$\sum_j x_{ij} = T_D$	all $j \epsilon D$, select the defenders
$\sum_i x_{ij} = T_A$	all $j \epsilon A$, select the attackers
$\sum_i s_i x_{ij} = V_j$	all j, calculate the total team skill rating
$V_j \leq VB$	all j, find the lowest skilled team
$x_{ik} + x_{i,k+1} + x_{i,k+2} \leq 2$	for $k \epsilon \{1, \ldots, d-2\}$

3.5.2.1 Allocating Workers to Projects

A group of n workers are available to work on m projects, but workers are not qualified to work on all projects, and each worker holds a skills rating relevant to each project (1 highest grade, 10 lowest grade), and ideally, the most skilful workers should be assigned to each project.

Let	$x_{ij} = 1$	indicates that worker i is to be a team member on project j
	$a_{ij} = 1$	indicates that worker i is able to work on project j
	r_{ij}	skill rating for worker i at project j
	D_j	number of workers required for project j

Model 1: each worker could be assigned to more than one project

$$\sum_i a_{ij} x_{ij} = D_j \qquad \text{all } j$$

$$\text{Minimise}\left(\sum_i \sum_j r_{ij} x_{ij}\right) \qquad \text{achieve the optimal skills rating}$$

Model 2: each worker could only be assigned to one project

$$\sum_i a_{ij}x_{ij} = D_j \qquad \text{all } j$$
$$\sum_j x_{ij} \le 1 \qquad \text{all } i$$
$$\text{Minimise}\left(\sum_i \sum_j r_{ij}x_{ij}\right)$$

Model 2a: each worker could only be assigned to one project and minimise the largest project skills rating

$$\sum_i a_{ij}x_{ij} = D_j \qquad \text{all } j$$
$$\sum_j x_{ij} \le 1 \qquad \text{all } i$$
$$\sum_i r_{ij}x_{ij} = v_j \qquad \text{all } j$$
$$v_j \le vv \qquad \text{all } j$$
$$\text{Minimise}(vv)$$

Model 3: all workers must be employed but on no more than 4 projects. Each project requires 6 workers.

$$\sum_i a_{ij}x_{ij} = D_j \qquad \text{all } j$$
$$\sum_j x_{ij} \le 4 \qquad \text{all } i$$
$$\sum_j x_{ij} \ge 1 \qquad \text{all } i$$
$$\text{Minimise}\left(\sum_i \sum_j r_{ij}x_{ij}\right)$$

Model 4: minimise the maximum score

$$\sum_i a_{ij}x_{ij} = D_j \qquad \text{all } j$$
$$\sum_j x_{ij} \le 4 \qquad \text{all } i$$
$$\sum_j x_{ij} \ge 1 \qquad \text{all } i$$
$$\sum_i r_{ij}x_{ij} = v_j \qquad \text{all } j$$
$$v_j \le vv \qquad \text{all } j$$
$$\text{Minimise}(vv)$$

Table 3.7 Staff allocations from each model

Solution Data Set 1: allocation of workers to tasks												
Model 1		*Cost/rating = 19*										
Project	Worker	1	2	3	4	5	6	7	8	9	10	*D*
	1									1	1	2
	2									1		1
	3										1	1
	4			1	1					1		3
	5						1					1
Model 2		*Cost/rating = 21*										
Project	Worker	1	2	3	4	5	6	7	8	9	10	*D*
	1							1			1	2
	2	1										1
	3		1									1
	4				1				1	1		3
	5						1					1
Data Set 2												
Model 3		*Cost/rating = 68*										
Project	Worker	1	2	3	4	5	6	7	8	9	10	
	1	1		1		1		1		1	1	
	2	1	1		1		1			1	1	
	3		1				1	1	1	1	1	
	4	1	1	1	1	1	1					
	5	1	1			1	1		1	1		
Total/worker		4	4	2	2	3	4	2	2	4	3	
Model 4		*Cost/rating = 78*										
Project	Worker	1	2	3	4	5	6	7	8	9	10	
	1	1				1		1	1	1	1	
	2	1	1	1			1			1	1	
	3		1	1			1	1		1	1	
	4	1	1		1	1	1			1		
	5		1		1	1	1		1		1	
Total		3	4	2	2	3	4	2	2	4	4	

Sample Data used in validating **Models 1 and 2** and results
Models were constructed where there are 12 workers and 5 projects (Table 3.7).

Availability matrix	*a*	1 available 0 not available
Ratings matrix	*r*	lower number indicates more able
Staffing requirement	*D*	staff required for each project

$$
\begin{array}{llll}
 & \text{Availabilities} & & \text{Ratings} \\
a :: & [1,1,1,0,0, & r :: & [4,2,8,1,3, \\
 & 1,0,1,0,1, & & 8,1,1,2,1, \\
 & 1,0,1,1,0, & & 7,7,7,7,8, \\
 & 0,0,1,1,1, & & 5,3,5,2,8, \\
 & 0,1,1,0,0, & & 1,4,7,3,1, \\
 & 1,1,0,0,1, & & 6,2,2,4,1, \\
 & 1,1,1,1,0, & & 3,4,2,8,9, \\
 & 1,0,1,1,1, & & 5,5,3,7,1, \\
 & 1,1,0,1,0, & & 1,2,3,4,2, \\
 & 1,0,1,0,1] & & 1,1,1,8,4]
\end{array}
$$

$$\text{Demands} \quad D :: \quad [2,\ 1,\ 1,\ 3,\ 1]$$

Models 3 and 4 alternative and additional data

$$
\begin{array}{ll}
\text{Demand}\, D :: & [6,\ 6,\ 6,\ 6,\ 6] : \\
\text{Workload restrictions} & 1 \leq x_{ij} \leq 4
\end{array}
$$

3.5.3 *Stock Cutting Problems*

This problem is used both to illustrate the need for good modelling, when the problem size increases the model does not increase exponentially in size, and the use of a Knapsack formulation in modelling a problem from a different area.

The roll of paper shown in Fig. 3.2 is to be cut into smaller rolls to satisfy customer demand.

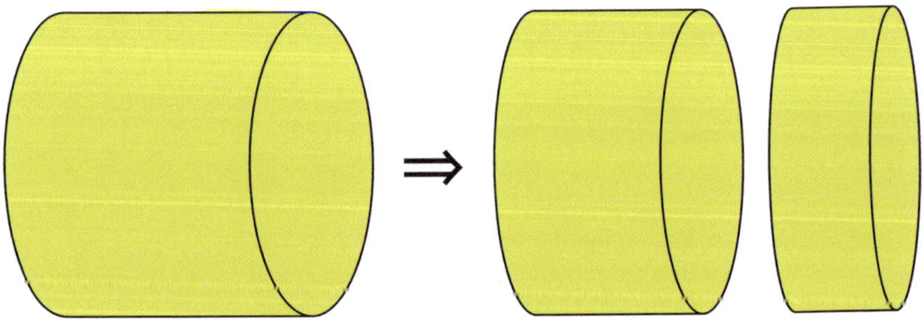

Fig. 3.2 The roll of paper in smaller rolls to satisfy customer demand

For example:

The width of a roll is W and there is a demand n_i rolls of width w_i.

This problem can be represented by a Knapsack model, let x_{ij} be the number of rolls of width w_i cut from roll j.

Leading to the model:

Constraints

$$\begin{array}{ll} \text{Width restriction} & \sum_i^r w_i x_{ij} \leq W_j \\ \text{Satisfy demand} & \sum_j^k x_{ij} \quad \geq n_i \\ \text{Roll status} & \sum_i^r x_{ij} \quad \leq M\delta_j \end{array}$$

Objective

Several objectives are again possible:

- Minimise number of rolls cut (equivalent to minimise the number of containers in a knapsack problem).

$$\text{Minimise} \sum \delta_j$$

- Minimise the cost of "material" used where C_j is the cost of using roll j.

$$\text{Minimise} \sum C_j \delta_j$$

Or when demand is greater than capacity, let u_i be the unsatisfied demand for rolls of width w_i the model becomes

Constraints

$$\begin{array}{ll} \text{Width restriction} & \sum_i^r w_i x_{ij} \quad \leq W_j \\ \text{Satisfy demand} & \sum_j^k x_{ij} + u_i = n_i \\ \text{Roll status} & \sum_i^r x_{ij} \quad \leq M\delta_j \end{array}$$

Objective

Several objectives are again possible:

- Minimise unsatisfied demand

$$\text{Minimise} \sum u_i$$

- Minimise cost of unsatisfied demand where c_i is the cost per unit of unsatisfied demand for customer i.

$$\text{Minimise} \sum c_i u_i$$

3.5.4 *More Knapsack "Type" Problems: Set Covering Problems*

A typical example is concerned with the reorganisation of fire services to provide a good coverage for all towns in the district.

○	Districts to be serviced
●	Possible sites for fire stations
-·-·-·-▶	Districts served (in an acceptable time) by a fire station

The aim is to determine and locate an appropriate number of fire stations so that all districts will receive an acceptable service. Whilst minimising costs or number of fire stations (Fig. 3.3).

Those models developed around the Knapsack problem can be used to determine the location of a suitable set of fire stations. To use these models, assume that each district is a "package" and that each fire station is a "container".

Thus

$x_{ij} = 1$ district i being allocated to fire station j,
$\delta_j = 1$ fire station site j being used, 0 otherwise
$s_{ij} = 1$ district i could be allocated to fire station j

The basic model is now given by:
For each district (for all i):

$$\sum_j s_{ij} x_{ij} = 1$$

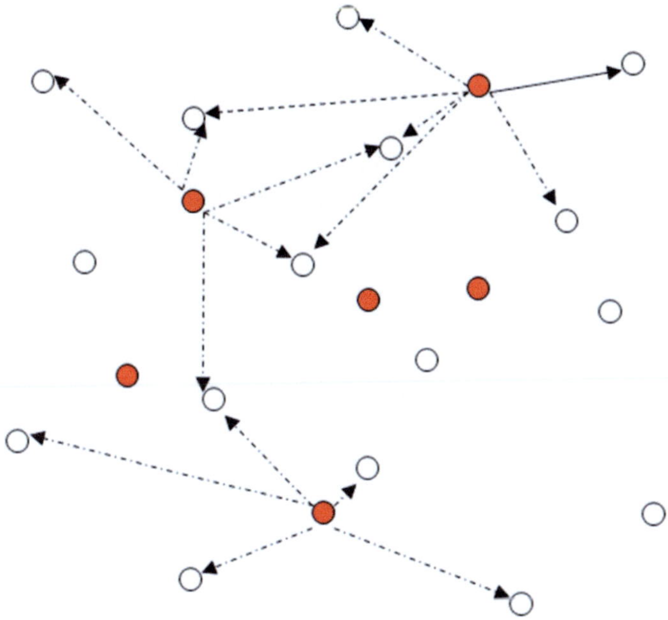

Fig. 3.3 Possible sites for fire stations

For each possible fire station:

$$\sum_i x_{ij} \leq M\delta_j$$

The objective function could be either to "Minimise the total number of fire stations used":

$$\text{Minimise} \sum \delta_j$$

or to minimise the cost of the fire stations used:

$$\text{Minimise} \sum C_j\delta_j$$

This approach also allows more complex modelling to represent more complex/realistic conditions.

For example, a fire station can service at most N districts, add the constraints

$$\sum_i x_{ij} \leq N \quad \text{for all}\, j,$$

or if the districts demands for service is known, add the constraints

$$\sum_i D_i x_{ij} \le K \quad \text{for all } j.$$

A final aim may be to combine objectives. For example, minimise the number of fire stations and the distance/time of travel to each district.

Thus combining the objectives

- Minimise the total number of fire stations, and

$$\text{Minimise} \sum \delta_j$$

- Minimise the total travel time/distance

$$\text{Minimise}\left(\sum_i \sum_j d_{ij} x_{ij}\right)$$

with the respective importance of each acting to determine the optimal solution.

Finally, it may be desirable to "even out" the demand at each fire station, so replace capacity constraint

$$\sum_i D_i x_{ij} \le K, \text{ with}$$
$$\sum_i D_i x_{ij} + b_j = K \text{ all } j$$

and add a third objective

$$\text{Minimise}\left(\sum_j b_j\right).$$

This model demonstrating that there may be more than one objective in a planning problem, and the solution process will need to be developed around the problem.

3.6 Network Models

These are an important class of problem where the form and structure of the model can lead to a simple solution technique.

The base models are as follows:

Transportation Problems, and
Network Flow Problems.

3.6.1 *Defining Transportation Problems*

This is a special type of Linear Programming problem. This problem type can be illustrated by the example:

A firm has two depots and three shops, and it wishes to deliver the goods from the depots (*A*, *B*) to the shops (*X*, *Y*, *Z*) as cheaply as possible.

Given the delivery network, 100 items available at *A* and a demand for 80 at *X* (Fig. 3.4; Table 3.8).

The problem can be modelled by considering each depot and shop in turn.

For example, at *A* it follows that the quantities $\{x_{11}, x_{12}, x_{13}\}$ dispatched are such that

$$x_{11} + x_{12} + x_{13} = 140$$

and at *X* the quantities $\{u, x\}$ received

$$x_{11} + x_{21} = 70$$

Considering all sources and destinations leads to the full formulation

$$\begin{aligned} x_{11} + x_{12} + x_{13} &= 140 \\ x_{21} + x_{22} + x_{23} &= 100 \\ x_{11} + x_{21} &= 70 \\ x_{12} + x_{22} &= 80 \\ x_{13} + x_{23} &= 90 \end{aligned}$$

with the cost function (to be minimised) given by

$$\text{cost} = 6x_{11} + 8x_{12} + 3x_{13} + 4x_{21} + 5x_{22} + 9x_{23}$$

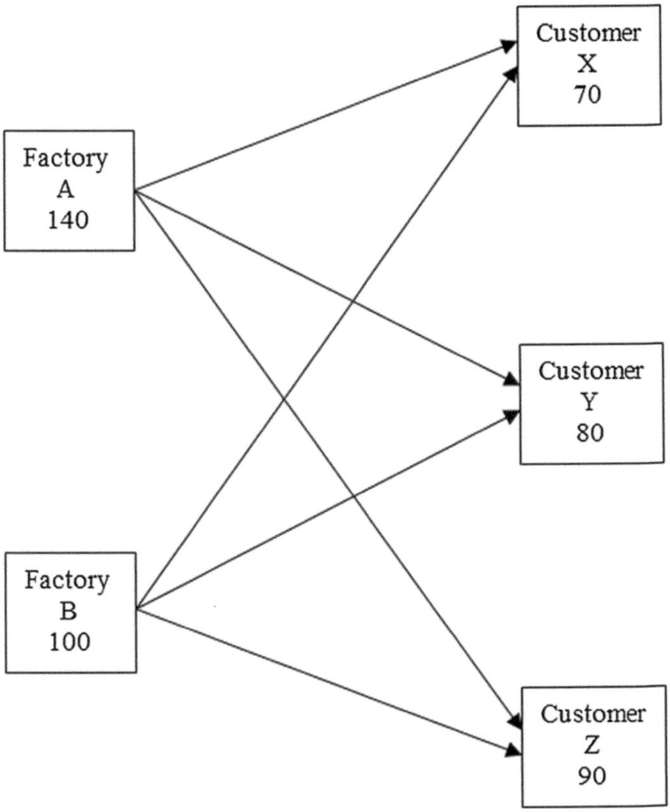

Fig. 3.4 Warehouse destination diagram with the cost data

Table 3.8 Shipping costs warehouse to destination

Route	Unit shipping cost	Quantity using the route
AX	6	x_{11}
AY	8	x_{12}
AZ	3	x_{13}
BX	4	x_{21}
BY	5	x_{22}
BZ	9	x_{23}

A linear programming problem where the solution has to have an all integer form with equalities rather than inequalities. But notice that there are only 4 independent equations and 6 unknowns.

$$x_{11}+x_{12}+x_{13}=140$$
$$x_{21}+x_{22}+x_{23}=100$$
$$x_{11}+x_{21}=70$$
$$x_{12}+x_{22}=80$$

Note that: if these are satisfied, then the other (5th) equation must be satisfied.

Such a set of equations can be solved by setting two of the variables to zero and solving the remaining set of simultaneous equations.

For example, if $x_{21}=0$ and $x_{12}=0$, then the solution is

$$x_{11}=70;\ x_{22}=80;\ x_{13}=70;\ x_{23}=20$$

Therefore, it seems as if there will be at most 15 possible solutions. A restricted search space each having 4 nonzero shipments. Therefore, it seems that there should be an alternative, simple, method to determine the optimal solution, than the Simplex Method.

{Note: The optimal solution uses no more than
Number of depots + number of shops – 1 routes}

x_{ij} quantity sent from warehouse i to shop j
c_{ij} cost of sending one item from warehouse i to shop j
A_i stock available at warehouse i
D_j stock available required at shop j

Giving the model:

$$\sum_i x_{ij} \geq D_j \quad \text{all } j$$
$$\sum_j x_{ij} \leq A_i \quad \text{all } i$$
$$\text{Minimise} \quad \text{Cost} = \sum_i \sum_j c_{ij} x_{ij}$$

3.6.2 Assignment Problems

These were originally observed when trying to allocate a set of jobs between a set of workers. For example, there are three jobs to be allocated between three workers so that each worker receives one of these jobs. The following cost Table 3.9 indicates the cost of allocating each job to each worker:

This is a special case of a Transportation Problem number of solutions n!.

$x_{ij}=1$ worker 1 is assigned to task j
c_{ij} cost of assigning worker I to task j

Table 3.9 Worker task cost data

	X	Y	Z
A	8	12	7
B	10	6	8
C	6	7	6

Giving the model:

$$\sum_i x_{ij} = 1 \qquad \text{all}\, j \,\text{assign all jobs}$$
$$\sum_j x_{ij} = 1 \qquad \text{all}\, i \,\text{assign all workers}$$
$$\text{Minimise Cost } = \sum_i \sum_j c_{ij} x_{ij}$$

Notice that if there are m workers and n tasks, where $m > n$ adding

E_i cost of worker i not being required,

the model becomes

$$\sum_i x_{ij} = 1 \quad \text{all j}$$
$$\sum_j x_{ij} = 1 \quad \text{all i}$$
$$\text{Minimise Cost } = \sum_i \sum_j \left(c_{ij} x_{ij}\right) + \sum_i (1 - x_i) E_i$$

3.6.3 Network Flow Models

Many problems can be represented in the form of a network and solutions found by applying suitable methods of solution.

Problems can involve the following:

- Finding the shortest path/route
- Connecting all points in a network in an optimum manner
- Maximising the flow through a network
- Planning and control of a network of activities finding the shortest completion time
- Areas reducible to network flows, for example transportation problems.

Examples of networks:

Nodes	*Branches*	*Flow*
Workstations	Cable	Information
Junctions	Roads	Traffic
Valves	Pipelines	Gas

3.6.3.1 Shortest Route

The shortest route problem is concerned with finding the shortest route from a source to a sink through a connecting network with nonnegative distances associated with respective branches of the network (Fig. 3.5).

The LP formulation of this problem would be as follows, send one item from node 1 to node N: let

$x_{ij} = 1$ if branch i to j is included in the shortest route
$x_{ij} = 0$ if branch is not used

The objective is to minimise the total cost of going from node 1 to node N, that is

$$\text{Minimise} \quad Z = \sum_i \sum_j d_{ij} x_{ij}$$

where d_{ij} is the direct distance between nodes i and j.

Subject to:

$$\sum_j x_{1j} = 1 \qquad \text{Source node}$$
$$\sum_i x_{iN} = 1 \qquad \text{Sink node}$$
$$\sum_k x_{ik} - \sum_k x_{kj} = 0 \quad \text{all other nodes } k.$$

3.6.3.2 Maximum Flow in a Network

Typically, there exists a quantity of gas at the source and the aim is to send as much as possible through a network pipes to supply a customer at the end of the network (Fig. 3.6).

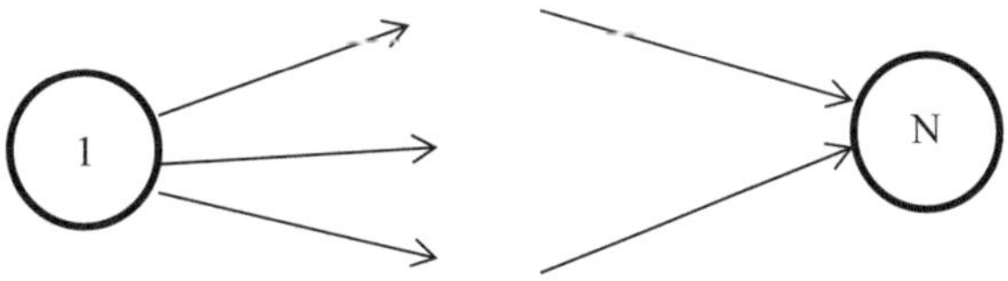

Fig. 3.5 Finding the shortest route from a source to a sink through a connecting network

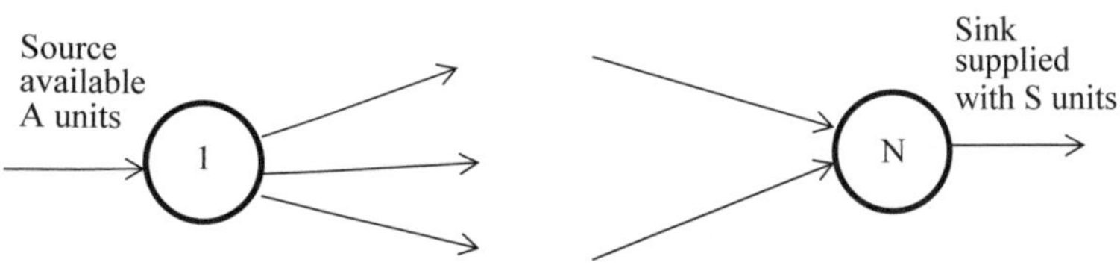

Fig. 3.6 Network pipes to supply a customer at the end of the network

The LP formulation of this problem would be as follows, let

x_{ij} be the quantity sent along the pipe joining nodes i and j,
C_{ij} be the carrying capacity of the pipe joining nodes i and j.

The objective is to maximise the total quantity reaching node N, that is

$$\begin{array}{lll} \text{Maximise} & Z = S & \\ \text{Subject to:} & \sum_j x_{1j} \leq A & \\ & \sum_i x_{iN} = S & \text{Sink node} \\ & \sum_k x_{ik} - \sum_k x_{kj} = 0 & \text{all other nodes } k \\ & \sum_i \sum_j x_{ij} \leq C_{ij} & \text{all pipes} \end{array}$$

This model can be extended through the addition of pumping stations, with capacity or/and cost at some of the nodes, for example node N has a capacity of C_N giving the additional constraint

$$\sum_k x_{Nk} \leq C_N$$

Note the objective could have been to satisfy demand or to send as much as possible or to minimise costs?

3.6.3.3 Network Planning Models (Critical Path Models)

The network in Fig. 3.7 can be represented by the Linear Programming model

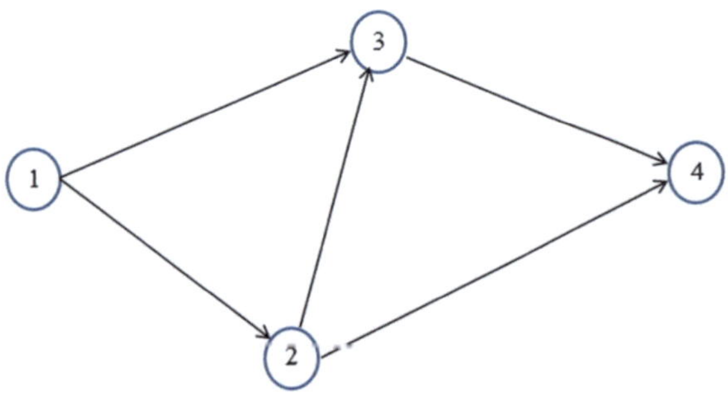

Fig. 3.7 Network planning models

$$
\begin{aligned}
x_2 &\geq x_1 + d_{12} \\
x_3 &\geq x_1 + d_{13} \\
x_3 &\geq x_2 + d_{23} \\
x_4 &\geq x_2 + d_{24} \\
x_4 &\geq x_3 + d_{34} \\
&\text{Minimise}(x_4)
\end{aligned}
$$

Following from this model, the case where some jobs could be completed using alternative technologies, with different costs and completion times, using job (1, 2) as an example with d_{12}^s as the alternative duration, can be modelled as follows:

$$
\begin{aligned}
x_2 &\geq x_1 + d_{12} - 100\delta_s \\
x_2 &\geq x_1 + d_{12}^s - 100\delta \\
\delta + \delta_s &= 1 \\
x_3 &\geq x_1 + d_{13} \\
x_3 &\geq x_2 + d_{23} \\
x_4 &\geq x_2 + d_{24} \\
x_4 &\geq x_3 + d_{34} \\
&\text{Minimise(costs)}
\end{aligned}
$$

3.6.3.4 Transhipment Models, Formulation and Problem Size

These are derived from transportation models through the addition of an extra stage (equivalent to keep the goods in a warehouse), for example Fig. 3.8:

If all factories were connected to each shop, connecting all sources with all destinations by way of all possible warehouses, this could be presented as a Transportation problem.

Here, there are 3 sources (m), 3 destinations (n) and 2 warehouses (w); therefore: there are $m \times n \times w = 18$ possible routes to be used by the company.

This can be modelled as a network flow problem, let

x_{ij} quantity shipped from factory i to warehouse j
y_{jk} quantity shipped from warehouse j to shop k
cx_{ij} cost of shipping an item from factory i to warehouse j
cy_{jk} cost of shipping an item from warehouse j to shop k
A_i available stock at factory i
C_j capacity of warehouse j
D_k demand at shop k.

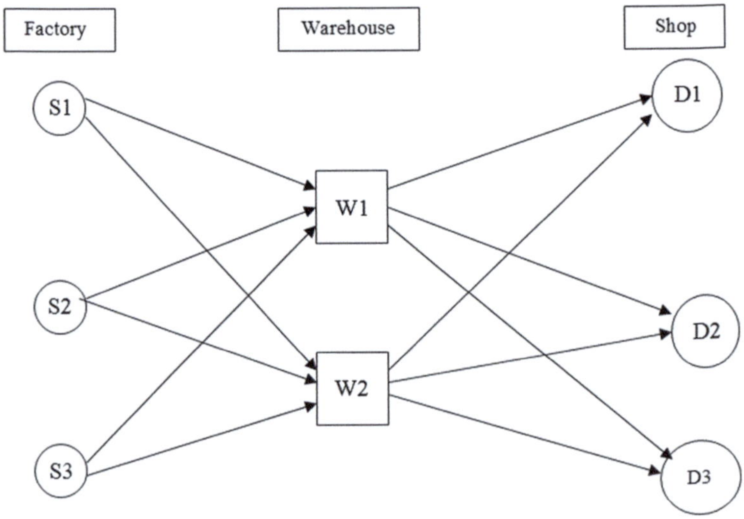

Fig. 3.8 Trans-shipment model, factory to warehouse to shop

Giving the constraints

$\sum_j x_{ij} \leq A_i$	sent from factory i
$\sum_i x_{ij} \leq C_j$	sent to warehouse j
$\sum_k y_{jk} \leq \sum_i x_{ij}$	balance stocks at warehouse j
$\sum_j y_{jk} \geq D_k$	sent to shop k
$\sum_i x_{ij} \leq M\delta_j$	indicates usage of warehouse j.

3.6.3.5 Production Planning Models

The objective is the determination of a production schedule so that the firm can satisfy demand whilst minimising costs. The costs to be considered are the production costs and holding costs (per unit produced).

Notation used are as foolows:

d_j the demand in month j
x_j the assigned production in month j
c_j the production capacity in month j
p_j the unit production cost in month j
h the unit holding cost per month.

Thus, it follows that the x_j need to satisfy the constraints:
demand in each month

$$\begin{aligned} &x_1 && \geq d_1 \\ &x_1 + x_2 && \geq d_1 + d_2 \\ &x_1 + \cdots && + x_n \geq d_1 + \cdots + d_n \end{aligned}$$

capacity constraint in each month

$$x_j \leq c_j \quad \text{all j}$$

whilst minimising the total cost, where cost is given by:
the production costs

$$p_1 x_1 + \cdots + p_n x_n$$

the holding costs, and as the month end stocks are given by

$$\begin{aligned} s_1 &= x_1 - d_1 \\ s_2 &= (x_1 + x_2) - (d_1 + s_2) \\ &\text{etc} \end{aligned}$$

the holding cost will be given by

$$hs_1 + hs_2 + \cdots + hs_n = h(nx_1 + (n-1)x_2 + \cdots + 1x_n)$$

to give the cost function, Production + Holding

$$\cos t = p_1 x_1 + \cdots + p_n x_n + h(nx_1 + (n-1)x_2 + \cdots + 1x_n)$$

Model Summary
Problem size: n planned production periods imply n variables
Planning for the daily production with a planning horizon of

one week implies	7 variables
one month	20 variables
per year	250 variables

Alternative Model
Additional Notation

x_{ij}	production in month i for use in month j
$x_{ij} = 0 \quad i > j$	no backlogging of orders

Constraints and Objective

Capacity constraint	$\sum_j x_{ij} \leq c_i$ all i
Demand requirement	$\sum_i x_{ij} \geq d_j$ all j
Production costs	$\sum_i \sum_j p_i x_{ij}$

Holding costs	$\sum_i \sum_j (j-i)hx_{ij}$
Total cost to be minimised	$\sum_i \sum_j ((j-i)h+p_i)x_{ij}$

Model Summary
Problem size: n planned production periods imply $n(n + 1)/2$ variables
Planning for the daily production with a planning horizon of

one week implies	30 variables
one month	190 variables
one day	31375 variables

Data requirement: production costs and holding costs needed to complete the model.

Both models are equally valid, represent the production planning process, but which model is best/most appropriate?

On the basis of number of variables, the first is more appropriate; but on the basis of management information, the second may be the best model. This question is addressed further within the case study.

3.7 Other Mathematical Modelling Applications

3.7.1 Data Envelopment Analysis

Typically Data Envelopment Analysis concerned with evaluating the performance of a department within an organisation:

- Is this department performing as well as others?
- Can it improve its performance by acting more like other departments?

To be able to compare a department with other like departments, there needs to be an agreed formal methodology for comparison.

For example, to demonstrate the need for an agreed methodology, consider the very straightforward sporting example "which team is the most effective/best association football team over a season".

Possible comparison methods used

- Points awarded $[\text{win, draw, loss}] = [2, 1, 0]$ or $[3, 1, 0]$
- Goals scored (S) and conceded (C) $[\text{goal measure}] = S{-}C \, or \, S/C$
- With the better team having gained the highest number of points and achieved the highest goal measure.

Alternatively

- Most wins, or win percentage
- Fewest losses.

The following example aims to show that the application of these rules can lead to very different results.

Consider a season (8 teams each playing 14 matches) where the results are presented in Table 3.10:

With the current points awarded, 3 points for a win and 1 for a draw, Haiton are relegated! and therefore can be said to be less efficient (not as good) than the other teams.

But could they claim to be better than the other teams if another criteria were to be used?

Using the alternative system awarding 2 points for a win and 1 for a draw, the final table would have given either:

Haiton equal top, if all goal differences and number of goals scored are equal, or Haiton "mid table" if all goal differences are not equal.

In either case, Haiton are not relegated! (Table 3.11).

In fact, Haiton could claim to be the "best team/most efficient" because they have not lost a match.

Note: In Association Football currently in UK leagues if two teams have equal points that team with the greater goal difference (scored-conceded) is judged to be the better team, but historically, they used to use goal ratio (scored/conceded)! this change can also alter the team's final standing when they have equal points.

Table 3.10 Association Football league table, current points system

	Played points	Won 3	Drawn 1	Lost 0	Points total	Goal difference
Ayton	14	6	2	6	20	15
Beeham	14	6	2	6	20	10
Ceeford	14	6	2	6	20	5
Deeton	14	6	2	6	20	0
Efax	14	6	2	6	20	−7
Fesley	14	6	2	6	20	−11
Geeton	14	6	2	6	20	−12
Haiton	14	0	14	0	14	0

Table 3.11 Association Football league table, previous points system

	Won 2	Drawn 1	Lost 0	Points total	Goal difference
Ayton	6	2	6	14	15
Beeham	6	2	6	14	10
Ceeford	6	2	6	14	5
Deeton	6	2	6	14	0
Haiton	0	14	0	14	0
Efax	6	2	6	14	−7
Fesley	6	2	6	14	−11
Geeton	6	2	6	14	−12

Data envelopment analysis (DEA), occasionally called frontier analysis, was proposed by Charnes, Cooper and Rhodes (1978). It is a performance measurement technique which can be used to enable the evaluation of the relative efficiency of decision-making units (DMUs) in organisations (above teams in a competition).

Examples of such units to which DEA has been applied are banks, police stations, hospitals, tax offices, prisons, defence bases (army, navy, air force), schools and university departments.

Often departments are compared by way of their efficiencies. A problem is how do you "fairly" compute efficiencies.

Data Envelopment Analysis proceeds by "challenging" each team to select the best weightings for their inputs (win or draw here) to demonstrate their efficiency (best overall result).

In this example,

Haiton could choose win 2 points, draw 1 point
Geeton could choose win 3 points, draw 2 points

DEA applied in business: An Illustrative Example

A company operates outlets in three regions each with two inputs and two outputs:

		Company			
		X	Y	Z	weights
INPUTS	A	20	40	30	v_1
	B	30	20	25	v_2
OUTPUTS	C	40	30	25	w_1
	D	30	50	20	w_2

To enable comparisons, the company's management challenges each department to use this data set to "prove" that they can be considered to be efficient, choosing the best weighting of their inputs and outputs.

Department X could use input A and output C to show that they are performing better than the other departments and thus demonstrating their efficiency,

Ratios Output/Input	X	Y	Z
	2.0	0.75	1.17
Efficiency %	100	37.5	58.5

Similarly department Y would could use input B and output D to demonstrate efficiency,

Ratios Output/Input	X	Y	Z
	1.0	2.5	1.6
Efficiency %	40	100	64

A formal approach for each company would require the determination of the weightings to be attached to each input and output quantity so that their efficiency rating is maximised. Notice also that each input and output must be used at an agreed specified weighted level, otherwise they can be assumed to have no importance.

For company X, this produces the model:

$$\begin{array}{ll} \max[(40w_1+30w_2)/(20v_1+30v_2)] & \text{optimise efficiency} \\ 40w_1+30w_2 \le 20v_1+30v_2 & \text{output less than input } X \\ 30w_1+50w_2 \le 40v_1+20v_2 & \text{output less than input } Y \\ 25w_1+20w_2 \le 30v_1+25v_2 & \text{output less than input } Z \end{array}$$

This can be represented as a linear programming problem by setting the input for company X to a value of 100, giving the model for company X as

$$\begin{array}{ll} \max[40w_1+30w_2] & \\ 20v_1+30v_2 = 100 & \text{set input at } 100 \\ 40w_1+30w_2\ 20v_1+30v_2 & \\ 30w_1+50w_2\ 40v_1+20v_2 & \\ 25w_1+20w_2\ 30v_1+25v_2 & \end{array}$$

Finally, a normal additional condition is that all inputs and outputs must influence the calculation, and all weights greater than zero, for example each input and output quantity, must contribute at least 15% towards the total quantity, adding the constraints:

$$\begin{aligned} 40w_1 &\ge 0.15(40w_1+30w_2) \\ 30w_2 &\ge 0.15(40w_1+30w_2) \\ 20v_1 &\ge 0.15(20v_1+30v_2) \\ 30v_2 &\ge 0.15(20v_1+30v_2) \end{aligned}$$

$$\begin{array}{lll} \text{Results} & \text{Efficiency} = 100 & \\ & \text{Weightings} & w=(0.375, 2.833) \quad v=(4.250, 0.500) \\ & \text{Relative Efficiencies} & \text{Firm}[X,Y,Z]=(100,73,26) \end{array}$$

The weights chosen by firm X show that X could be considered to be operating efficiently.

The model for company Y will be

$$\begin{aligned}
&\max[30w_1 + 50w_2]\\
&40v_1 + 20v_2 = 100\\
&40w_1 + 30w_2\, 20v_1 + 30v_2\\
&30w_1 + 50w_2\, 40v_1 + 20v_2\\
&25w_1 + 20w_2\, 30v_1 + 25v_2\\
&40w_1 \geq 0.15(40w_1 + 30w_2)\\
&30w_2 \geq 0.15(40w_1 + 30w_2)\\
&20v_1 \geq 0.15(20v_1 + 30v_2)\\
&30v_2 \geq 0.15(20v_1 + 30v_2)
\end{aligned}$$

Results Efficiency = 100

Weightings	$w = (2.833, 0.300) \quad v = (0.375, 4.250)$
Relative Efficiencies	$\text{Firm}[X, Y, Z] = (87, 100, 59)$

The weights chosen by firm *Y* show that *Y* could be considered to be operating efficiently.

The model for company *Z* will be

$$\begin{aligned}
&\max[40w_1 + 35w_2]\\
&30v_1 + 25v_2 = 100\\
&40w_1 + 30w_2\, 20v_1 + 30v_2\\
&30w_1 + 50w_2\, 40v_1 + 20v_2\\
&25w_1 + 20w_2\, 30v_1 + 25v_2\\
&25w_1 \geq 0.05(25w_1 + 20w_2)\\
&20w_2 \geq 0.05(25w_1 + 20w_2)\\
&20v_1 \geq 0.05(20v_1 + 30v_2)\\
&30v_2 \geq 0.05(20v_1 + 30v_2)
\end{aligned}$$

Results Efficiency = 67

Weightings	$w = (2.282, 0.503) \quad v = (0.852, 2.798)$
Relative Efficiencies	$\text{Firm} = (100, 100, 67)$

The results indicating that firm *Z* is inefficient, because in comparison with the other departments' inputs and outputs, it is unable to demonstrate that it could be considered to be efficient. They also demonstrate that firm *Z* could seem to become more efficient by acting like either firm *X* or firm *Y*.

3.7.2 Goal Programming

In many situations when a model is being constructed to represent a particular application, there can exist:

- More than one objective, for example in a diet problem minimise cost and maximise "taste"
- Conflicting constraints, in a manufacturing context, spend no more than the allowed budget and make at least the required number of items.

These objectives and constraints are often in conflict:

- the "tasty" diet could include expensive foodstuffs and the economic diet very non"tasty" foodstuffs, and in general, the cost of the tasty diet will be (much) greater than the economic diet and the economic diet will not be "tasty",
- the set budget may not be sufficient to be able to operate the factory (direct and indirect, fixed and variable costs).

Goal programming aims to reconcile these aims, often satisfying neither, producing a compromise solution.

Illustrative Example: production example

$$\begin{array}{ll} \text{Capacity constraints} & \sum_j a_{ij}x_j \leq b_i \quad \text{all } i \\ \text{Income target} & \sum_j r_j x_j \geq T \\ \text{Maximise Profit} & \sum_j p_j x_j \end{array}$$

Where a conflict between the capacity constraints and the income target may not allow a feasible solution.

Goal programming proceeds by the addition of "deviational variables" to give the enhanced model where the constraints have been replaced with equations

$$\begin{array}{ll} \text{Capacity constraints} & \sum_j a_{ij}x_j = d_i^- - d_i^+ + b_i \quad \text{all } i \\ \text{Income target} & \sum_j r_j x_j = t^- - t^+ + T \end{array}$$

Minimise deviations from capacities/targets, here minimise overcapacity and under income generation, measured by d_j^- and t^+ (assuming that unused capacity d_j^+ and higher than target revenue t^- are acceptable) to give the objective

$$\text{Minimise} \quad \sum_j w_j d_j^- + wt^+$$

The values for p_j and p are weights representing the relative importance of each capacity and the income target constraint.

The problem now is the determination of suitable costs, consider the extreme cases:

w very large the objective is simplified to Minimise (wt^+) generate the target income at the expense of the addition of extra capacity.

w very small the objective would seem to become Minimise $\left(\sum_j p_j d_j^-\right)$ but in fact, the deviational model is replaced with

$$\text{Capacity constraints} \quad \sum_j a_{ij} x_j \le b_i \quad \text{all } i$$

$$\sum_j r_j x_j \ge T$$

$$\text{Maximise Profit} \quad \sum_j p_j x_j$$

The difficulty in formulating this goal programming model is the determination of the deviational weights. There are many approaches to goal programming; however, heuristic approaches can often be applied with more success to this type of problem.

References

1. Stigler G (1945) Cost of subsistence. J Farm Econ 25:303–314
2. Dantzig GB (1990) The diet problem: interfaces. In: The practice of mathematical programming vol 20(4), pp 43–47

Chapter 4
Heuristic Techniques in Optimisation

Val Lowndes and Stuart Berry

These approaches can be useful when the problem to be solved is large and/or complex, when a timely good solution is considered to be better than a late optimal solution.

Many problems are such that the model to represent them is too large and complex to be solved in a reasonable time or the search space is too large, for example, the travelling salesman problem or scheduling air traffic controllers.

Note that

- in an n city travelling salesman problem, there are $(n-1)!$ routes to be (potentially) evaluated; if the route $\{a, b, c, d\}$ is deemed to be the same as route $\{d, c, b, a\}$, there are $(n-1)!/2$ routes.
- while scheduling m tasks between n air traffic controllers implies, in the worst case, $(n!)^m$ possible solutions, and
- when n jobs are to be processed through a flow shop, there are $n!$ possible schedules, more in the case of a job shop.

However, in reality, travelling salesmen have determined good routes, air traffic controllers are scheduled, and manufacturing work in both flow and job shop scenarios is scheduled.

In all cases, the planners have used heuristic approaches to determine good feasible schedules.

There are many approaches used to develop heuristics methodologies and algorithms which lead to the "solution" of optimisation problems. In general, each

V. Lowndes
University of Derby, Kedleston Road, Derby DE22 1GB, UK
e-mail: V.P.Lowndes@derby.ac.uk

S. Berry (✉)
College of Engineering and Technology, University of Derby, Kedleston Road, Derby DE22 1GB, UK
e-mail: s.berry@derby.ac.uk

S. Berry et al. (eds.), *Guide to Computational Modelling for Decision Processes*,
Simulation Foundations, Methods and Applications,
DOI 10.1007/978-3-319-55417-4_4

problem type tends to require a "tailor-made" heuristic or variation on existing methodologies. Here, two methodologies are presented: one is stochastic (genetic algorithms), and the other is deterministic (tabu search), to indicate their possible usages as approaches to problem-solving.

Defining Heuristics

A heuristic contributes to a reduction of search space in a problem-solving activity. A heuristic approach attempts to understand the mental operations in the thinking process, i.e. a study of the cognitive processes. Heuristic problem-solving takes a heuristic approach or uses heuristics in a problem-solving or decision-making situation.

Heuristic programming

A program of work can be defined to be a scheduled procedure of tasks. A heuristic program is a scheduled procedure of heuristics, that is a formal ordered presentation of aids to discovery.

Rationale for the use of heuristics?

The heuristic approach is ideally suited to two types of problems:

- *problems too large* for traditional OR methods; or
- *problems too loosely structured* or *ill structured* to be expressed in the mathematical terms necessary for the traditional algorithmic models.

If too large, a heuristic approach can provide a shortcut to the process of deriving/developing a solution, for example a travelling salesman problem. If the problem is too loosely structured, a heuristic approach can provide an orderly approach to a solution. However, the solution is not necessarily the optimal solution or even close to the optimal solution.

4.1 Genetic Algorithms

This approach to the determination of a good solution of an optimisation problem uses concepts drawn from biology, Holland [1].

Typically, it progresses from an initial set of trial solutions (created randomly) to a new set/generation of trial solutions by combining selected members of the initial population, thereby improving the state of the known solution.

This approach can be useful when the problem to be solved is large and/or complex often with multiple objectives when a timely good solution (compared with a late optimal solution) is both appropriate and useful.

The concept underlying genetic algorithms is that of natural evolution. In evolution, the problem each species faces is one of searching for beneficial adaptations to a complicated and changing environment. The *knowledge* that each species has gained is embodied in the make-up of the chromosomes of its members.

Operations that alter this chromosomal make-up are applied when parents reproduce. Examples of such operations are *random mutation* and *crossover* of chromosomal material between two parents' chromosomes. Random mutation provides background variation and occasionally introduces beneficial material into a species' chromosomes. Crossover exchanges the corresponding genetic material from two parent chromosomes, allowing beneficial genes of different parents to be combined in their offspring.

Holland [1] is the founder of the field of genetic algorithms. Holland's research evolved around the ability to encode complicated structures as simple representations (bit strings) and the power of simple transformations to improve such structures.

He showed that with the proper control structure, rapid improvements of bit strings could occur so that a population of bit strings could be made to "evolve" like a population of animals.

He described a genetic algorithm as a control structure with which these representations and operations could be managed in order to evolve bit strings that well-represented solutions to the problem to be solved.

Holland suggested that even in large complicated search spaces, given certain conditions on the problem domain, genetic algorithms would tend to converge on solutions that were globally optimal or nearly so.

The design and implementation of a ***Genetic Algorithm*** to a particular problem requires the following components/stages:

- A chromosomal representation of solutions to the problem;
- A way to create an initial population of solutions;
- An evaluation function that plays the role of the environment, rating solutions in terms of their *fitness*;
- Genetic operators that generate and alter the composition of offspring during reproduction;
- Values for the parameters that the genetic algorithm uses (population size, probabilities of applying genetic operators, etc.
  Holland [1], Goldberg [2] and Mitchell [3].

Chapter 17 contains descriptions and examples of methods for selection, crossover and mutation.

4.1.1 Implementation Examples

The genetic algorithmic implementation tends to be particular to each application. The following examples aim to show that these stages tend to be particular to each problem. In each case, selection of parents is based on both Tournament and Roulette procedures to be able to compare the effectiveness of each approach.

4.1.1.1 Knapsack Problems Genetic Algorithmic Approaches

First consider the problem where there are n items but just one container; here, the objective is the selection of the items to be packed.

Define the variables

w_j weight of item j
v_j value of item j
C capacity

and

$S_{ij} = 1$ item j is packed, string i
$S_{ij} = 0$ item j is not packed

with

W_i weight of packing defined by string i
V_i value of packing defined by string i

Thus, the GA string for an 8 variable problem can have the form:

$$S = [1\ 1\ 0\ 1\ 0\ 1\ 1\ 0]$$

Indicating a packing list with the weight and value given by

$$\begin{array}{llll} \text{Weight} & & W_i = \sum S_{ij} w_j & \text{and} \\ \text{Value} & \text{if} & W_i \leq C & V_i = \sum S_{ij} v_j \quad \text{however} \\ & \text{else} & W_i > C & V_i = 0 \quad \text{(an infeasible solution)} \end{array}$$

Example: Suppose $C = 160$ and there are two strings $(S_1,\ S_2)$

$$\begin{array}{llll} S_1 & [1\ 1\ 0\ 1\ 0\ 1\ 1\ 0] & \text{Weight } W_1 = 142 & \text{Value } V_1 = 200 \\ S_2 & [1\ 1\ 1\ 1\ 1\ 1\ 1\ 0] & \text{Weight } W_2 = 202 & \text{Value } V_2 = 330 \end{array}$$

Imposing the weight restriction gives

$$\begin{array}{llll} S_1 & & \text{Weight } W_1 = 142 & \text{Value } V_1 = 200 \\ S_2 & \text{not allowed} & \text{Weight } W_2 = 202 & \text{Value } V_i = 0 \end{array}$$

Crossover and mutation can be employed to obtain a good solution; crossover single or multiple will always provide a "reasonable" possible solution:

Crossover: choose two parent strings and a crossover point to generate new strings

	parents		new strings
S_1	[1 1 0 1 | 0 1 1 0]	→	[1 1 0 1 1 1 0 0]
S_2	[1 0 1 1 | 1 1 0 0]		[1 0 1 1 0 1 1 0]

Mutation: when a location is selected for mutation, then the new value can be given by:

$S_{ij} = \text{mod}_2(S_{ij} + 1)$ altering the status of item i, or
$S_{ij} = \text{Random from}\{0, 1\}$ randomly setting the status of item i.

4.1.1.2 Multiple Container Problem

Now consider the problem where the aim is to minimise the number of containers, or total cost of containers used to pack n items, all items to be packed.

Here, the GA string for an 8 variable problem (assuming that there are 4 containers available for use) will have the form:

$$S_i = [1\ 1\ 2\ 1\ 3\ 1\ 4\ 2]$$

check capacities for feasibility and then calculate the cost of this solution; the individual container usages (s_i) are as follows:

$$\begin{aligned}
s_1 &= [1 \quad 1 \quad 0 \quad 1 \quad 0 \quad 1 \quad 0 \quad 0] \\
s_2 &= [0 \quad 0 \quad 1 \quad 0 \quad 0 \quad 0 \quad 0 \quad 1] \\
s_3 &= [0 \quad 0 \quad 0 \quad 0 \quad 1 \quad 0 \quad 0 \quad 0] \\
s_4 &= [0 \quad 0 \quad 0 \quad 0 \quad 0 \quad 0 \quad 1 \quad 0]
\end{aligned}$$

With the individual container, weights given by

$$w_i' = \sum_j s_{i_j} w_j$$

The solution is valid if

$$w_i' \leq C_i$$

n, the cost of the solution, is given by the total cost of the used containers.

Crossover: choose two parent strings and a crossover point to generate new strings

	parents		new strings
S_1	[1 1 2 1 | 3 1 4 2]	→	[1 1 2 1 1 1 3 3]
S_2	[1 4 4 2 | 1 1 3 3]		[1 4 4 2 3 1 4 2]

check for feasibility, 4 containers used

Mutation: when a location is selected for mutation, then a possible implementation would be to replace the existing container number S_{ij} with

$$S_{ij} = \text{random}\{1, 2, 3, 4\}, \text{ or}$$

This model/implementation can be used to minimise either the number of containers or the cost of the containers.

Selection for crossover could be by way of either the Tournament or the Roulette methodologies (see Sect. 17.1 for examples of Tournament and Roulette selection)

Both approaches have been applied to the sample problem with the typical set of results:

Data:

15 items available to be packed into a container with capacity of 59 units to maximise the value of items packed.

$$\begin{array}{ll} \textit{Weights } W = & [3 \;\; 7 \;\; 2 \;\; 4 \;\; 6 \;\; 11 \;\; 9 \;\; 3 \;\; 2 \;\; 15 \;\; 8 \;\; 6 \;\; 12 \;\; 15 \;\; 4] \\ \textit{Values } \;\; V = & [2 \;\; 6 \;\; 1 \;\; 3 \;\; 5 \;\; 8 \;\; 2 \;\; 5 \;\; 22 \;\; 7 \;\; 3 \;\; 1 \;\; 4 \;\; 6 \;\; 11] \end{array}$$

The results from three iterations of the genetic algorithm were as follows:

Best value	weight
64	59
64	59
68	59

4.1.1.3 Travelling Salesman Problem Genetic Algorithmic Approach

This example is used to demonstrate that the crossover mechanism is dependent on the problem.

Defining the variables

d_{ij} distance from location i to location j.
c_{ij} cost of travelling from location i to location j.

The genetic algorithm string will indicate the order in which the cities are to be visited, for example in a 10 city problem it could have the form:

$$S_1 = [7\ 8\ 2\ 9\ 3\ 4\ 1\ 0\ 5\ 6\ 1], \text{or}$$
$$S_1 = [8\ 2\ 7\ 6\ 1\ 0\ 9\ 1\ 5\ 3\ 4]$$

Point crossovers, using two genetic algorithm strings, not be appropriate here, it could cause multiple visits to the same location.

A simple alternative is to select a single string and a single crossover point

$$S_1 = [7\ 8\ 2\ 9\ 3\ 4|\mathbf{10\ 5\ 6\ 1}]$$

and reverse the order of those jobs after the crossover point

$$S_1 = [7\ 8\ 2\ 9\ 3\ 4\ \mathbf{1\ 6\ 5\ 10}]$$

This crossover causes two changes to the order of cities visited

$$\{4,\ 1\} \text{ replaces } \{4,\ 10\} \text{ and}$$

$$\{10,\ 7\} \text{ replaces } \{1,\ 7\}$$

This process can be repeated to perform multiple crossovers

$$S_1 = [7\ 8\ 2\ 9|3\ 4\ 1\ 6\ 5\ 10]$$

Giving

$$S_1 = [7\ 8\ 2\ 9\ 10\ 5\ 6\ 1\ 4\ 3]$$

An alternative crossover is to select two random points giving

$$S1 = [7\ 8\ 2|9\ 3\ 4\ 10|5\ 6\ 1]$$

extract

$$M_1 = [9\ 3\ 4\ 10]$$

Now select another random point in R_I and replace the extracted string

Remainder $R_1 = [7\ 8|2\ 5\ 6\ 1]$

To give the new string

$$S_1 = [7\ 8\ 9\ 3\ 4\ 10\ 2\ 5\ 6\ 1]$$

This crossover has made three changes to the original (visiting order) genetic algorithm string:

{8, 9} replaces {8, 2};
{10, 2} replaces {10, 5} and
{2, 5} replaces {2, 9}

Mutation can be achieved by: select two positions and exchange the cities
Selection for crossover: use Tournament or Roulette selection

The following example has been "solved" using both a standard genetic algorithmic approach (TSP_GA) and a mini genetic algorithm approach (TSP_miniGA).

In a "mini genetic algorithm" approach, several small (few strings) genetic algorithms are processed, each generating one good solution. These good solutions are then used as the starting population for a final genetic algorithm.

Both approaches were applied to a 30 city problem then to a 40 city problem, applying one crossover on 99% of occasions and 2 crossovers on 1% of occasions. Comparing the solution times for the 30 city problems

TSP_GA	80 strings and 1500 iterations		100 strings 2500 iterations
	6.70 seconds	best route cost	=374
	17.53 seconds	best route cost	=374
	6.23 seconds	best route cost	=374
	6.70 seconds	best route cost	=374
	6.21 seconds	best route cost	=375
	6.70 seconds	best route cost	=374

with

TSP_miniGA	40 mini genetic algorithm populations		
	40 strings each with 30 iterations, and final string with 250 iterations		
	3.20 seconds	best route cost	=370
	3.26 seconds	best route cost	=392
	3.21 seconds	best route cost	=377

Comparing the solution times for the 40 city problems			100 strings 2500 iterations
TSP_GA	8.60 seconds	best route cost	=506
	17.53 seconds	best route cost	=489
	8.50 seconds	best route cost	=494
	17.06 seconds	best route cost	=499
	8.60 seconds	best route cost	=503
	16.82 seconds	best route cost	=499

with

TSP_miniGA	4.42 seconds	best route cost	=530
	4.41 seconds	best route cost	=501
	4.38 seconds	best route cost	=513

Comparing the solution times for the 20 city problems

TSP_GA	4.35 seconds	best route cost	=251
	4.48 seconds	best route cost	=255
	4.36 seconds	best route cost	=248

Combining the results obtained using TSP_GA suggested that the solution time is given by:

$$\text{Time} = 0.1928 + 0.2085\ \text{Cities} \quad \text{or} \quad \text{Time} \approx \frac{\text{Cities}}{5}$$

Continuing the investigation using one crossover on 1% of occasions and 2 crossovers on 99% of occasions indicated that mostly using one crossover gave better results.

Comparing the solution times for the 40 city problems

TSP_GA	8.58 seconds	best route cost	=548
	8.53 seconds	best route cost	=535
	8.66 seconds	best route cost	=556

with

TSP_miniGA	4.37 seconds	best route cost	=573
	4.39 seconds	best route cost	=560
	4.44 seconds	best route cost	=566

4.1.1.4 Flow Shop Scheduling Genetic Algorithmic Approach

Define the variables

$$T_{ij} \quad \text{time required on process } i \text{ by job } j.$$

The GA string will indicate the order in which the jobs are to be processed, for example in a 10 job problem, it can have the form:

$$S_k = [7\ 8\ 2\ 9\ 3\ 4\ 10\ 5\ 6\ 1]$$

Essentially, the same problem with the same problems as a travelling salesman problem except here, there are $n!$ possible solutions compared with $(n-1)!/2$ possible solutions with the travelling salesman problem, and the fact that the resultant schedule following crossover is more changed than with the travelling salesman problem.

Replacing S_k with $[7\ 8\ 2\ 9\ 3\ 4\ 1\ 6\ 5\ 10]$ has made 4 changes to the schedule $\{4,1\},\{1,6\},\{6,5\},\{5,10\}$ compared with only two changes in a travelling salesman problem using the same genetic algorithm string, these were as follows:

$$\{4,1\} \text{ and } \{10,7\}.$$

Other possible crossovers methodologies retain more of the existing order typical crossover mechanisms are as follows:

- One point crossover
- Two (multiple point) crossover
- Position-based crossover

All are illustrated using the genetic algorithm string

$$S_k = \{1,2,3,4,5,6,7,8\}$$

<u>One Point Crossover</u>
One point is randomly selected from Parent 1,
Jobs selected from one side (random) are copied to the child, (Jobs 1, 2 and 3 below); the other jobs are then placed in the order given by the 2nd parent. (i.e. 5, 8, 4, 7 and 6).

```
1 2 3 | 4 5 6 7 8     Parent 1
1 2 3 |               Taken from Parent 1

5 8 1 4 2 3 7 6       Parent 2
      5 8 4 7 6       Remainder removing jobs "taken from parent 1"

1 2 3 5 8 4 7 6       Child, new string.
```

<u>Two Point Crossover</u>

Two points are randomly selected from Parent 1, Jobs outside the 2 crossover points are mapped to the child (A, B and G, H).

The remaining jobs are copied in the order given by Parent 2 (E, D, C and F).

```
       Cross over points
1 2 | 3 4 5 6 | 7 8   Parent 1

5 8 | 1 4 2 3 | 7 6   Parent 2
      5 4 3 6         Remainder removing jobs "taken from parent 1"

1 2   5 4 3 6   7 8   Child
```

In an alternative approach, the parents inside the two crossover points are mapped to the child (i.e. C, D, E and F). The remaining jobs (H, A, B and G) are then copied in the order given by Parent 2.

```
1 2 | 3 4 5 6 | 7 8   Parent 1

5 8   1 4 2 3   7 6   Parent 2
  8   1   2     7     Remainder removing jobs "taken from parent 1"

8 1   3 4 5 6   2 7   Child
```

<u>Position-based Crossover</u>

N positions are chosen randomly from Parent 1, and these jobs are inherited directly by the child (B, C, E and H). The other jobs are then placed in the order given by Parent 2 (A, D, G and F).

```
1 2 3 4 5 6 7 8   Parent 1
  2 3   5     8   Parent 1

5 8 1 4 2 3 7 6   Parent 2
    1 4     7 6   Remainder removing jobs "taken from parent 1"

1 2 3 4 5 7 6 8   Child
```

4.1.1.5 Quadratic Assignment Problem Genetic Algorithmic Approach

Define the variables

$T_{ij} = 1$ worker i is assigned to carry out task j
$D_{ij} = 1$ duration/cost of task j when assigned to task i
M_i allowable working time for worker i

Initially, assuming that each worker is assigned to a single task and each task is assigned to a single worker, the problem formulation becomes as follows:

$$\sum_j T_{ij} \le 1$$

$$\sum_i T_{ij} = 1$$

$$\text{Minimise}\left\{\sum_i \sum_j f(D_{ij}, T_{ij})\right\}$$

where the objective function has a quadratic form.

If more than one task can be allocated to a worker, the formulation becomes as follows:

$$\sum_j D_{ij} T_{ij} \le M_i$$
$$\sum_i T_{ij} = 1$$
$$\text{Minimise}\left\{\sum_i \sum_j f(D_{ij}, T_{ij})\right\}$$

For the basic model with $n = 8$ jobs and $m = 15$ workers, the genetic algorithm string will have the form:

$$S_k = [7\ 8\ 12\ 9\ 3\ 14\ 10\ 5]$$

For the extended model, each worker could appear more than once.

Crossover: any crossover could be appropriate the simplest (single point) leads to valid solution strings, for example:

$$S_1 = [7\ 8\ 12\ 9\ 3\ 14\ 10\ 5]$$
$$S_2 = [3\ 5\ 7\ 11\ 6\ 12\ 13\ 2]$$

Can give the new strings

$$S_3 = [7\ 8\ 12\ 9\ 6\ 12\ 13\ 2]$$
$$S_4 = [3\ 5\ 7\ 11\ 3\ 14\ 10\ 5]$$

Mutation, for a randomly chosen location *j*, the existing value is replaced with

$$S_{jk} = 1 + \bmod_{15}\left(S_{kj} + \text{ceil}(15 \times \text{rand})\right)$$

alternatively, exchange two values chosen at random.

4.1.1.6 Fire Station Location Problem Genetic Algorithms and Fuzzy Logic

The object can be to determine the best locations for a fire station or to determine the best allocation of districts to be serviced by each, existing, fire station.

Using the notation

$x_{ij} \in \{0, 1\}$	district *i* being allocated to fire station *j*,
d_{ij}	distance from district *i* to fire station *j*,
r_i	expected demand from district *i*,
R_j	capacity of fire station *j*
C_j	cost of fire station *j*
$\delta_j \in \{0, 1\}$	fire station site *j* being used, 0 otherwise
$s_{ij} \in \{0, 1\}$	district *i* could be allocated to fire station *j*

The basic model is now given by:
for each district (for all *i*):

$$\sum_j s_{ij} x_{ij} = 1$$

$$\sum_i r_i x_{ij} \le R_j$$

for each possible fire station, determine its usage

$$\sum_i x_{ij} \le M\delta_j$$

With the objective function aiming to minimise some function of distance travelled using the decision model becomes

$$\text{Minimise}\left(\sum_i \sum_j d_{ij}^n x_{ij}\right)$$

$$\sum_j x_{ij} = 1$$

$$\sum_i r_i x_{ij} \leq R_j$$

Or aiming to minimise both distance and cost the model becomes

$$\text{Minimise}\left(\sum_i \sum_j d_{ij}^n x_{ij}\right) \text{and} \quad \text{(and)}$$

$$\text{Minimise}\left(\sum_j C_j \delta_j\right)$$

$$\sum_j x_{ij} = 1$$

$$\sum_i r_i x_{ij} \leq R_j$$

$$\sum_i x_{ij} \leq M\delta_j$$

In both cases, the genetic algorithm string can have the form

$$\{4, 3, 5, 2, 2, 1, 4, 3, 3, 1\}$$

that is fire station 1 is allocated to cover cities 6 and 10.

In the second model, fuzzy logic can be used to combine the two objectives, for example starting with optimal solutions to the two problems:

$$\text{Minimise}\left(C_1 = \sum_j C_j \delta_j\right)$$

$$\sum_j x_{ij} = 1$$

$$\sum_i r_i x_{ij} \leq R_j$$

$$\sum_i x_{ij} \leq M\delta_j$$

$$\text{Minimise}\left(C_2 = \sum_i \sum_j d_{ij}^n x_{ij}\right)$$

$$\sum_j x_{ij} = 1$$

$$\sum_i r_i x_{ij} \leq R_j$$

The values C_1 and C_2 can be used to enable the implementation of a fuzzy logic-based objective.

Technical Notes (see Sect. 17.1 for an example of selection methods).
Selection methods
Roulette
The selection procedure randomly picks out two parent chromosomes, based on their fitness values, which are then used by the crossover and mutation operators (described below) to produce two offspring for the new population.

The selection process:

1. sum the fitness of all population members to give the *total fitness*;
2. generate, from a uniform distribution, a random integer, *r, between* 0 *and total fitness*;
3. select the first chromosome whose cumulative fitness is greater than or equal to *r*.

Thus, the better strings tend to be selected more often than the others, and thus, the average fitness of the population (of strings) tends to improve as the process advances towards convergence (of the fitness values).

Tournament
The selection procedure has two stages:

Stage 1: select at random m strings and from these retain the best, repeat this process until n strings have been selected.
Stage 2: select two parents from these *n* strings and perform crossover and mutation to generate two new strings. Note at this stage, each string is used only once.

4.2 Tabu Search

This is a deterministic method. This search technique operates by stepping through the solution space constructing a tabu list (of visited solutions) so that the search does not (repeatedly) return to the same solution.

Glover [4, 5] "invented" the procedure and processes of a tabu search. This is a heuristic method that allows a search procedure to explore the solution space beyond local optimality. Application areas include scheduling, resource allocation, planning and telecommunications.

Tabu search is based on the premise that problem-solving, in order to qualify as intelligent, must incorporate adaptive memory and responsive exploration. The use of adaptive memory by tabu search is in contrast to memoryless techniques such as simulated annealing and genetic algorithms. The emphasis on responsive exploration derives from the supposition that poor strategic choice may produce more information than a good random choice.

Practical implementations of tabu search seek to exploit the properties of:

(1) Flexible memory—to drive the search into new regions of the solution space;
(2) Responsive exploration—to identify and explore "good" regions of solution space.

Unlike some other techniques (for example simulated annealing and genetic algorithms), randomisation is not normally employed to avoid termination at a local optimum; thus, *most* tabu search implementations are largely or wholly deterministic.

Tabu search implementation, consider the first problem

$$\begin{aligned}&\text{Maximise} f(x_1,\ldots,x_n)\\&\text{Subject to}\\&\sum x_i = M\\&x_i^k \text{ the } k\text{th iteration for } x_i\end{aligned}$$

From an initial solution, step 1,

$$\sum x_i^1 = M$$

Investigate all (valid) neighbouring points to determine the best neighbouring point and move to this point

$$\begin{aligned}&\sum x_i^2 = M\\&\boldsymbol{x_i^2 = x_i^1 \qquad i \neq k, i \neq j}\\&\boldsymbol{x_k^2 = x_k^1 + h}\\&\boldsymbol{x_j^2 = x_j^1 - h}\end{aligned}$$

adding the previous point to a tabu list, restricting a return to this point.

Note: the tabu list normally has a fixed finite length, so a solution could be investigated more than once.

4.2.1 *Basic Financial, Investment Problem, Tabu Search Approach*

A sum of money is available for investment into several (n) projects, let

x_i^s number of units of investment into project i at iteration s
M number of units available to be invested
$f(x_1^s, \ldots, x_n^s)$ return from the investments at iteration s

Initial solution, let

$x_i^1 = (M/n)$ number of units of investment into project i at iteration s
$T(1) = \{x_1^1, x_2^1, \ldots, x_n^1\}$ initial tabu list

after selecting the new solution $\{x_1^2, \ldots, x_n^2\}$, this solution is added to the tabu list

$$T(2) = \{x_1^2, x_2^2, \ldots, x_n^2\} \qquad \text{the initial tabu list}$$

Note that the tabu list can either have an unlimited size or a maximum size. In the latter case, the more "historic" points are successively removed from the tabu list, allowing them to be revisited.

4.2.2 *Extended Financial, Investment Problem, Tabu Search Approach*

This problem can be extended through the addition of limits on the size of the investment in each project. This adds the constraints $x_i^s \leq M_i$ to be checked at each iteration.

The minimum and maximum allowed investments into each project are 16 and 24, and the available investment fund is 80 (Table 4.1).

The first two stages in a tabu search, step size 1, are as follows:

Table 4.1 Investment opportunities points

Cash flows, unit of investment 10, available 80				
Time	Project W	Project X	Project Y	Project Z
Cost/unit	10	9	11.1	9.6
Return/unit	12	13	12.7	12.4

Stage 1	W	X	Y	Z		
Current point	20	20	20	20	208	
Neighbours	21	19	20	20	206	
	21	20	19	20	208.4	
	21	20	20	19	207.2	
	19	21	20	20	210	
	20	21	19	20	210.4	
	20	21	20	19	209.2	Best new point
	19	20	21	20	207.6	
	20	19	21	20	205.6	
	20	20	21	19	206.8	
	19	20	20	21	208.8	
	20	19	20	21	206.8	
	20	20	19	21	209.2	
Stage 2	W	X	Y	Z		
Tabu list	20	20	20	20		
Current point	20	21	20	19	209.2	
Neighbours	21	20	20	19	207.2	
	21	21	19	19	209.6	
	21	21	20	18	208.4	
	19	22	20	19	211.2	
	20	22	19	19	211.6	Best new point
	20	22	20	18	210.4	
	19	21	21	19	208.8	
	20	20	21	19	206.8	
	20	21	21	18	208	
	19	21	20	20	210	
	20	20	20	20	208	Tabu
	20	21	19	20	210.4	

At the next stage, the tabu list will contain the two points

W	X	Y	Z
20	20	20	20
20	21	20	19

4.2.3 Travelling Salesman Problem

Here, the objective is to produce a route visiting each city only once starting and finishing at a given city.

A typical route in a 6 city problem could be defined by

$$[4, 6, 1, 2, 5, 3]$$

The search procedure operates by investigating the effect of switching all allowed pairs of cities, and the best option with respect to cost change from this search is selected as the next move generating a new route, for example exchanging the 6 and the 3 to give

$$[4, 3, 1, 2, 5, 6]$$

The tabu list is implemented by prohibiting the move of these cities {6, 3} for a certain number of iterations.

Thus, giving the routine:

Investigate all possible allowed exchanges
Select the best exchange
Generate the new schedule
Update the tabu list
Repeat.

Typical result from the tabu search: random city positions 40 and 50 city problem solution plots (Fig. 4.1).

Note in an n city problem ($n > 2$), there are $((n-1)!)/2$ possible valid tours, many possible tours.

Given 40 city problem, for comparison, solutions were obtained to problems where the cities were placed randomly onto a grid using both a greedy algorithm and a tabu search.

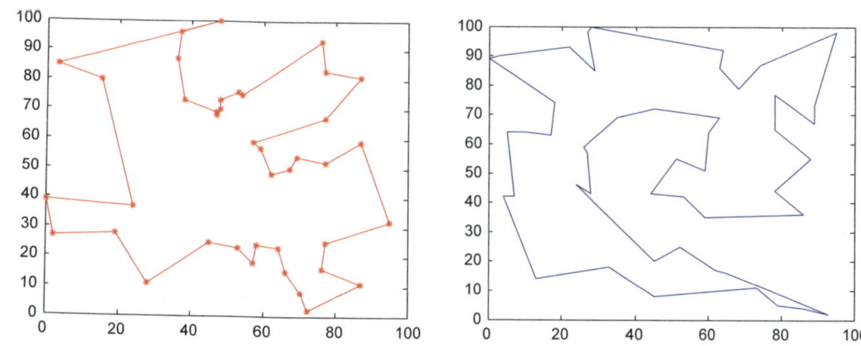

Fig. 4.1 Sample travelling salesman solution plots

The results from this investigation were as follows:

Greedy Algorithm	solution cost	= 476,
Tabu Search	solution cost	= 455;

The tabu search giving a more economic solution than that obtained from the use of a greedy algorithm.

Flow shop scheduling can be achieved using the same approach. A similar, but harder, problem when scheduling n jobs, there are $n!$ possible schedules to be considered.

4.3 Review Questions

Question 1

Q1.1 A knapsack problem can be defined as:
A container is to be packed with several items so that the value of its contents is maximised.
For example, six items are available for packing, but the available container has a maximum capacity of only 100 kg:

Item i	1	2	3	4	5	6
Value v_i	13	13	15	17	11	10
Weight w_i (kg)	20	20	60	50	10	15

Describe how a genetic algorithm approach could be used to determine how to pack this container, as an example generate 4 genetic algorithm strings and carry out at least two iterations.

Q1.2 Show how your model could be adapted so that items 1 and 2 may not both be packed into the container.

Q1.3 Produce a genetic algorithm model to determine the most appropriate containers to be used if all items have to be packed. There are five containers available with capacities and costs.

Container	1	2	3	4	5
Capacity	100	40	50	90	90
Cost	60	10	12	18	50

Question 2
This question is concerned with traffic flow and the control of the traffic flow through a network.

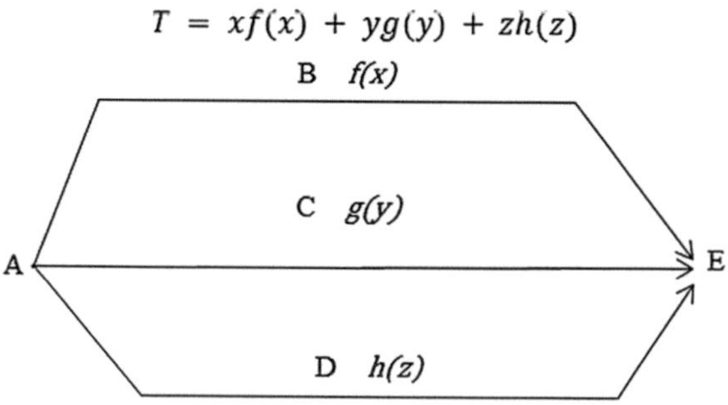

Fig. 4.2 Routes from source A to destination E

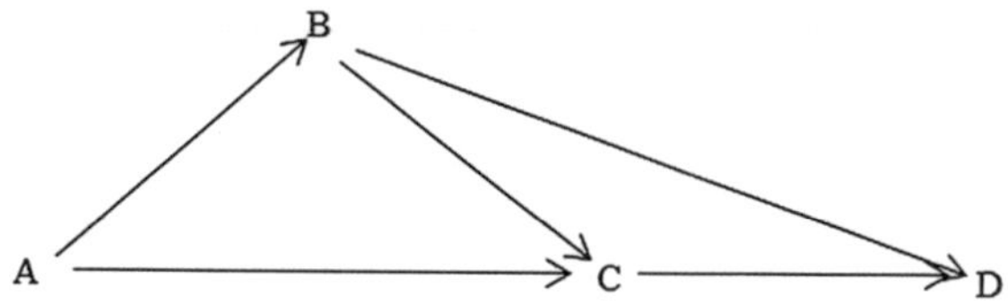

Fig. 4.3 Routes from source A to destination D

An example of a network is given in Fig. 4.2 where the aim of the morning traffic is to progress from the start (node A) to the finish (node E) as economically as possible.
The total traffic flow on a typical day is N units, and there are three routes through the network

ABE,
ACE and
ADE.

The aim is therefore to determine the flows (x, y and z, respectively) along each route to determine the best solution minimising costs, where the total cost T is given by

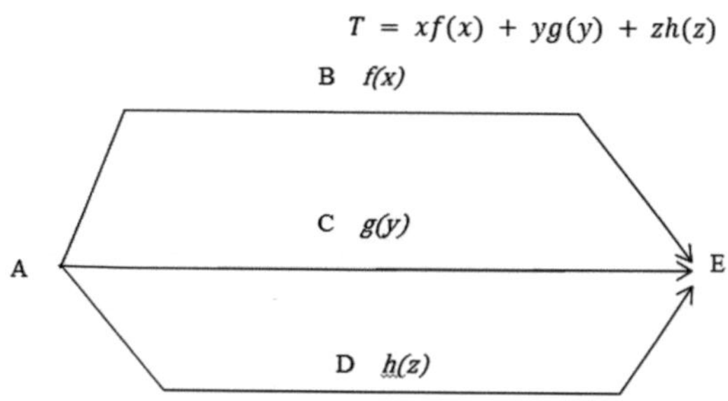

Currently, traffic is free to choose its own route and as a result the flows are at an equilibrium, each costs/times along each route, that is

$$f(x) = g(y) = h(z) \qquad x + y + z = N$$

Example:
Cost functions

$$\begin{aligned} &f(x) = x^2 + 1; \\ &g(y) = y + 8; \\ &h(z) = 4z + 3 \\ &\text{and} \\ &N = 12 \\ &x + y + z = 12 \end{aligned}$$

the equilibrium (cost) flows are as follows:

$$\begin{aligned} x &= 3.57 \quad f(3.57) = 13.74 \\ y &= 5.74 \quad g(5.74) = 13.74 \\ z &= 2.69 \quad h(2.69) = 13.76 \end{aligned}$$

Q2.1 Assuming a starting traffic flow (4, 4, 4), discuss the form of the neighbourhood and the nature of the tabu list used to obtain a solution to this problem and carry out three steps to demonstrate the operation of your implementation.

Q2.2 How could a tabu search be applied to networks shown in Figs. 4.3 and 4.4 where each connection will have an associated cost,

for example: Link AB unit cost $f_{AB}(x)$ when x units access this link.

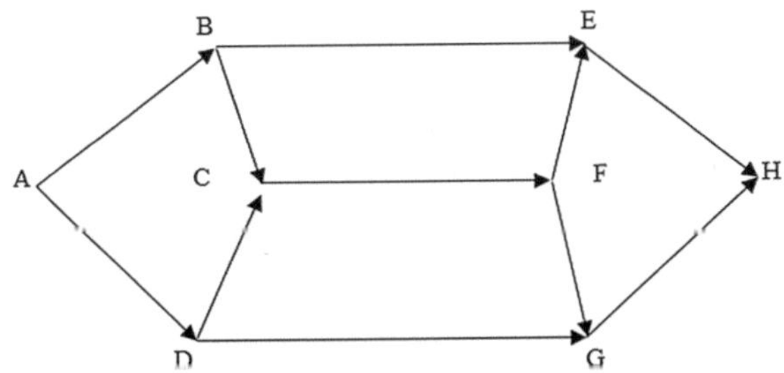

Fig. 4.4 Routes from source A to destination H

Referencces

1. Holland JH (1992) Adaption in natural and artificial systems. MIT Press, Cambridge
2. Goldberg D (1989) Genetic algorithms in search optimisation and machine learning. Addison Wesley, MA
3. Mitchell M (1998) An introduction to genetic algorithms. MIT Press, Cambridge
4. Glover F (1989) Tabu search—Part I. ORSA J Comput 1(3):190–206
5. Glover F (1990) Tabu search—part II. ORSA J Comput 2:4–32

Chapter 5
Introduction to the Use of Queueing Theory and Simulation

Val Lowndes and Stuart Berry

Queueing theory can be used to provide information about systems that can be represented by a queue/server or by series of queues/servers enabling queueing theory results to provide a means of estimating the expected performance of a manufacturing system.

Some of the common examples where queues are encountered are as follows:

- patients in a dentist's waiting room
- customers in the Post Office
- supermarket checkouts
- road junctions
- aeroplanes waiting to land at an airport
- at a work station in a factory

These applications can be extended to describe

- factory production systems, queues in both series and parallel
- traffic control in a city, queues in series

There are three basic elements in a queueing system.

- The arrivals (need to know the statistical distribution of arrivals)
- The service process (number of servers, the statistical distribution of service time)
- The queue discipline (FIFO, priority, …)

V. Lowndes
University of Derby, Kedleston Road, Derby DE22 1GB, UK
e-mail: V.P.Lowndes@derby.ac.uk

S. Berry (✉)
College of Engineering and Technology, University of Derby,
Kedleston Road, Derby DE22 1GB, UK
e-mail: s.berry@derby.ac.uk

S. Berry et al. (eds.), *Guide to Computational Modelling for Decision Processes*,
Simulation Foundations, Methods and Applications,
DOI 10.1007/978-3-319-55417-4_5

The purpose of queueing theory is to enable us to study (and hence model) queues in order to be able to predict the effect of changes on a system before we actually implement those changes. (Implementing change can be very costly).

For example, the investigation might want to consider changes to

- Pattern of arrivals
- Mean length of service
- Number of servers

These changes could then affect

- the average time customers have to wait
- the average number of customers waiting
- the proportion of time the service facility is in use.

Kendall Notation for Queueing Systems

There is a standard notation for classifying queueing systems into different types; this was proposed by D.G. Kendall.

Systems are described by the notation: **A/B/C/D/E.**

Where

A Distribution of "Interarrival" times of customers
B Distribution of service times
C Number of servers
D Maximum total number of customers allowed in the system
E Population size

and **A** and **B** can take any of the following distribution types:

M Exponential Distribution (random)
D Deterministic
G General distribution

General Results

The derivation of the standard queueing results is outlined in Chap. 16. The following results aim to demonstrate the appropriateness of queueing theory applied to modelling and evaluating alternative manufacturing systems.

M/G/1 queueing model

average queue length is given by $\frac{\lambda^2\sigma^2+\rho^2}{2(1-\rho)}$

where σ is the variance of the service times.

M/M/1 queueing model

Variance of service time $\sigma = \frac{1}{\mu}$

leading to the standard results

average queue length $\frac{\rho^2}{(1-\rho)}$

average number in system $\frac{\rho^2}{(1-\rho)} + \rho = \frac{\rho}{(1-\rho)}$

average time in system $\frac{\rho}{(1-\rho)}\frac{1}{\mu} + \frac{1}{\mu} = \frac{1}{(1-\rho)}\frac{1}{\mu}$

M/U/1 Uniform (Rectangular)

Here, the service time distribution will be symmetric about the mean, $\left[\left(\frac{1}{\mu} \pm a\right)\right]$,

and the variance is given by $\sigma^2 = \frac{a^2}{3}$

With the extreme (largest possible) variance when $a = \frac{1}{\mu}$

To give the maximum variance $\sigma^2 = \frac{1}{3\mu^2}$

giving an average queue length $\frac{2}{3}\frac{\rho^2}{(1-\rho)}$

M/T/1 Triangular

Here, the service time distribution will be symmetric about the mean, $\left[\left(\frac{1}{\mu} \pm a\right)\right]$,

with the variance given by $\sigma^2 = \frac{a^2}{6}$

Thus the "extreme" variance is given by $\sigma^2 = \frac{1}{6\mu^2}$

with the average queue length now given by $\frac{7}{12}\frac{\rho^2}{(1-\rho)}$

Finally for a deterministic service time **M/D/1,** the variance is 0

The average queue length is given by $\frac{1}{2}\frac{\rho^2}{(1-\rho)}$

Summary

These results can be used to place limits on the length of the queue:

$M/M/1$	UPPER LIMIT	average length	$= L$
$M/U/1$		average length	$< 0.67L$
$M/T/1$		average length	$< 0.58L$
$M/G/1$	LOWER LIMIT	average length	$= 0.50L$

indicating the rationale for the use of queueing theory where for an application where jobs arrive randomly the upper and lower limits on the service time can be calculated from the M/M/1 and M/G/1 results.

Now incorporate average service time in the system, giving estimates for delivery times.

$$M/M/1 \quad \frac{1}{(1-\rho)}\frac{1}{\mu}$$
$$M/D/1 \quad \left(\frac{1}{2}\frac{\rho}{(1-\rho)}\frac{1}{\mu}\right)+\frac{1}{\mu}$$

These results can be extended to cases where there is limited waiting space, for example the system can hold at most n customers.

Here $P_0 = (1-\rho)/(1-\rho^n)$
And $P_n = \rho^n(1-\rho)/(1-\rho^{n+1})$
note as $n \to \infty$ then $P_0 \to (1-\rho)$ and $P_n \to \rho^n(1-\rho)$

5.1 Evaluating Manufacturing Systems Using Queueing Theory (Random Arrivals)

5.1.1 Manufacturing Systems 1: n Work Stations in Series

A manufacturing system can be considered to be constructed from a series of queue, workstation pairs, for example (Fig. 5.1).

Consider the two extreme cases:

All service times described by exponential distributions, random $\lambda < \mu_i$ all i.
All service times described by deterministic distributions.

In the first case, the arrival rate at each work station will follow the same random distribution, and hence the average times and numbers in the system can be obtained by a repeated application of the standard results.

Thus it follows that the *Average time in system* $= \sum \frac{1}{(1-\rho_i)}\frac{1}{\mu_i}$ provides an upper limit.

To calculate a lower limit, first assume that the work stations are such that the work stations are progressively quicker, that is:

$$\mu_1 < \mu_2 < \cdots < \mu_n$$

Fig. 5.1 *n* work stations in series

and that all work times are deterministic in which case queues can only occur at the first work station and it follows that

$$\textit{Average time in system} = \textit{Average time queueing} + \sum \frac{1}{\mu_i}$$

and *Average time queueing* can be approximated by

$$\frac{1}{2}\frac{\rho^2}{(1-\rho)}\frac{1}{\mu_1} + \frac{1}{2}\rho\frac{1}{\mu_1} = \frac{1}{2}\frac{\rho}{(1-\rho)}\frac{1}{\mu_1}$$

Giving the lower estimate for *Average time in system* as

$$\frac{1}{2}\frac{\rho}{(1-\rho)}\frac{1}{\mu_1} + \sum \frac{1}{\mu_i}$$

Notice that simulation modelling can be used to demonstrate that when the job times are deterministic, then the overall duration is given by

$$\text{Lower estimate for average time in system} = \frac{1}{2}\frac{\rho_D}{(1-\rho_D)}\frac{1}{\mu_D} + \sum \frac{1}{\mu_i}$$

where ρ_D and μ_D relate to the dominant (longest no necessarily the first) stage in the production process. (Simulation, Example 1, gives a validation of this result).

5.1.2 Evaluating Manufacturing Systems 2: n Work Stations in Series with Rework

The final work station represents final inspection at which point items are either accepted (and shipped to the customer) or sent back for reworking, see Fig. 5.2, note that an item could return to the first work station many times.

Although λ new jobs arrive per time period, adding the number reworked (assuming an infinite number of reworks are possible) gives the correct system loading as

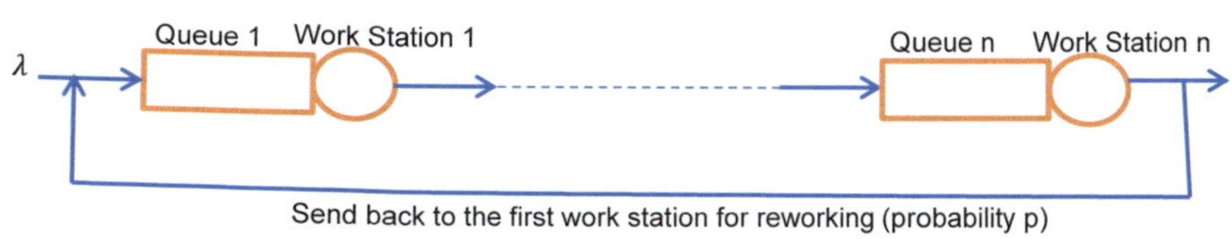

Fig. 5.2 *n* work stations in series with rework

$$\text{Arrivals } = \lambda + p\lambda + p2\lambda + \cdots \text{ or } \quad \lambda_A = \lambda/(1-p)$$

Therefore, $\frac{\lambda}{1-p} < \mu$ or $\frac{\lambda\left(1-p^{r+1}\right)}{1-p} < \mu$, r reworks allowed all i for finite queueing.

Consider the two extreme cases: all service times described by exponential distributions, random, and all service times described by deterministic distributions.

Each job could pass through the system more than once, so that the average times in the system need to be calculated as

Expected number of "journeys" through the system	$\frac{1}{(1-p)}$
Expected time in system (exponential service times)	$\frac{1}{(1-p)}\sum\frac{1}{(1-\rho)}\frac{1}{\mu}$
Expected time in system (deterministic service times)	$\frac{1}{(1-p)}\left(\frac{1}{2}\frac{\rho_D}{(1-\rho_D)}\frac{1}{\mu_D} + \sum\frac{1}{\mu_i}\right)$

5.1.3 *Evaluating Manufacturing Systems 3: Splitting the Jobs*

Here, the probability that "routeing a" is followed is p, and the probability that "routeing b" is followed is $(1-p)$ (Fig. 5.3).

The arrival rate at Queue 2a is therefore $p\lambda$ and at Queue 2b $(1-p)\lambda$; thus, the average times for both types of job can be estimated using

arrivals	λ	at work station 1 and	$\rho = \lambda/\mu_1$
	$p\lambda$	at work station 2a and	$\rho_{2a} = p\lambda/\mu_{2a}$
	$(1-p)\lambda$	at work station 2b and	$\rho_{2b} = (1-p)\lambda/\mu_{2b}$

Thus, the upper limits can be calculated using the M/M/1 results as follows:

$$\frac{1}{(1-\rho)}\frac{1}{\mu_1} + \frac{1}{(1-\rho_{2a})}\frac{1}{\mu_{2a}} \quad \text{and} \quad \frac{1}{(1-\rho)}\frac{1}{\mu_1} + \frac{1}{(1-\rho_{2a})}\frac{1}{\mu_{2a}}$$

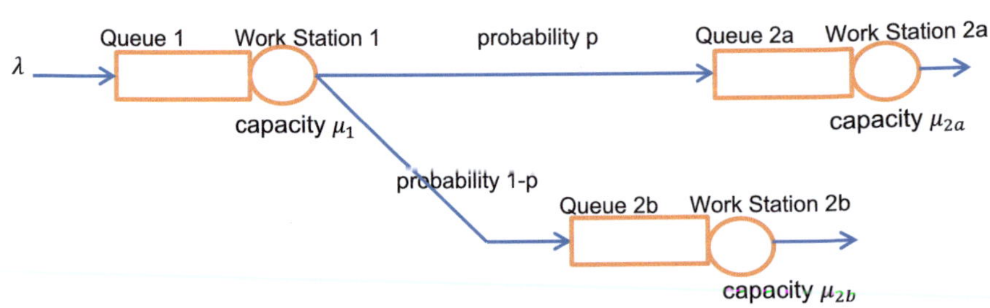

Fig. 5.3 Splitting work

Lower estimates for average times in system by M/D/1

$$\frac{1}{2}\frac{\rho_{Da}}{(1-\rho_{Da})}\frac{1}{\mu_{Da}}+\sum\frac{1}{\mu_i} \quad \text{and} \quad \frac{1}{2}\frac{\rho_{Db}}{(1-\rho_{Db})}\frac{1}{\mu_{Db}}+\sum\frac{1}{\mu_i}$$

5.1.4 Evaluating Manufacturing Systems 4: Jobs Processed in Batches

Here, jobs arrive in batches of n-like items, are processed individually and then left as a batch of n items.

The interarrival rate for the batches is λ, and the service rate for each item is μ, $\lambda<\mu/n$ (Fig. 5.4).

As all jobs are processed on a single machine, this could be modelled either as an M/M/1 system where the service rate is or as an M/D/1 system where the batch is treated as if it were a single job.

Thus, the upper and lower times in the system can be calculated from

Upper Limit	M/M/1	$\sum\frac{1}{(1-\rho)}\frac{n}{\mu}$
Lower Limit	M/D/1	$\frac{1}{2}\frac{\rho}{(1-\rho)}\frac{n}{\mu}+\sum\frac{n}{\mu}$ where $\rho=n\lambda/\mu$

5.1.5 Evaluating Manufacturing Systems 5

Manufacturing processes when all jobs move to the next stage at the same time, for example a KANBAN system and/or a production assembly line (Fig. 5.5).

All stage times are deterministic, and the time between stage movements $T=\text{Max}(T_1,\ldots,T_n)$

Thus time for job in system	nT
Time between successive finished items	T
Uniformly distributed with stage n dominant	$\mu=T,\sigma=\frac{range}{2\sqrt{3}}$ (minimum time at stage n > maximum time at all other stages)

Therefore, time for job in system is

Normally distributed (n large),	$\mu=nT,\sigma_s=\sigma\sqrt{n}$
Time between successive finished items	Uniform $\mu=T,\sigma=\frac{range}{2\sqrt{3}}$

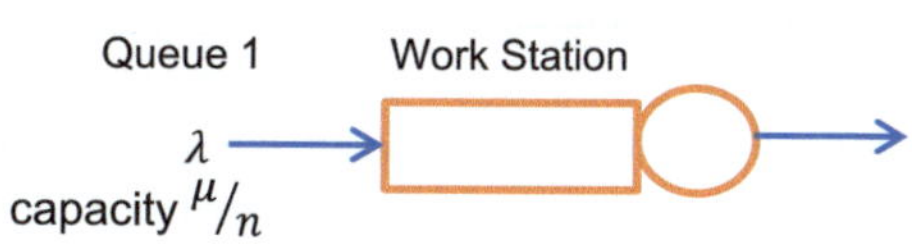

Fig. 5.4 Processing batches of jobs

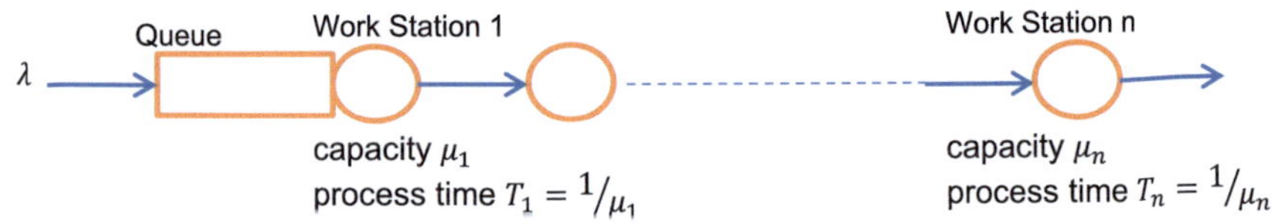

Fig. 5.5 KANBAN system

5.1.6 Changing Service Capacities

When there are parallel servers, for example in a bank, not all service points may be staffed at all times, but when demand is high more servers are used to improve the service to the customer.
If, for example, when there are 4 customers present additional service capacity becomes available (doubling the service capacity) the steady state probabilities become

$$P_1 = \rho P_0, P_2 = \rho^2 P_0, P_3 = \rho^3 P_0, P_4 = \frac{\rho^4}{2} P_0, P_5 = \frac{\rho^5}{2^2} P_0, \ldots$$

note: $P_4 = \frac{\rho^4}{2} P_0 = \rho^3 \frac{\rho}{2} P_0$

the effect, on the expression for P_0 of the doubling of the service capacity. Giving

$$P_0\left(1 + \rho + \rho^2 + \rho^3 + \frac{\rho^4}{2} + \frac{\rho^5}{2^2} + \cdots\right) = 1, \text{ or}$$

$$P_0\left((1 + \rho + \rho^2) + \rho^3\left(1 + \frac{\rho}{2} + \frac{\rho^2}{2^2} + \cdots\right)\right) = 1, \text{ or}$$

$$P_0\left((1 + \rho + \rho^2) + \rho^3\left(\frac{2}{2 - \rho}\right)\right) = 1$$

5.2 Using Simulation to Evaluating Planning and Control Systems in Flow Shops

To justify the need for alternative approaches to PP&C, consider first a flow shop consisting of 4 work stations where the job times at each station are deterministic, in the first case the first work station is dominant and in the second case the final work station is dominant.

In both cases, there are (always) jobs waiting to be processed before the first stage; Gantt charts showing the status of the flow shop if all jobs were to start at as early a date as possible are shown in Fig. 5.6a.

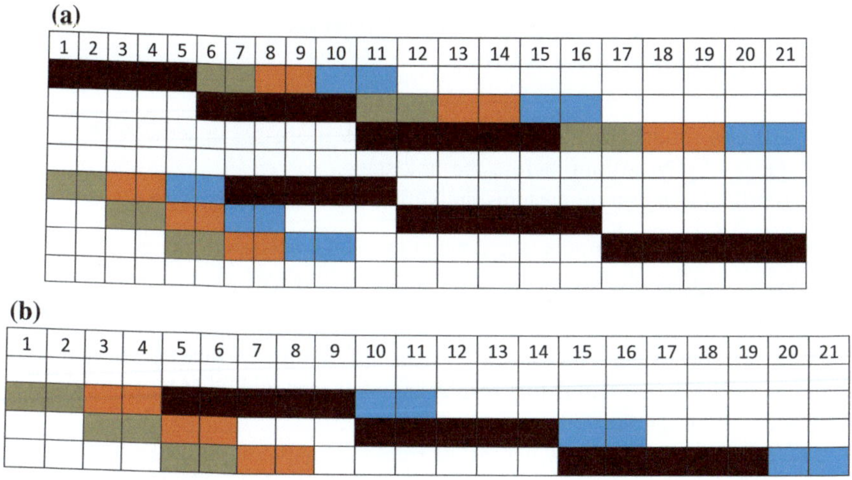

Fig. 5.6 **a** Illustrating the effect on total time of JIT, the dominant stage. **b** Illustrating the effect on total time of JIT, the dominant stage

Completion times are the same for both cases although more control will be needed in the second case; here, there will be a constant increasing work in progress, and hence both increased the costs (material and stock holding) and possibly an increased requirement for storage space. This problem (work in progress) will occur whenever the dominant stage is anywhere other than at the first stage; for example, Fig. 5.6b shows the same result when the third stage is dominant.

Simulation modelling can be used to compare and contrast the use of alternative methods for planning and control in flow shops, and the methods considered are as follows:

KANBAN,
CONWIP (**CON**stant **W**ork **I**n **P**rogress),
Production lines
A greedy system for comparison—start a job when the production stage is ready.

In each case, the following parameters will be calculated

Distribution of job times
Time to produce M items

5.2.1 Simulating to Compare and Contrast Production Line Systems

There exists a production system consisting of N processes in series with a nonempty queue of jobs waiting before the first stage.

Two production control systems are simulated:

- A job can move to the next stage when this stage becomes free, no control, and
- Production line where all jobs move at the same time

In each of the simulations, there existed one (absolutely) dominant stage with the times at all other stages being smaller than the minimum time at this dominant stage. In each plot, the simulation results are compared with the plot from a normal distribution with the same mean and variance.

The first pair of plots show the results from the "move when the next machine is free" simulation, and the second pair the results from the production line simulation. In each case, as the number of stages is increased, the results from the simulation tend towards being normally distributed (Fig. 5.7).

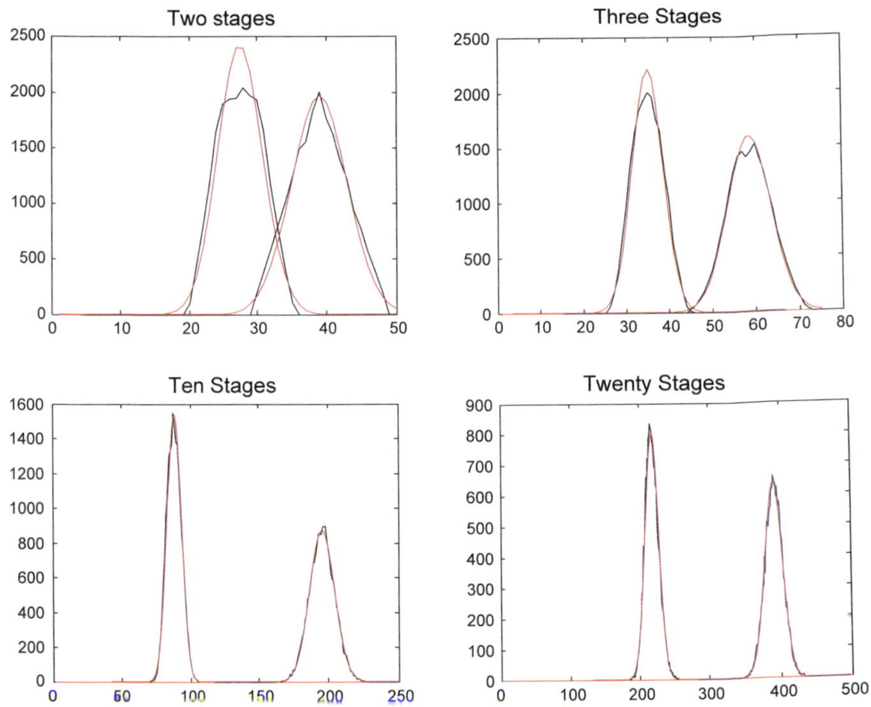

Fig. 5.7 Comparing production line and greedy control systems

In each simulation, the job duration on the first machine followed a uniform duration, between 15 and 25, whilst the job times at all other machines followed a uniform duration, between 5 and 15.

So that in each case, the first machine is absolutely dominant; the results (mean, standard deviation and coefficient of variation, v) from these simulations were as follows:

Machines	Move when ready			Production line		
	Mean	sd	v (%)	Mean	sd	v (%)
2	30.05	4.13	13	38.95	4.02	10
3	40.10	4.95	12	58.52	5.02	9
10	112.8	7.79	7	194.98	9.03	5
20	218.02	10.07	5	390.42	12.27	3

5.2.1.1 Extending the Investigation

The investigation was extended through the simulation of the following cases:

- 20 stages: first [15, 25], rest [5, 15]; first strictly dominant; the minimum time for a job at the first process is greater than the maximum time at any other process.
- 20 stages: first [15, 25], rest [10, 20]; first dominant; the average time at the first process is greater than the average times at all other processes, and the greatest time at the first process is greater than the greatest time at any other process.
- 20 stages: first [15, 25], rest [10, 25]. First slightly dominant; the average time at the first process is greater than the average times at all other processes but the maximum time is not greater than all other maxima.

Note: the variability in process duration is large to emphasise its effect.

The first results are derived from the simulation of a production line where all jobs move onto the next stage at the same time. The second results are obtained from the simulation of a KANBAN control system where a job can move onto the next process when that process becomes free.

Set 1 Results: Production line
20 stages: first [15, 25], rest [5, 15]; first strictly/strongly dominant; here, the first stage (alone) determines the movement along the production line. Therefore by the central limit theorem, expect that the total process time will be normally distributed. The following plots compare the results from a simulation of this configuration with a normal distribution with the same mean and standard deviation (Fig. 5.8).

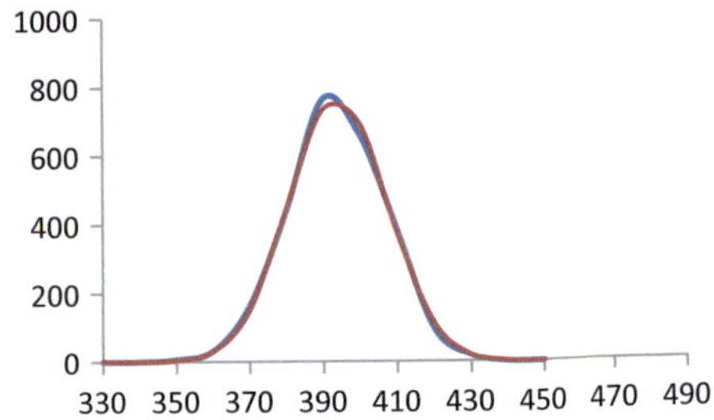

Fig. 5.8 Simulation results *blue*, Normal distribution with same mean and variance *red*

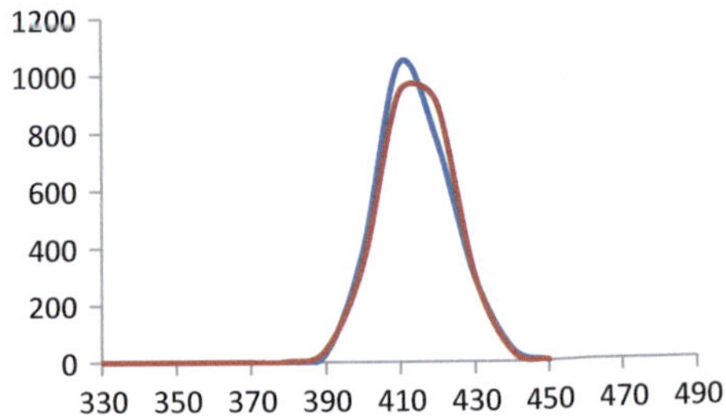

Fig. 5.9 Simulation results *blue*, Normal distribution with same mean and variance *red*

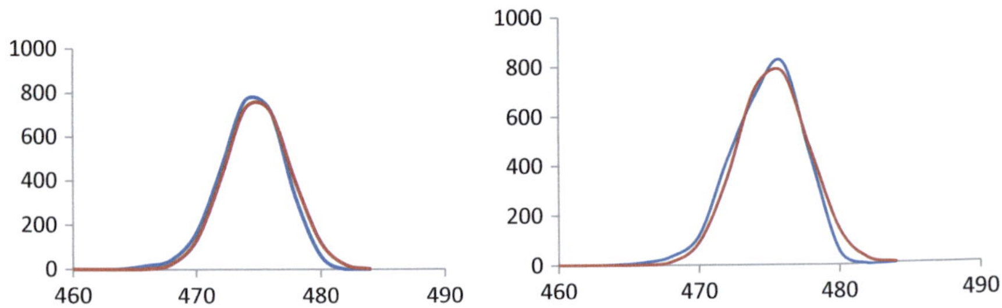

Fig. 5.10 Simulation *blue*: Normal same mean and standard deviation *red*

20 stages: first [15,25], rest [10,20]; first weakly dominant

In this case, although the first stage is most often dominant on 76.1% of occasions, the durations are generated by this distribution (Fig. 5.9).

20 stages: first [15,25], rest [10,25]. First slightly dominant; on (only) 34.4% of occasions, the durations are sampled from this distribution, and other stages are dominant on 65.6% of occasions (Fig. 5.10).

Set 2 Results: Move When Next Stage Available

20 stages; first [15, 25] rest [5, 15]; first stage is strictly dominant but here each job moves when its following stage becomes free. The plot shows that the results from the simulation are similar to those from a normal distribution with the same mean and variance (Fig. 5.11).

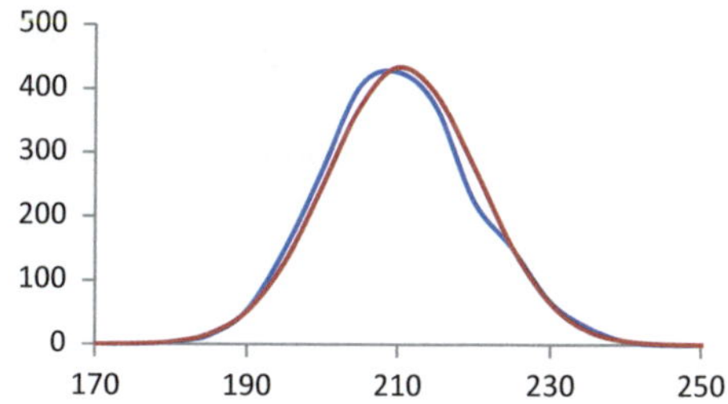

Fig. 5.11 Simulation results *blue*, Normal distribution with same mean and variance *red*

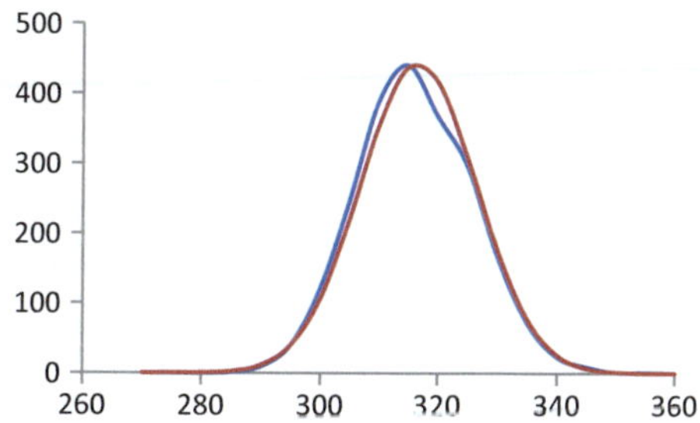

Fig. 5.12 Simulation results *blue*, Normal distribution with same mean and variance *red*

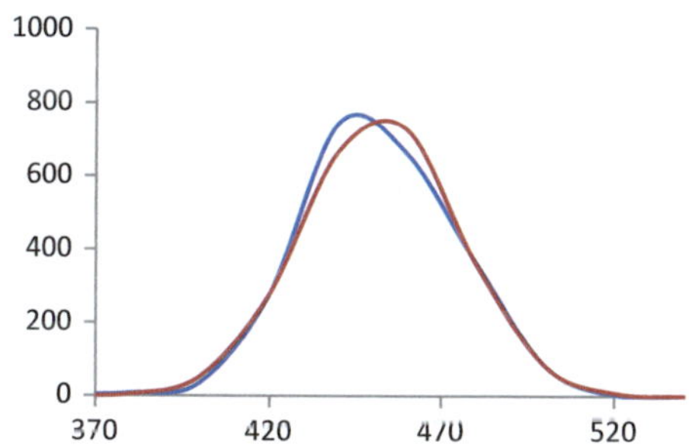

Fig. 5.13 Simulation results *blue*, Normal distribution with same mean and variance *red*

20 stages: first [15, 25], rest [10, 10]; first stage dominant

Again the distribution of the resultant job times is similar to a normal distribution with the same mean and variance the plots are (Fig. 5.12).

20 stages: first [15, 25], rest [10, 25]; first slightly dominant.

Again the plots are similar (Fig. 5.13).

The results indicating that when a "move-when-able" system is employed within a flow shop environment, the process time (from starting at the first stage to completing the final stage) will be (approximately) normally distributed.

5.2.2 *Simulating a KANBAN System*

As an introductory example, a flow shop consisting of five stages was simulated; the process times at each stage were described by uniform distribution with parameters, where stage 4 is weakly dominant:

Stage	1	2	3	4	5	
Minimum time	6	4	4	8	4	
Maximum time	10	8	8	12	8	
Average times	8	6	6	10	6	Total Average Times = 36
Dominance	N	N	N	D	N	

Simulating the operation of the KANBAN control system gave the result:

Mean duration = 45.9
Standard Deviation = 2.6

Repeating with the parameters rearranged so that the final stage is dominant

Stage	1	2	3	4	5	
Minimum time	4	4	6	4	8	
Maximum time	8	8	10	8	12	
Average times	6	6	8	6	10	Total Average = 36
Dominance	N	N	N	N	D	

Gave the results Mean duration = 49.9
Standard Deviation = 2.6

Plotting the results together with plots of a normal distribution with the same parameters gave Fig. 5.14, indicating that in this example the time in system is approximately normally distributed.

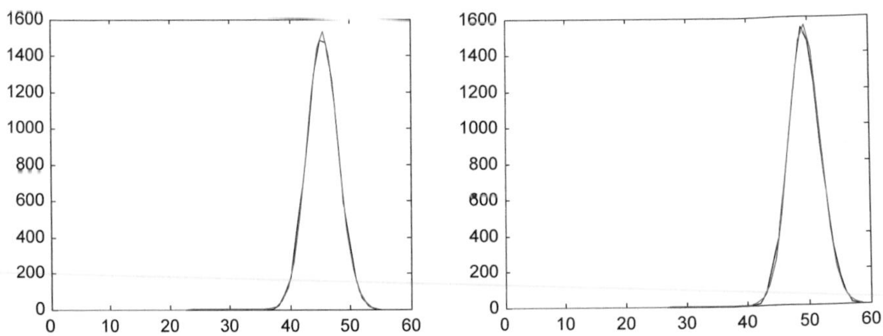

Fig. 5.14 Simulation results *blue*, Normal distribution with same mean and variance *red*

Finally, simulating with the machines ordered by way of their (descending) dominance

Stage	1	2	3	4	5	
Minimum time	8	6	4	4	4	
Maximum time	12	10	8	8	8	
Average times	10	8	6	6	6	Total = 36
Dominance	D	N	N	N	N	

Gave the results Mean duration = 36.3
Standard Deviation = 2.4

Plotting the results with plots of a normal distribution with the same parameters gave the results as shown in Fig. 5.15.

Repeating to generate extreme values when a flow shop has 5 stages with the job times described by :

	Simulation 1					Simulation 2				
Stage	1	2	3	4	5	1	2	3	4	5
	15	5	5	5	5	5	5	5	5	15
	25	20	20	20	20	20	20	20	20	25
	D	N	N	N	N	N	N	N	N	D

Total Average times = 70

Results	Mean	= 73.5	Mean	= 100.2
	Standard Deviation	= 7.8	Standard Deviation	= 100.2

Both plots (Fig. 5.16) indicate that the job times (within the flow shop) are, approximately, normally distributed.

Note in the second simulation, each job at each stage "occupies" on average a 20-time unit slot, size governed by the job time at the final stage.

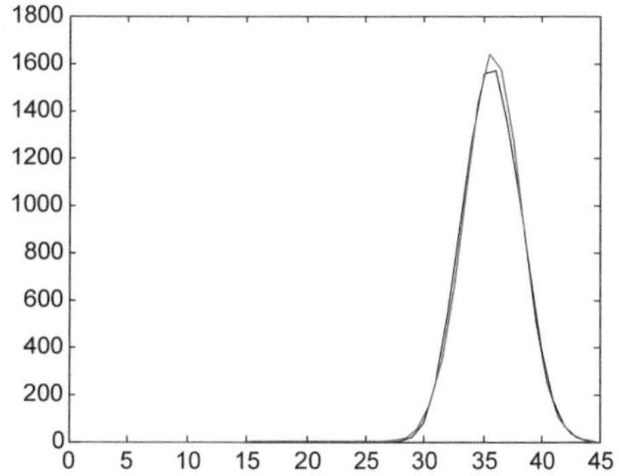

Fig. 5.15 Comparing production line and greedy control systems

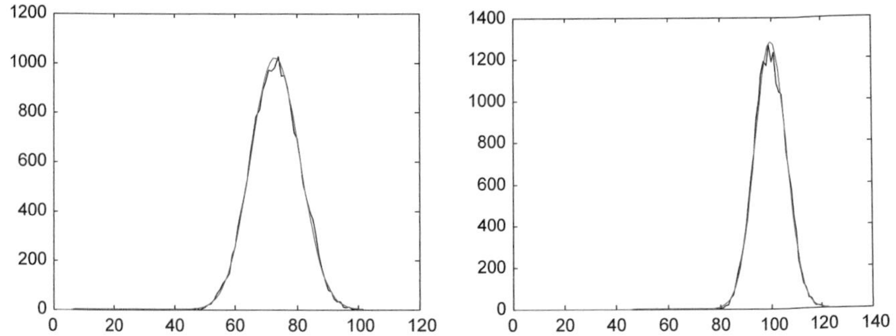

Fig. 5.16 Comparing the effect of the location of the dominant stage

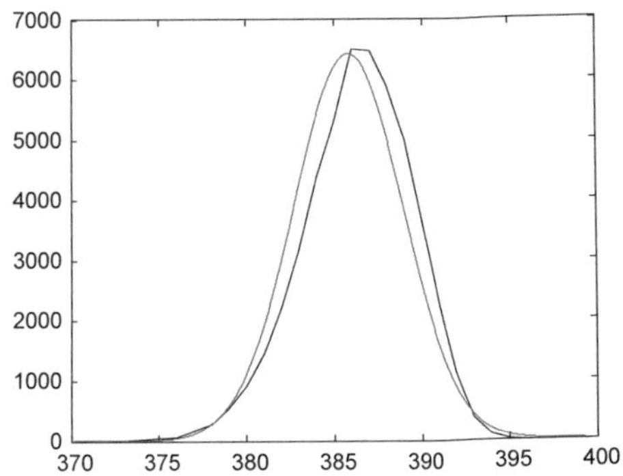

Fig. 5.17 20 stage production process

Extending to flow shop with 20 stages indicated a more skewed distribution for the job times; Fig. 5.17 shows the production time distribution and a normal distribution with the same mean and variance.

Figure 5.18 shows the result where there are 100 then 200 stages, together with a plot of the normal distribution with the same mean and variance.

5.2.3 CONWIP Control System

Production control using a CONWIP, or "Limiting the quantity of work in progress", methodology has the greatest effect when the final stage in the process is dominant, similar to the effect of a KANBAN system.

To illustrate the effect of this methodology on a flow, shop a production process consisting of 20 production stages was simulated; the process times at these stages were

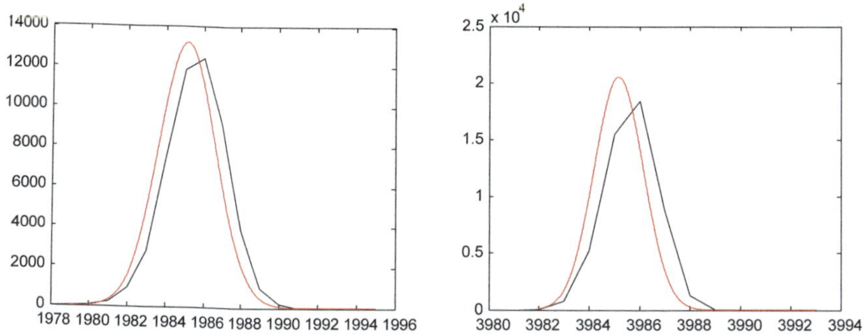

Fig. 5.18 100 and 200 stage production processes

Stage 1 to 19 uniform $[5, 20\}$
Stage 20 uniform $[15, 25]$,
the final stage was weakly dominant, worst case.

The average time in production (no queueing) is therefore the sum of the "stage distribution" averages = 257.5.

The results from these simulations were as follows:

Allowed WIP CONWIP parameter (x)	Average time in system	Standard deviation of time in system	Average time to produce 100 items	Maximum time in system
1	257.5	19	25,750	257.5
5	266	19	5335	344
10	279	19	2797	355
15	297	18	1995	378
18	314	17	1763	390
19	321	17	1710	398
20	329	18	1665	410
22	346	19	1595	432
25	375	20	1527	469
28	407	22	1483	496
30	429	24	1462	532
35	487	26	1460	588
40	544	30	1402	665

Suggesting a model for the average time in the system (TIS) with the maximum allowed work in progress, Fig. 5.19:

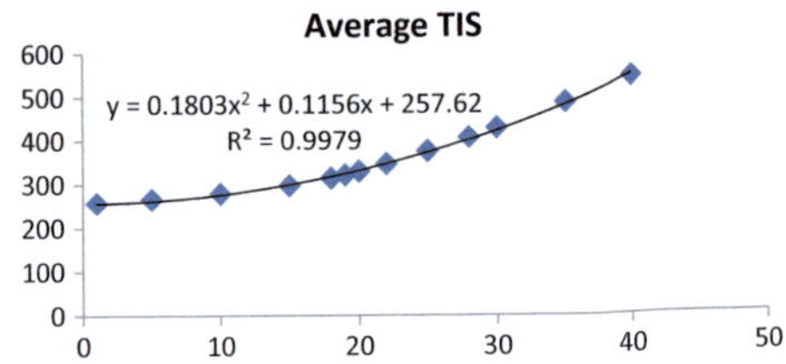

Fig. 5.19 Modeling time in system

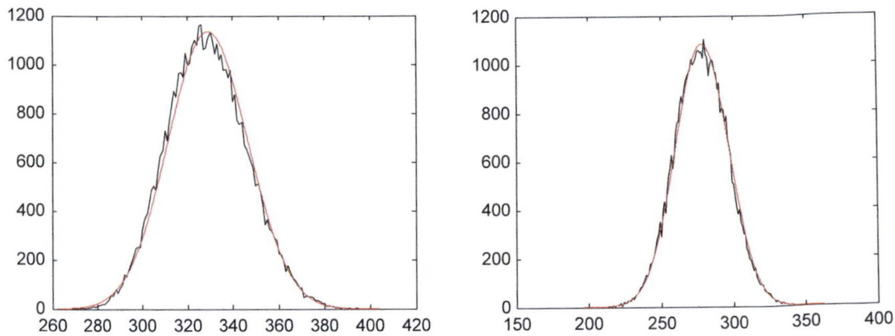

Fig. 5.20 Comparing production line and greedy control systems

$$TIS = 0.180x^2 + 0.116x + 257.6$$

Note: $Average\ Queueing\ Time = 0.1803x^2 + 0.1156x + 0.1181$.

Figure 5.20 shows that with a constant work in progress control system, the times in the system are approximately normally distributed.

Capacity	= 20	Capacity	= 10
mean	= 329	mean	= 278
sd	= 17.6	sd	= 18.5

Similarly, the simulations estimated the time between successive jobs leaving the system, and these results are shown in Table 5.1.

These results seem to indicate that an appropriate value for the CONWIP parameter is greater than the number of stages (x) but probably less than 50% more than the number of stages.

Table 5.1 Effect of CONWIP parameter on completion times

Allowed WIP CONWIP parameter (x)	Average time between completed jobs	Standard deviation of time
1	250	19.3
10	27.0	16.3
15	19.3	9.7
20	16.1	7.0
25	14.8	6.0
30	14.1	5.5
35	13.8	5.3
40	13.6	5.1
50	13.3	4.9

5.2.4 *Summary*

When the dominant stage is the first stage within the production process, all planning and control systems behave in the same way, leading to the same results (time in system, work in progress and time to produce N items).

Thus, it would seem that when the first stage is not dominant the planning and control system should act to make this stage dominant.

5.3 Simulating Manufacturing Systems, to Define PP&C Systems Requirement When Arrivals Are Random

A four-stage production system was simulated with customers arriving randomly with an interarrival time of 10 and the random stage service times {8, 6, 4, 2} in the first simulation then reversing the order to {2, 4, 6, 8} in the second simulation.

The results were:

Model	Average	Queueing Time				
Job order	Time	Total	Stage1	Stage2	Stage3	Stage4
$\{2,4,6,8\}$	36.47	16.24	0.24	1.09	3.23	11.24
$\{8,6,4,2\}$	36.36	16.20	16.20	0.00	0.00	0.00

Thus demonstrating that when arrivals are random and service times deterministic, the location of the dominant (slowest) stage does not alter the total time in the system, but it does indicate that the location of the dominant machine does influence the need for a formal approach to production planning and control; when the first stage is dominant, the system is self controlling, and when the final stage is dominant active control is necessary.

Same result occurs for a production line system with a nonzero queue of jobs waiting at the start.

Regardless of the location of the dominant stage, if all other times are the same time to complete n jobs is the same.

5.4 Review Questions

(A) Introductory Queueing Questions

1. In a supermarket, the arrival of customers at the cash desk is random at an average rate of 15 every 30 min. The average time it takes to scan and calculate the cost of customer purchases is 1½ min, and this time is exponentially distributed.

   (i) How long will a customer expect to wait before being served?
   (ii) What is the chance that queue length will exceed 5?
   (iii) What is the probability that the cashier is working?

   [Ans. (i) 4.5 min, (ii) 0.133, (iii) 0.75]
2. A TV repairer finds that the time spent on his jobs has an exponential distribution with mean 30 min. If he repairs TV sets in the order in which they come in, and if the arrival of sets is approximately Poisson with an average rate of 10 per 8 h day,
   What is the repairer's expected idle time each day?
   How many jobs are ahead of the average set just brought in?
   [Ans. 3 h, $1^2/_3$ jobs]
3. An average of 10 customers per hour arrive at a one window drive-in "take-away". Service time per customer is exponential with mean 5 min. The space in front of the window, including that for the vehicle/driver just being served, can accommodate a maximum of 3 cars. Other cars can wait outside this space.

   (i) What is the probability that an arriving customer will have to wait outside the indicated space?
   (ii) How long would an arriving customer expect to wait before starting service?
   (iii) How many spaces should be provided in front of the window so that all arriving customers can wait in front of the window at least 70% of the time?
   [Ans: (i) 0.579, (ii) 25 min, (iii) 7]

(B) Developing Modelling with Queueing Theory

Q1

A technician fixes broken laptops. The repair time is exponentially distributed with a mean of 30 min. Broken laptops arrive according to a Poisson stream, on average 10 broken laptops per day (8 h).

(i) What is the fraction of time that the technician has no work to do?
(ii) How many laptops are, on average, at his repair shop?
(iii) What is the mean throughput time (waiting time plus repair time) of a laptops?

Q2

In a service station, there is one petrol pump. Cars arrive at the service station according to the Poisson process. The arrival rate is 20 cars per hour. Cars are served in order of arrival. The service time (i.e. the time needed for filling and paying) is exponentially distributed with a mean service time of 2 min.

(i) What is the fraction of cars that has to wait longer than 2 min?
(ii) How does this change if the distribution is no longer exponential?

Q3

A service station has two pumps, one for petrol and the other for LPG. For each pump customers arrive according to a Poisson process, on average 20 customers per hour for petrol and 5 customers for LPG. The service times are exponential. For both pumps, the mean service time is 2 min.
Determine the distribution of the number of customers at the petrol pump, and at theLPG pump, assuming that there are two queues one for petrol and another for LPG.

Q4

Consider an M/M/1 queue with an arrival rate of 60 customers per hour and a mean service time of 45 s. A period during which there are 5 or more customers in the system is called crowded, and when there are less than 5 customers it is quict.

(i) What is the mean number of crowded periods per day (8 h), and
(ii) How long do they last on average?

(C) Simulation questions

Q1 Construct a simulation model for the application defined in question 1.3. Extend the model to represent the case where there is a single two-queue system servicing both pumps and with the customers moving to the first free pump.

Q2 Comparing manufacturing systems

N&F plc operates a small workshop in Ayton producing specialist items. The workshop currently contains 6 machines and has had a workforce of 6, one worker for each machine. Currently jobs arrive randomly with, on average, an arrival every 12 min.

The firm is about to reorganise their production facilities either by buying new machines or by establishing production lines using their existing machines.

The current system uses the machines organised as the flow shop shown in Fig. 5.21.

Alternative proposals:

Proposal 1: Replace the machines at stages B and C with two multipurpose machines (M), job time between 10 and 15, replacing two production stages with one stage (Fig. 5.22).

Proposal 2: Set up two production lines each (assume that there are now two machines of type C and 2 machines of type A), and buy an extra machine type C (Fig. 5.23).

With one machine at each stage and jobs allocated at random to each of these production lines.

Proposal 3: Set up two production units with a common final stage where 66% of the jobs are allocated to production unit *X* (Fig. 5.24).

Production data (Table 5.2).

Using results from queueing theory and by constructing simulation models, evaluate these alternative systems.

Q3 Production planning at a furniture manufacturing company

The firm produces a range of fireplace surrounds from softwood, hardwood and stone at its Beeham factory. The production process used varies with the type of product made; however, the equipment used is given in Table 5.3, and the layout is shown in Fig. 5.25.

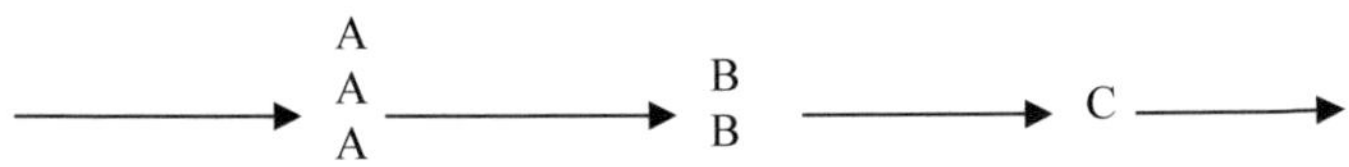

Fig. 5.21 Existing workshop

Fig. 5.22 Proposal 1

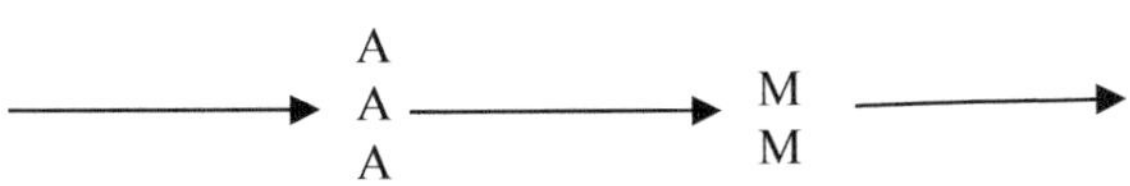

Fig. 5.23 Proposal 2

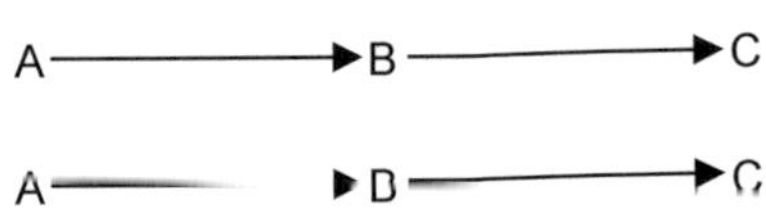

Fig. 5.24 Proposal 3

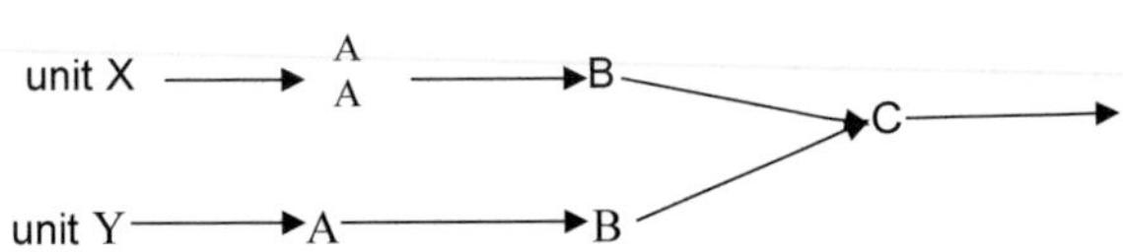

Table 5.2 Job time parameters at each stage, random or uniform distributed

Machine type	Job times random job average times in minutes	Job times uniform job distribution times in minutes
A	15	Between 10 and 20
B	6	Between 5 and 7
C	5	Between 3 and 6
M	9	Between 8 and 10

Table 5.3 Current manufacturing process

	Time (minutes)		
	Softwood	Hardwood	Stone
Prepare wood	30 ± 10	120 ± 15	240 ± 60
Make decorations	10 ± 2	15 ± 5	Not relevant
Construct frame	20 ± 10	40 ± 15	15
Finish frame	15 ± 1	30 ± 1	Not relevant
Add decorations	10 ± 1	10 ± 1	Not relevant

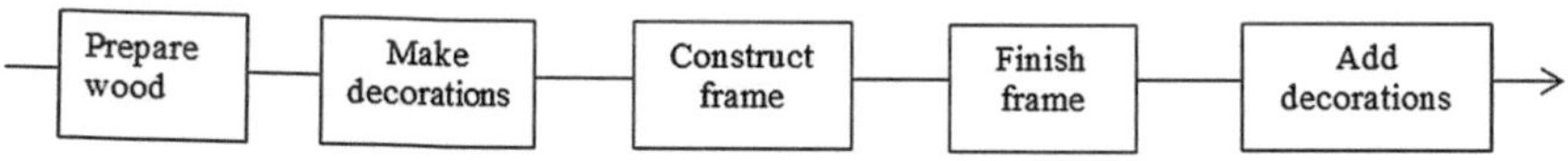

Fig. 5.25 Current production system

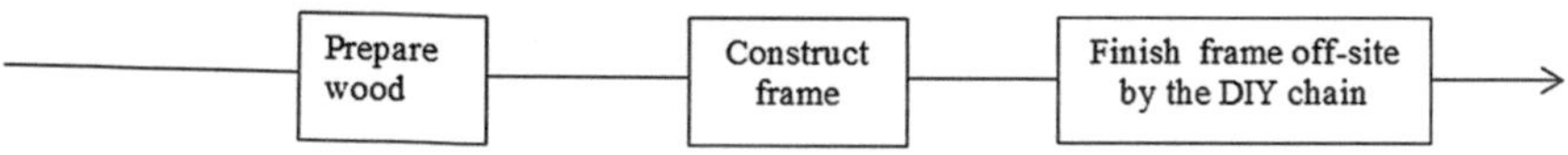

Fig. 5.26 Proposed production

Currently, during the peak demand seasons, each day the manufacture of 10 units is planned, normally 6 softwood, 3 hardwood, and 1 stone surround using the factory layout defined by table and Fig. 5.25.

However, the firm has just gained a new contract to supply a DIY chain with on average 25 softwood kits a week; the production and installation of these surrounds is described by table and Fig. 5.26, where the finishing tasks are carried out by the DIY chain's workers. Note if this new product is successful, the DIY chain intends to make hardwood and stone surrounds available through their outlets.

Part 1 Proposed system for the DIY kits

Assuming that all workers are multi-skilled, construct a simulation model to represent the proposed production system as described in Table 5.4 and use this model to determine the number of machines at each stage and the staffing required to satisfy this DIY demand.

Table 5.4 Proposed manufacturing process

	DIY chain product time (minutes)		
	Softwood	Hardwood	Stone
Wood and decorations	45 ± 15	140 ± 10	240 ± 60
Construct frame	20 ± 10	40 ± 15	15 ± 5
Finish frame	Off site	–	–

Table 5.5 Breakdown probabilities and repair times

	Breakdown probability and repair times (minutes)		
	Softwood	Hardwood	Stone
Prepare	0.075; 5 ± 1	0.05; 10 ± 3	Not relevant
Make frame	0.1; 8 ± 1	0.1; 8 ± 2	Not relevant

Part 2

For each of the model developed in part 1, investigate the effect of machine breakdowns and repairs on your recommended model.

The breakdown and repair information is given in Table 5.5.

Q4 Evaluation of Alternative service systems at Swifts bank

Swifts Bank operates its retail banking services through a chain of high street outlets. To improve customer's service and to reduce costs, Swift's are considering the reorganisation of their branch counter services.

Customers arrive randomly, on average one every 1.5 min. About 60% of these customers are classified as "short-service-time customers", 30% are "medium-service-time customers", 9% are "long-service-time customers" and the remaining 1% require in-depth advice.

The service times for these customers are as follows:

Short service	between 1 and 3 min
Medium service	between 2 and 6 min
Long service	between 5 and 15 min
In depth advice	between 10 and 30 min

Construct appropriate queueing and/or simulation models to evaluate the performance of current service system, considering customer waiting time, queue length, server idle time and number of servers needed to provide the customers with a good service.

Part 4.1 Evaluating the current system

Currently, there are four service counters in use with queues of waiting customers forming in front of each server, see Fig. 5.27, you may assume that there is no queue switching.

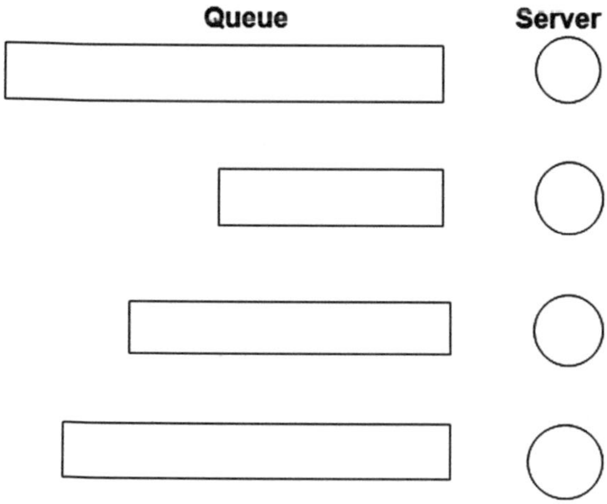

Fig. 5.27 4 servers and 4 queues

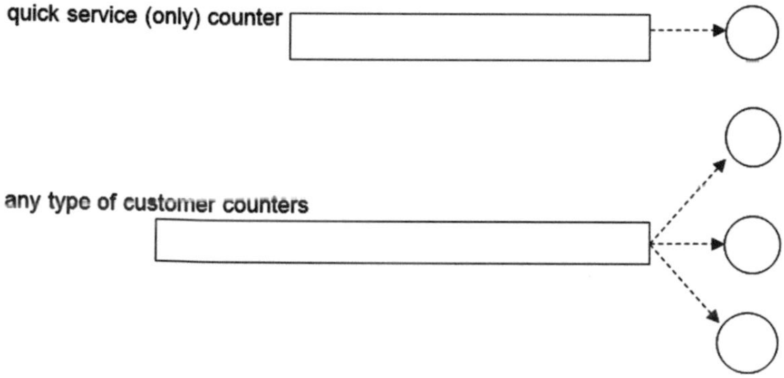

Fig. 5.28 2 queues quick service counter

Part 4.2 Evaluating first alternative system

A first possible reorganisation is to establish two queues: one serving a new quick service counter (only quick jobs) and the other serving the other three counters, see Fig. 5.28.

Part 4.3 Evaluating second alternative systems

The second possible reorganisations are

- to establish two queues each serving two counters,
- to establish one queue serving all counters, see Figs. 5.29 and 5.30.

Part 4 Implement a reception (triage) counter to direct customers

The receptionist directs the customers to the queue in front of the most appropriate server, see Fig. 5.27, duration 90% 1 min, 10% 2 min (Fig. 5.31).

Simulation Example 1

A process with 4 work stations in series was simulated.

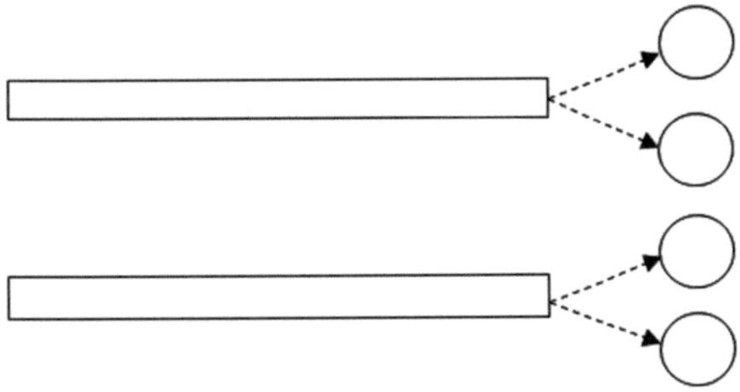

Fig. 5.29 Two-queue system

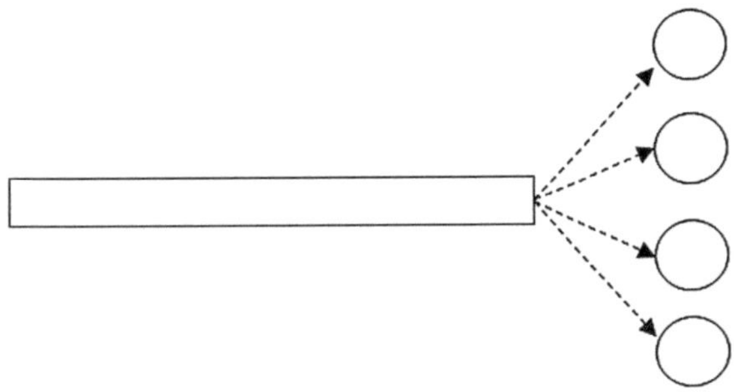

Fig. 5.30 Single queue four servers

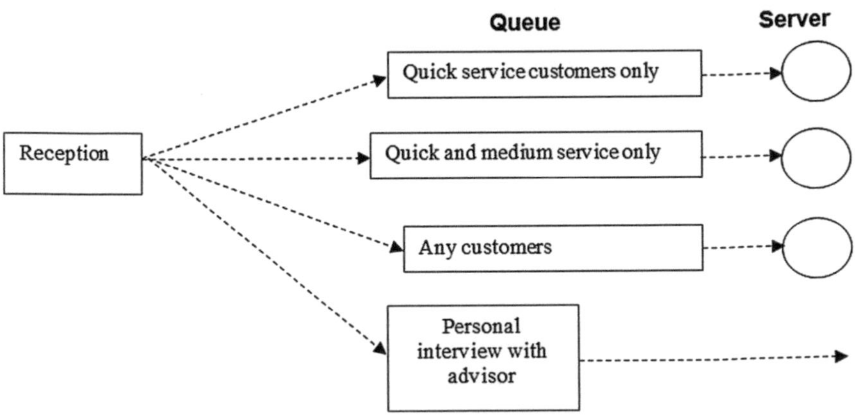

Fig. 5.31 Multiple queues and advisor

In the first simulation, the work station capacities were $\{10, 5, 3, 2\}$,
In the first simulation, the work station capacities were $\{2, 3, 5, 10\}$

The results from simulating these cases were

Work stations, IAT = 20. Job durations				Simulations	Average time in system	Estimating formula result
1	2	3	4			
10	5	3	2	30,000	25.06	25
2	3	5	10	30,000	24.89	25

Simulation Example 2

A process with 4 work stations in series was simulated. The capacities were $\{10, 5, 3, 2\}$,

The probability of rework and true arrival rates/time unit were

Arrival rate + reworked items	Dominant stage ρ	Average time in system
3	0.50	25.0
3.33	0.55	29.5
3.75	0.63	37.0
4.3	0.71	51.5
5.0	0.83	94.5

Simulation 3 Validating calculations for total time in system (TTIS)

Given the formula $\text{TTIS} = \left(\frac{1}{2}\frac{\rho}{(1-\rho)}\frac{1}{\mu_1}\right) + \sum \frac{1}{\mu}$ with $\rho = 0.6$ at dominant machine/stage.

Then, TTIS = (0.5 * 1.5 * 12) +(12 + 8 + 4) = 9 + 24 = **33**, as obtained from the simulation.

Arrivals random average interarrival Time = 20				Average times for machine layouts CDF and FDC			
Process	Service time	One stage	Averages	Machine-order CDF	Times	Machine-order FDC	Times
C	12	Time in system	21.11	Time in system	33.11	Time in system	33.11
		WIP	1.06				
D	8	Time in system	10.65	Waiting times in queues			
		WIP	0.54	Queue1	9.11	Queue1	0.49
F	4	Time in system	4.49	Queue2	0	Queue2	2.16
		WIP	0.23	Queue3	0	Queue3	6.46
		Total individual times	36.25	Individual total 3.15 greater than actual value		Total =	9.11

Part II
Case Studies

These case studies aim to investigate the application of various solution methodologies to problems demonstrating that the "best" approach is problem specific and that the production and analysis of a mathematical model can lead to an efficient and effective solution methodology.

Chapter 6
Case Studies: Using Heuristics

Val Lowndes, Ovidiu Bagdasar and Stuart Berry

The aim of these case studies is to demonstrate how large and complex decision-making problems can be "solved" using heuristic methods. These methods are derived and developed from the known structure of the problem to enable an efficient solution.

6.1 Using Heuristics to "Solve" Case Studies from Mathematical Programming

This section shows how the process of producing a mathematical model can lead to an understanding of the complexity of a problem. Here, travelling salesman problems and production planning problems are considered to show that one falls into the category of NP and solved using a heuristic method to give a good solution (not necessarily optimal) and the other will (and is) always be easily solvable using a simple heuristic to give an optimal solution.

V. Lowndes
University of Derby, Kedleston Road, Derby DE22 1GB, UK
e-mail: V.P.Lowndes@derby.ac.uk

O. Bagdasar · S. Berry (✉)
College of Engineering and Technology, University of Derby,
Kedleston Road, Derby DE22 1GB, UK
e-mail: s.berry@derby.ac.uk

O. Bagdasar
e-mail: O.bagdasar@derby.ac.uk

S. Berry et al. (eds.), *Guide to Computational Modelling for Decision Processes*,
Simulation Foundations, Methods and Applications,
DOI 10.1007/978-3-319-55417-4_6

6.1.1 Modelling the Travelling Salesman Problem

The aim here is to show that it is possible to construct a linear programming model to determine the optimal solution to the travelling salesman problem. The suitability of this model is then discussed showing that the number of constraints increases greatly as the number of cities increases.

This analysis is initially based around developing a model to solve the small 4-city problem,

where

$x_{ij} = 1$ indicates that the salesman will travel from city i to city j

$x_{ij} = 0$ indicates that the salesman will not travel from i to j

Thus, the basic model would seem to require the equations

$$\sum_{j,j \neq i} x_{ij} = 1 \quad \text{leave city } i \text{ only once, all } i.$$

and

$$\sum_{i,j \neq i} x_{ij} = 1 \quad \text{arrive at city } j \text{ only once all } j.$$

However, this formulation is not satisfactory (will not always produce an acceptable solution) because these equations can be satisfied by the *not valid* solution with "subtours"; for example, in a four town problem, these equations could be satisfied by:

$$x_{12} = 1, x_{21} = 1 \text{ and } x_{34} = 1, x_{43} = 1 \text{ two non-connected sub tours.}$$

Thus, this model will not always produce an acceptable solution, a full tour visiting each town only once, and has therefore to be extended to prevent the production of such unacceptable solutions.

Extended Model, to remove subtours

Add the new variables y_1, y_2, y_3, y_4 to represent the four towns in this problem. These variables are used to generate additional constraints (6 extra constraints here) which will cause the formulation to become infeasible if the model tries to generate solutions containing subtours.

In this formulation, the tour is assumed to start and end at town 1, and a constraint is added to represent each possible link between towns.

$$y_i - y_j + (n-1)x_{ij} \leq (n-2) \quad \text{all} \quad i,j > 1,\ i \neq j$$

As an illustration to show that these extra constraints will not allow subtours whilst allowing valid tours:

- Consider the effect of subtour 2–3 and 3–2 on these extra constraints

$$(x_{23} = 1, x_{32} = 1)$$

The constraints generated for route 2–3 and route 3–2 are

$$\begin{aligned} y_2 - y_3 + (n-1) &\le (n-2) \\ y_3 - y_2 + (n-1) &\le (n-2) \end{aligned}$$

Adding these constraints would give the infeasible condition

$$2(n-1) \le 2(n-2) \quad \text{FALSE, route prohibited}$$

Hence, these additional constraints have act (here) to prevent subtours occurring in the solution.

- Now, consider the effect of the invalid tour 2–3, 3–4, and 4–2 on these extra constraints

$$(x_{23} = 1, x_{34} = 1, x_{42} = 1)$$

giving

$$\begin{aligned} y_2 - y_3 + (n-1) &\le (n-2) \\ y_3 - y_4 + (n-1) &\le (n-2) \\ y_4 - y_2 + (n-1) &\le (n-2) \end{aligned}$$

Add to give an infeasible statement preventing this subtour.

$$3(n-1) \le 3(n-2) \quad \text{FALSE, route prohibited}$$

- Finally, consider the valid tour 1 2, 2–3, 3–4, and 4–1 ($x_{12} = 1, x_{23} = 1, x_{34} = 1, x_{41} = 1$) and substitute into the full set of constraints to give

$$\begin{array}{ll} y_2 - y_3 + (n-1)x_{23} \le (n-2) & y_2 - y_3 + (n-1) \le (n-2) \\ y_3 - y_4 + (n-1)x_{34} \le (n-2) & y_3 - y_4 + (n-1) \le (n-2) \\ y_2 - y_4 + (n-1)x_{24} \le (n-2) & y_2 - y_4 + 0 \le (n-2) \\ y_3 - y_2 + (n-1)x_{32} \le (n-2) & y_3 - y_2 + 0 \le (n-2) \\ y_4 - y_2 + (n-1)x_{42} \le (n-2) & y_4 - y_2 + 0 \le (n-2) \\ y_4 - y_3 + (n-1)x_{43} \le (n-2) & y_4 - y_3 + 0 \le (n-2) \end{array}$$

$$\begin{array}{ll} \text{add to give} & 2(n-1) \le 6(n-2) \\ \text{or} & 10 \le 4n \quad \text{which can be TRUE} \end{array}$$

This can be satisfied for n, hence showing that this model allows valid tours whilst not allowing invalid tours.

Notice that in an m town problem, the equivalent result for a valid tour would be

$$(m-2)(n-1) \leq \frac{1}{2}m(m-1)(n-2)$$

or

$$n \geq \left(\frac{2m^2 - 4m + 4}{m^2 - 3m + 4}\right)$$

This can be solved for a given value of m. Therefore, this model is successful provided that n is greater than 3 as m becomes large.

Problem Size: For n city travelling salesman problem, an exhaustive search would require the evaluation of $(n-1)!/2$ possible tours. A 1000 town problem would have $999!/2$ possible tours, and an exhaustive search would not be completed in a reasonable time.

This, linear programming, model could determine the optimal tour for any number of cities; however, if there are n cities, then there will be:

Arrive once	n equations
Leave once	n equations
Prohibit sub tours	$(n-1)(n-2)$ equations,

City variables	$(n-1)$ cities
City Links variables	$(n-1)(n-2)$ links

giving

$2n + (n-1)(n-2) = n^2 - n + 2$	constraints and
$((n-1) + (n-1)(n-2)) = (n-1)^2$	variables, both $O(n^2)$.

Thus, a 1000 town problem (a small problem) will have approximately 1,000,000 constraints and 1,000,000 variables and may not be solvable in a reasonable time.

6.1.2 Applying Heuristic Methods to Travelling Salesman Problems

This section indicates how a travelling salesman problem can be "solved" using both genetic algorithmic and Tabu search methodologies.

6.1.2.1 Genetic Algorithm Implementation

A genetic algorithm approach will use:

String	A rearrangement of the cities
Selection for crossover	Tournament selection
Crossover	Select a point and reverse the order of the cities after this point
Mutation	Select two positions and exchange the cities

Examples, using randomly placed cities, have been "solved" using both a standard genetic algorithmic approach and using a mini-genetic algorithm approach.

Mini Genetic Algorithm approach, several small (few strings) genetic algorithms are processed, each generating one good solution. These solutions are then used as the starting population for a final genetic algorithm, again employing only a few runs.

Both approaches were applied first to a 30 city problem and then to a 40 city problem,

with one crossover on 99% of occasions and 2 crossovers on 1% of occasions.

		Time	Best solution
30 city problem			
TSP_GA	100 strings and 2500 iterations	6.51	374
TSP_miniGA	40 mini-GAs. Each 40 strings and 30 iterations. Final population 250 strings	3.22	380
40 city problem			
TSP_GA	100 strings and 2500 iterations	12.9	498
TSP_miniGA	40 mini-GAs. Each 40 strings and 30 iterations. Final population 250 strings	4.41	515
20 city problem			
TSP_GA	100 strings and 2500 iterations	4.4	251

For comparison with one crossover on 1% of occasions and 2 crossovers on 99% of occasions.

		Time	Best solution
40 city problem			
TSP_GA	100 strings and 2500 iterations	8.6	547
TSP_miniGA	40 mini-GAs. Each 40 strings and 30 iterations. Final population 250 strings	4.4	566

Summary of GA approaches

The results from these randomly generated problems suggest that better (cheaper) solutions have been obtained using a genetic algorithm (rather than a set of mini-genetic algorithms) with mostly one crossover and two crossovers infrequently and randomly. These results suggested that the solution time could be modelled by:

$$\text{GA Solution time model} \quad \text{Time} = 0.1928 + 0.2085\,\text{Cities}$$

6.1.2.2 Tabu Search Implementation

Within this implementation, a typical route in a 6 city problem could be represented by the tour

$$[4, 6, 1, 2, 5, 3]$$

The implemented Tabu search procedure operates by investigating the effect of switching all allowed pairs of cities, and the best option with respect to cost change from this search is selected as the next move generating a new route, for example exchanging the 6 and the 3 to give

$$[4, 3, 1, 2, 5, 6]$$

The tabu list is implemented by prohibiting the move of these cities {6,3} for a certain number of iterations, for example

$$[0, 0, 4, 0, 0, 4]$$

Here, at the next 4 iterations, no “swaps” using cities 3 or 6 are allowed, and after each iteration, the tabu list is updated by adding two new tabu cities and reducing the tabu time for the existing tabu cities.

Giving the Tabu search implementation routine:

Investigate all possible allowed exchanges, $n(n-1)/2$ possible pairs.
Select the best exchange
Generate the new schedule
Update the tabu list
Repeat.

Typical result from the Tabu search: random city positions and sample solution plots (Fig. 6.1)

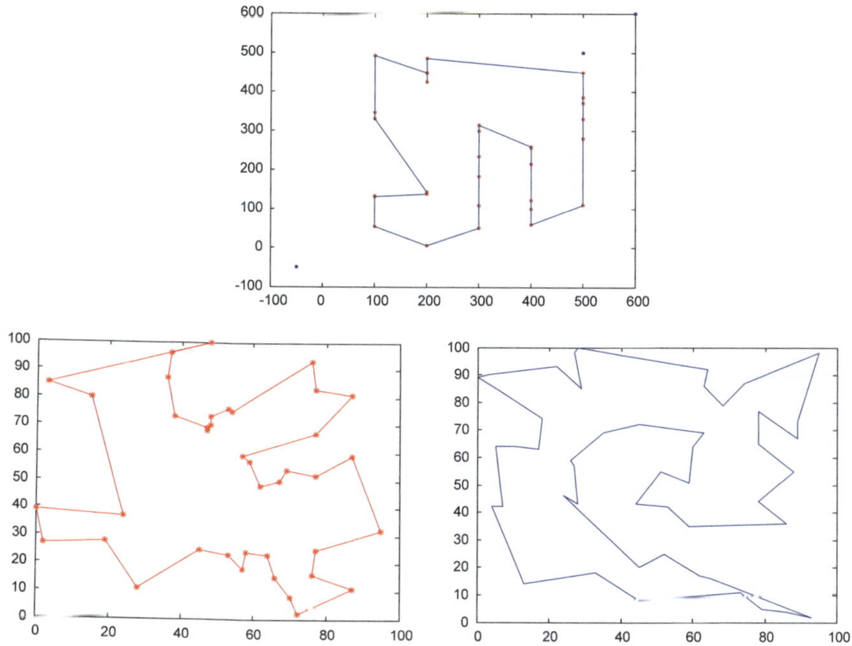

Fig. 6.1 Plots of solutions to travelling salesman problems using a Tabu search

$$\text{Results}: \quad \text{greedy} = 476; \quad \text{tabu search} = 455;$$

Note: Regardless of the size of the tabu period or tabu list, the search could start to cycle around the same results. The possibility of cycling occurring can be reduced through the use of a stochastic tabu period, the "tabu period" being randomly generated.

6.2 Extended Travelling Salesman Problem: Garbage Collection

A related problem is concerned with determining a number of routes/vehicles needed to collect a given set of garbage. In these problems, there exists a set of customers represented: for domestic collection, a set of arcs (Fig. 6.2), and for commercial customers, a set of nodes in a network (Fig. 6.3).

In Fig. 6.2, the quantity of refuse to be collected on each road is shown by the thickness of the line and the collectors have to traverse the full length of the road (similar to a postman problem), whereas Fig. 6.3 shows the locations of the commercial customers; here, the collectors are not required to traverse the full length of

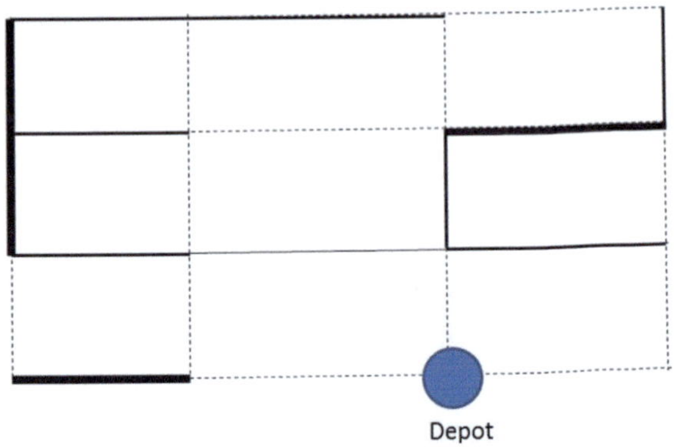

Fig. 6.2 Garbage collection by road

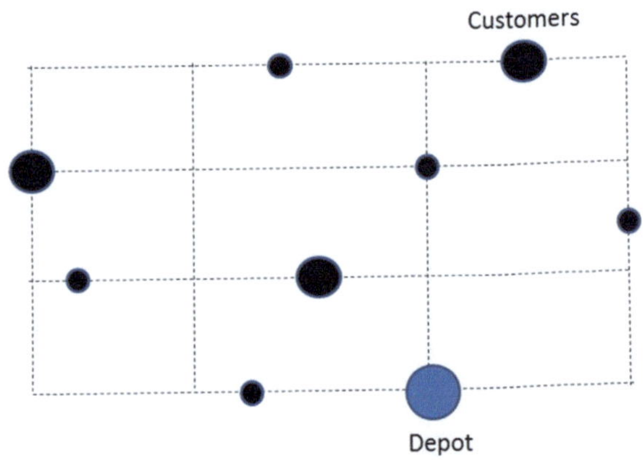

Fig. 6.3 Garbage collection by customer

the road, and the quantity of refuse to be collected is indicated by the size if the location symbol.

The required solution is a set of routes from and returning to the depot such that both travel time and load restrictions are not violated; notice that if the collections could be split into "route sets", this problem is reduced to a set of travelling salesman problems, thus suggesting the use of a heuristic approach to determine good solutions.

Methods Investigated

GA	Full string break down into loads for best solution testing
GA + TS	Combining methodologies, TS to improve the best GA solution
TS	To generate half loads, iterative set of loads best order and then exchange between orders; back to best order and so on. After loads best order construct full routes and save the best costs
TS + GA	To determine the "half" loads, use a greedy approach to construct a set of half loads, and then, using a Tabu search or genetic algorithm-based approach, convert these into full loads

6.2.1 Genetic Algorithm Implementation

Figure 6.4 shows the routes selected by the genetic algorithm applied to a 20 city problem on a 100×50 grid

Figure 6.5 shows the routes selected by the genetic algorithm applied to a 16 city problem on a 20×20 grid.

6.2.2 Half Loads Sample Results

This approach starts with a set of "half load" routes from the depot, each of these routes is such that it requires no more than half of the wagons capacity and then uses an approach using genetic algorithms to select the best routes for the wagons (joining half routes). If for example the depot is located at (0, 0) with 22 collection points (Fig. 6.6a); the total load (here) requires 3 wagons and thus 6 half-wagons; the generated half wagon schedules are shown in Fig. 6.6b, c.

The "half route" end points from this depot location are as follows:

Half route	End point
1	$(8, 14)$,
2	$(0, 12)$,
3	$(12, 8)$,
4	$(0, 15)$,
5	$(3, 15)$,
6	$(9, 9)$

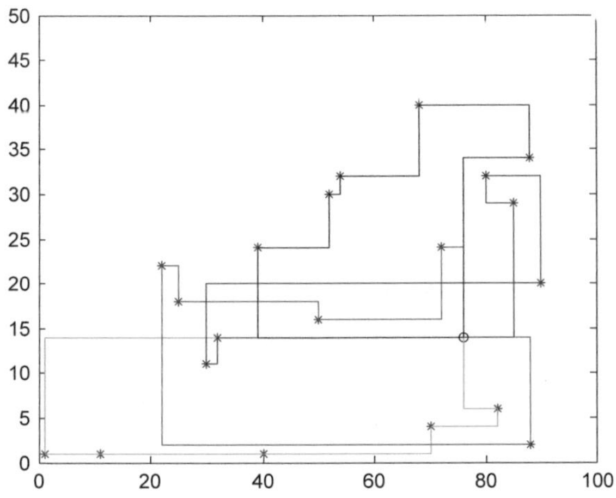

Fig. 6.4 Garbage collection routes generated by a genetic algorithm

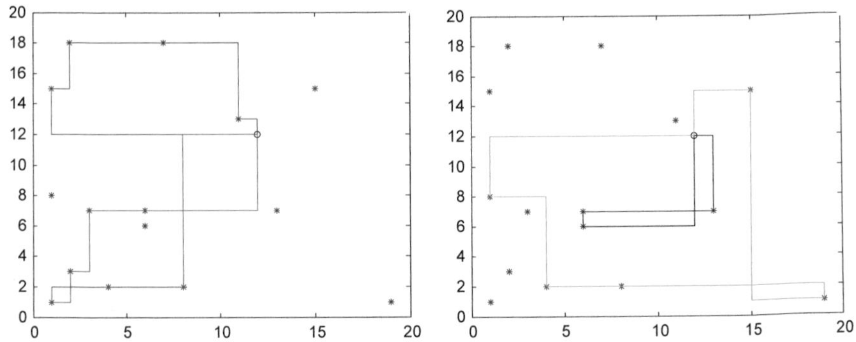

Fig. 6.5 Routes for wagons 1 and 2 and Routes for wagons 3 and 4

Leading to the full routes

1	Half routes $1+2$	joining distance	$=10$
2	Half routes $3+6$	joining distance	$=4$
3	Half routes $4+5$	joining distance	$=3$
		Total joining distance	$=17$

In a second example, the depot is located at the point (15, 7), and the ending pickup, for the half load schedule, is shown by the symbol "$*$"; the three full route can now be constructed by searching the alternative pairings using an exhaustive search when the number of loads is low and a Tabu search when the number of alternatives is very high (Fig. 6.7).

Here, the half route end points for this depot location are as follows:

Half route	End point
1	$(0, 15)$,
2	$(1, 6)$,
3	$(1, 8)$,
4	$(3, 4)$,
5	$(2, 11)$,
6	$(8, 1)$

Leading to the full routes

1	Half routes $2+3$	joining distance	$=2$
2	Half routes $1+5$	joining distance	$=6$
3	Half routes $4+6$	joining distance	$=8$
		Total joining distance	$=16$

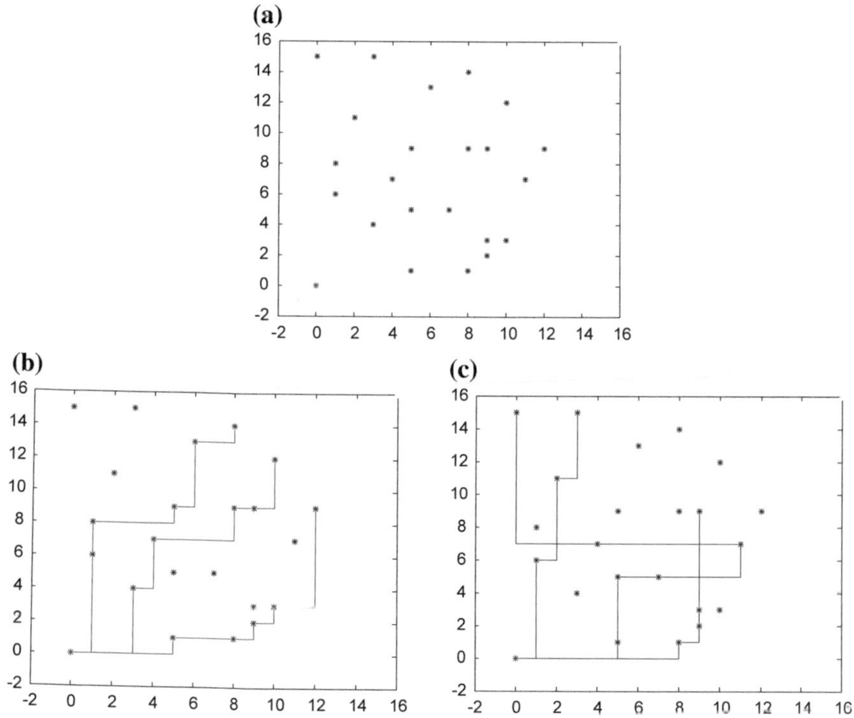

Fig. 6.6 **a–c**. Half routes 1, 2 and 3 and half routes 4, 5 and 6

Fig. 6.7 Generated "half routes"

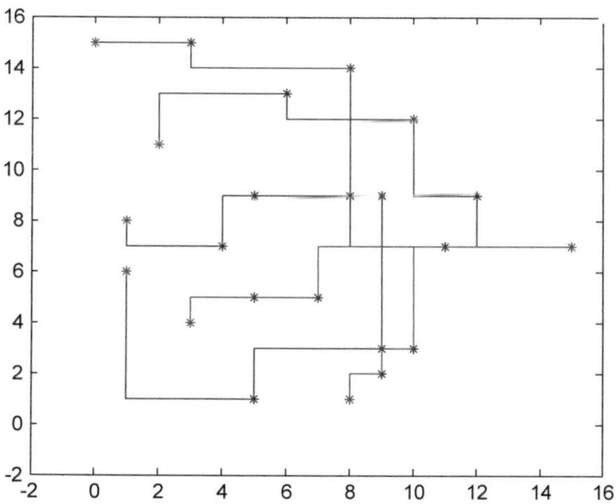

6.2.3 *Using Genetic Algorithms to Generate Half Loads*

To increase the effective search space, genetic algorithms can be used to determine the half loads and half load end points.

Here the GA string gives a "wagon picking order", for example the string {2, 3, 5, 1, 4} leads to the routine

Stage 1: Greedy choice picking orders using the order given by

$$\{2, 3, 5, 1, 4\}$$

Wagon 2 picks the nearest collection point
Wagon 3 then picks the closest available (valid) collection point
And so on until
Wagon 4 picks the closest available (valid) collection point

Stage 2: The picking order is now given by

$$\{3, 5, 1, 4, 2\}$$

Wagon 3 then picks the closest available (valid) collection point
And so on until
Wagon 2 picks the closest available (valid) collection point

and so on cycling through the wagons in the order defined by the genetic algorithm string until all loads have been allocated to a half route. Then, employ a tabu or exhaustive search to produce full strings, from these half strings, calculating costs, and then employ crossover mutation to generate a new set of GA strings and so on.

Genetic algorithm string solutions: in four sample sets of results, the end collection points were as follows:

Sample set, sets of half loads

	1	2		3		4	
Point	Job	Point	Job	Point	Job	Point	Job
1, 8	*E*	*8, 1*	*A*	*0,15*	*G*	*0,15*	*G*
8, 1	*A*	*2,11*	*F*	*3, 4*	*C*	*8, 1*	*A*
0,15	*G*	*0,15*	*G*	*1, 8*	*E*	*1, 6*	*D*
1, 6	*D*	*1, 8*	*E*	*5, 1*	*B*	*3,15*	*H*
2,11	*F*	*3,15*	*H*	*8, 1*	*A*	*2,11*	*F*
5, 1	*B*	*3, 4*	*C*	*2,11*	*F*	*1, 8*	*E*
Half routes total costs							
	90		90		90		90

Constructing routes by joining these end points gives the solutions

1	2	3	4
AB	AC	AB	AD
DE	EF	CE	EF
FG	GH	FG	GH
Joining costs, joining two half loads			
11	15	15	19
Total costs			
101	105	105	109

6.3 Production Planning Problems Network Models

This case study aims to show how that production planning problems can be represented by a network models; this fact suggests that there may be an effective heuristic means of deriving optimal, or near optimal, solutions to the problem.

The initial models consider a case where there is a single product, all working is carried out during the normal working hours, and all the demanded items move from the source (start) to the sink (despatch).

Each arc has an associated cost and each node a capacity (production capacity) or requirement (demand).

Production planning problems can be represented as transportation or trans-shipment problems; note (here) that late deliveries are not allowed! Thus, the demand from month 2 can only be produced in either month 1 or month 2.

This problem can be formulated as a network flow problem and hence there probably exists a simple solution technique, see Fig. 6.8b for an example.

Proof of this approach is obtained from an evaluation of the implied linear programming model.

6.3.1 Production Planning Problems (Mathematical Programming Techniques)

This case study aims to show how the Mathematical Programming Model for this problem is used to develop, and validate, a heuristic means of deriving optimal solutions to such a problem.

The initial models consider a case where there is a single product, and all working is carried out during the normal working hours. This model is then extended to investigate cases where overtime working or backlogging orders is allowed.

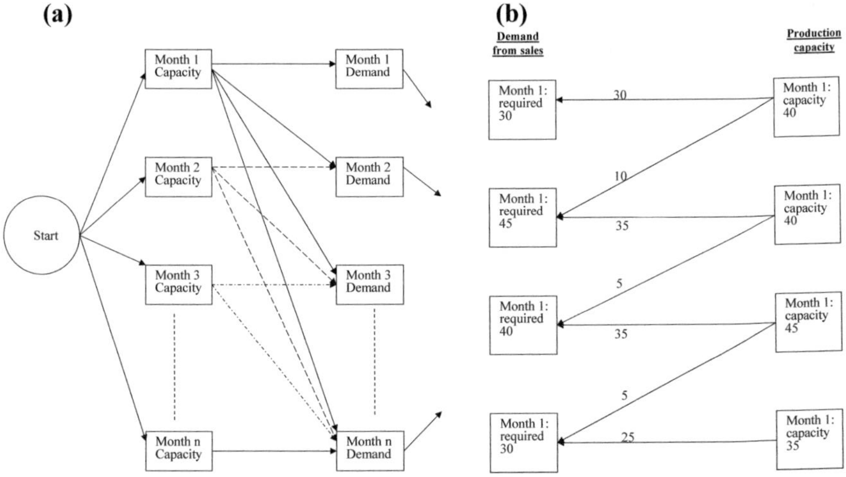

Fig. 6.8 **a** and **b** Example allocating production to satisfy demand

Notation,

d_j the demand in month j
c_j production capacity in month j
p_j production cost per unit in month j
h unit holding cost per unit per month

Model 1, notation

x_j production in month j

Thus, it follows that the x_j need to satisfy the constraints:

$$\sum_{i=1}^{j} x_i \geq d_j \quad \text{all}\, j$$

$$x_i \leq c_i \quad \text{all}\, i$$

whilst minimising the total cost, where cost is given by:

$$C = \sum_i p_i x_i + h \sum_i (n + 1 - i) x_i$$

See Sect. 3.6.5.3 for the development of this model.
Model 2, this uses the alternative notation

x_{ij} production in month i for use in month j;
$x_{ij} = 0,\ i > j$ no backlogging of orders

Thus, it follows that the x_j need to satisfy the constraints:

$$\sum_j x_{ij} \leq c_i \quad \text{all } i$$

$$\sum_i x_{ij} \geq d_j \quad \text{all } j$$

whilst minimising the total cost, where cost is given by:

$$C = \sum_i \sum_j ((j - i)h + p_i)x_{ij}$$

Data requirement for this model: production costs and holding costs and demand forecasts.

6.3.2 Evaluation the Model: Simplifying the Cost Function

For both models as

$$p_1 = p_2 = \cdots = p_n$$

the unit production cost will be constant for production during normal working time (not normally time period dependent) and as the total production will be (just) enough to satisfy demand when the aim is to minimise costs, then it follows that

$$\sum p_i x_i = p \sum x_i = pX \quad \text{where} \quad X - \sum x_i \text{ is a constant}$$

has a constant value, and as a consequence, the cost expression in model 1 can be reduced to

$$\text{cost} = h(nx_1 + (n - 1)x_2 + \cdots + 1x_n)$$

and finally as h is common to all terms, then h can be removed to give the simplified cost expression to be optimised:

$$\text{cost} = (nx_1 + (n - 1)x_2 + \cdots + 1x_n)$$

Similarly, in model 2, the cost expression reduces to

$$\sum_i \sum_j ((j-i)h)x_{ij}$$

Notice that both formulations imply that there is no requirement for any precise costing!

Thus, the production planning process can be modelled by the linear programming formulation:

Model 1

Select x_i so that the total cost is minimised:

$$\text{cost} = (nx_1 + (n-1)x_2 + \cdots + 1x_n)$$

Whilst satisfying the constraints:

$$\sum_{i=1}^{j} x_i \geq d_j \quad \text{all}\, j$$

$$x_i \leq c_i \quad \text{all}\, i$$

Model 2

Select x_{ij} so that the total cost is minimised

$$\text{cost} = \sum_i \sum_j ((j-i)h)x_{ij}$$

Whilst satisfying the constraints:

$$\sum_j x_{ij} \leq c_i \text{ all } i$$

$$\sum_i x_{ij} \geq d_j \text{ all } j$$

In both cases, the objective function has been reduced to "minimise stockholding time".

The optimal solution to this problem, both models (which can be verified using LP package), is to produce as late as possible producing early and storing only when there is insufficient capacity, thus minimising stockholding time.

These models explain how in particular small manufacturing firms with a low resource base have been able to operate effectively/profitably; the obvious production planning technique is optimal.

This approach, and model 1, will be extended to demonstrate how it can be applied when there are many items using the same production facilities and where overtime working or subcontracting could be used by the firm.

The approach has been employed by a company producing 30 types of item in a single production unit.

Illustrative example: the (optimal) production schedule will be built backwards as demonstrated in the example.

Month	1	2	3	4	5	6	7
Capacity	40	40	40	40	40	40	40
Demand	20	20	30	40	30	60	20

This must be the optimal solution; all items are produced as late as possible, minimising holding costs (Fig. 6.9).

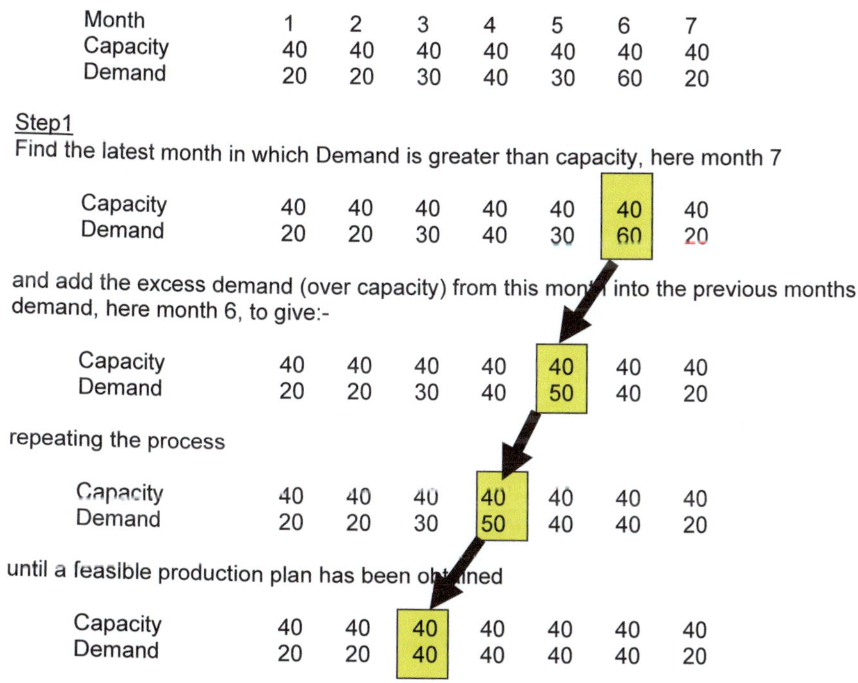

Fig. 6.9 Optimal allocation of production time

6.3.3 Extending the Model to Include Overtime Working

The extended model requires the additional notation, let

y_j overtime worked in month j
O_j overtime capacity in month j
rp cost of producing an item using overtime working.

Then, it follows that the planning process can be modelled by:

$$\begin{array}{llll} \text{Demand} & x_1 & +y_1 & \geq d_1 \\ & x_1+x_2 & +y_1+y_2 & \geq d_1+d_2 \\ & x_1+\ldots+x_n & +y_1+\ldots+y_n & \geq d_1+\ldots+d_n \end{array}$$

$$\text{Capacity} \quad x_j \leq c_j, y_j \leq C_j \quad \text{all}\, j$$

$$\text{Production cost} \quad p_1x_1+\ldots+p_nx_n+rp_1y_1+\ldots+rp_ny_n$$

and as the month end stocks are given by

$$\begin{array}{ll} s_1 = x_1+y_1 & -d_1 \\ s_2 = (x_1+y_1)+(x_2+y_2) & -(d_1+d_2) \\ s_n = \sum\limits_j (x_j+y_j) - \sum\limits_j d_j & \end{array}$$

holding cost will be given by

$$\sum_j hs_j = h(n(x_1+y_1)+\ldots+1(x_n+y_n)) - h(nd_1+\ldots+1d_n)$$

and the variable holding cost by

$$h(n(x_1+y_1)+\ldots+1(x_n+y_n))$$

to give the cost function

$$\text{cost} = p_1x_1+\ldots+p_nx_n+rp_1y_1+\ldots+rp_ny_n+h(n(x_1+y_1)+\ldots+1(x_n+y_n))$$

However, as $p_1 = p_2 = \ldots = p_n$ and $\sum\limits_j (x_j+y_j) = \sum\limits_i d_j$

the cost expression will reduced to

$$\text{cost} = (r-1)p_1y_1+\ldots+(r-1)p_ny_n+h(n(x_1+y_1)+\ldots+1(x_n+y_n))$$

Considering the extreme cases shows that:
if $r = 1$ (no overtime premium), the problem becomes minimised

$$\text{cost} = h(n(x_1 + y_1) + \ldots + 1(x_n + y_n))$$

The optimal solution is to produce as late as possible (the same as in the basic problem),

if r is very large, the problem becomes minimised

$$\text{cost} = (r - 1)py_1 + \ldots + (r - 1)py_n \equiv y_1 + \ldots + y_n$$

Thus, produce as little as possible on overtime, and when overtime has to be used, produce as late as possible.

In a more general problem, overtime costs more than normal working but not infinitely more; there will be the options:

(a) Produce in month n using overtime working, or
(b) Produce in month $(n - k)$ using normal working, k number of month storage

The relevant costs for comparison are as follows:

(a)	rp	overtime production
(b)	$p + kh$	production plus holding cost

from which it follows that the produce early and store policy will be used when this option is cheaper than overtime working:

$$rp \geq p + k \quad \text{or} \quad k \leq (r - 1)/h$$

Thus, when demand in a month exceeds, capacity determines whether the excess should be produced early or using overtime working using this relationship.

Example: Given the data $r = 1.5$, $p = 5$, $h = 0.5$

Capacity	40	40	40	40	40	40	40
Overtime Capacity	10	10	10	10	10	10	10
Demand	20	20	30	40	30	50	20

Here, the excess demand in period 6 could be produced in month 6 on overtime or in month 5 and stored for one month, $k = 1$.

Here, $\frac{(r-1)p}{h} = \frac{(1.5-1)10}{0.5} = 10$ therefore as $k = 1 < 10$

The optimal policy is to produce early and store (Fig. 6.10).

Capacity	40	40	40	40	40	40	40
Overtime Capacity	10	10	10	10	10	10	10
Demand	20	20	30	40	30	50	20
Normal Production	20	20	30	40	40	40	20
Overtime	0	0	0	0	0	0	0

Fig. 6.10 optimal use of overtime working

6.3.4 *Extending the Model to Include Subcontracting Production*

The extended model requires the additional notation, let

z_j quantity subcontracted for use in month j
S_j maximum number of subcontracted items in month j
sp cost of a subcontracted item.

$$\begin{array}{llll} \text{Demand} & x_1 & +z_1 & \geq d_1 \\ & x_1 + x_2 & +z_1 + z_2 & \geq d_1 + d_2 \\ & x_1 + \ldots + x_n & +z_1 + \ldots + z_n & \geq d_1 + \ldots + d_n \end{array}$$

$$\text{Capacity} \quad x_j \leq c_j, z_j \leq B_j \quad \text{all } j$$

Production and subcontracted costs

$$p_1x_1 + \ldots + p_nx_n + sp_1z_1 + \ldots + sp_nz_n$$

and as the month end stocks are given by

$$\begin{aligned} s_1 &= x_1 + z_1 \qquad\qquad - d_1 \\ s_2 &= (x_1 + z_1) + (x_2 + z_2) - (d_1 + d_2) \\ s_n &= \sum_j (x_j + z_j) - \sum_j d_j \end{aligned}$$

the variable holding cost will be represented by

$$h(nx_1 + \cdots + 1x_n)$$

to give the cost function

$$\text{cost} = (s-1)pz_1 + \cdots + (s-1)pz_n + h(nx_1 + \ldots + 1x_n)$$

as

$$p_1 = \cdots = p_n$$

Extreme cases:
Notice that

$s = 1$ (no additional cost from subcontracting) the optimal plan is to subcontract all excess (of capacity) production requirements; in effect, the optimal solution is to obtain stock as late as possible.

$$\text{cost} = h(nx_1 + \cdots + 1x_n)$$

If s is very large, the problem reduces to minimising

$$\text{cost} = z_1 + \cdots + z_n$$

Produce as late as possible using only the existing production capacity.

In a more general problem, subcontracting costs more than normal working but not infinitely more, there will be the options:

Subcontract for use in month n, or
Produce in month $(n-k)$ using normal working, store for k months.

The relevant costs for comparison are as follows:

sp subcontracting cost per item production
$p + kh$ production plus holding cost

from which it follows that the produce early and store policy will be used when this option is cheaper than subcontracting production:

$$sp \geq p + kh \quad \text{or} \quad k \leq \frac{(s-1)p}{h}$$

6.3.5 Extending the Model to Include Backlogging Deliveries

The extended model requires the additional notation, let

z_j quantity backlogged in month j
bp cost of a backlogged item

$$\begin{array}{llll} \text{Demand} & x_1 & +z_1 & \geq d_1 \\ & x_1+x_2 & +z_1+z_2 & \geq d_1+d_2 \\ & x_1+\ldots+x_n & +z_1+\ldots+z_n & \geq d_1+\ldots+d_n \end{array}$$

$$\text{Capacity} \quad x_j \leq c_j, z_j \leq B_j \quad \text{all}\, j$$

$$\text{Cost function} \quad \text{cost} = (b-1)pz_1 + \ldots + (b-1)pz_n + h(nx_1 + \ldots + 1x_n)$$

Extreme cases:

Notice that if $b = 1$ (no additional cost from subcontracting), the optimal plan is to backlog sales so that there are no stockholding costs, and if s is very large, no backlogging and the problem reduce to minimising

$$\text{cost} = h(nx_1 + \ldots + 1x_n)$$

Therefore, produce as late as possible. In a more general problem, backlogging costs more than normal working but not infinitely more, there will be the options:

Backlog for use in month n, or
Produce in month n using normal working, store for k months.

The relevant costs for comparison are as follows:

bp subcontracting cost per item production
$p + kh$ production plus holding cost

from which it follows that the produce early and store policy will be used when this option is cheaper than subcontracting production:

$$bp \geq p + kh \quad \text{or} \quad k \leq \frac{(b-1)p}{h}$$

6.3.6 One Production Line More Than One Item

Here, the process will be modelled by:

$$\begin{array}{lll} \text{Item 1} & \text{Item 2} & \\ \sum_{i=1}^{j} x_{i1} \geq d_{j1} & \sum_{i=1}^{j} x_{i2} \geq d_{j2} & \text{all}\, j \end{array}$$

capacity constraints $t_1 x_{i1} + t_2 x_{i2} \leq C_i$ all i

variable costs are given by $\sum \sum (n - i + 1)x_{ji}$

Therefore, produce all items as late as possible. However if, because of capacity constraints, if some items have to be produced early produce the cheapest lines early and the more expensive lines as late as possible.

6.3.7 Summary

These models demonstrate that:

The optimal production plan can be produced without the need for a complex model, and often without detailed costing.
The understanding gained from the derivation of a model to represent can lead to an efficient approach to the derivation of an optimal (or very good) solution.

6.4 Overall Summary: Two Problems

Travelling salesman problem, a known hard problem

Visiting n cities (n very large)

$(n-1)^2$	variables	$O(n^2)$
(n^2-n+2)	constraints	$O(n^2)$

Production Planning Problem (model 2)
Planning for n months, m products (m large)

$mn(n+1)/2$	variables	$O(mn^2)$
$nm+m$	constraints	$O(mn)$

Production Planning Problem (model 1)

nm	variables	$O(mn)$
$nm+m$	constraints	$O(mn)$

An initial assumption, had production planning model 2 been produced, may have been that production planning problems were like travelling salesman problems, hard problems, but in general, they have been shown to be an easy problem.

Chapter 7
Further Use of Heuristic Methods

Val Lowndes, Stuart Berry, Chris Parkes, Ovidiu Bagdasar and Nicolae Popovici

The next section shows not only how a heuristic approach can be employed to solve hard problems but also how an investigation of the mathematical model can lead to a simpler solution technique.

7.1 Flow Shop Scheduling

In these problems, there are n jobs to be processed through m machines, in the same order on each machine, there are $n!$ possible work schedules and the aim is to determine the optimal or a very good schedule, where the optimal schedule minimises the total process time for these jobs.

Often, the number of possible schedules is such that an exhaustive search is not feasible, solution takes too long and good solutions can be obtained using genetic algorithms.

As an example, do establish suitable genetic algorithm operators consider an example containing jobs to be scheduled:

$$\{1, 2, 3, 4, 5, 6, 7, 8\}$$

V. Lowndes
University of Derby, Kedleston Road, Derby DE22 1GB, UK
e-mail: V.P.Lowndes@derby.ac.uk

S. Berry · C. Parkes · O. Bagdasar (✉)
College of Engineering and Technology, University of Derby,
Kedleston Road, Derby DE22 1GB, UK
e-mail: o.bagdasar@derby.ac.uk

S. Berry
e-mail: s.berry@derby.ac.uk

N. Popovici
Babes-Bolyai University, Cluj-Napoca, Romania

S. Berry et al. (eds.), *Guide to Computational Modelling for Decision Processes*,
Simulation Foundations, Methods and Applications,
DOI 10.1007/978-3-319-55417-4_7

The genetic algorithm strings are rearrangements of these jobs, for example,

$$\{3, 1, 2, 8, 4, 5, 7, 6\}, \text{ and } \{8, 7, 6, 5, 1, 2, 4, 3\}$$

Here, strings are selected for "crossover" using tournament selection, initially select two strings at random that retain that string with the best (lowest) cost.

Here, crossover is performed by removing a random-length substring from the selected string

$$\{8, 7, 6, 5, 1, 2, 4, 3\}, \text{ giving } \{8, 7, 6, 4, 3\} \quad \{5, 1, 2\}$$

and then replacing the substring to give, as an example, the new string

$$\{8, 5, 1, 2, 7, 6, 4, 3\}$$

Finally, mutation is performed by interchanging the positions, in the string, of two randomly selected jobs.

Example 1: 8 jobs through 4 machines, 8! or 40,320 possible schedules
Implementation: 50 GA strings and a maximum of 50 iterations, at most 2500 schedules tested

The results from 5 runs of the GA routines are shown in Table 7.1. The optimal solution has a cost of 177 thus demonstrating that the approach using genetic algorithms is able to generate good work schedules, generating 2500 work schedules gave similar results to the GA solutions (costs = 181, 178, 178, 177, 179).

Example 2: 12 jobs through 5 machines, 12! or 479,001,600 possible schedules.
Implementation: 100 GA strings and a maximum of 200 iterations, at most 20,000 schedules tested.

The results from 5 runs of the GA routines as shown in Table 7.2.

A fuller search determined a better solution with the cost 225 again demonstrating that the GA approach can determine good solutions to large problems.

Table 7.1 Schedules generated by the genetic algorithm 8 jobs 4 machines

Cost	Job processing schedule							
181	1	8	3	6	2	5	4	7
177	8	3	1	2	5	6	4	7
178	1	8	3	6	5	4	2	7
177	8	3	1	5	2	6	4	7
178	1	8	3	5	2	6	4	7

Table 7.2 Schedules generated by the genetic algorithm 12 jobs 5 machines

Cost	Job processing schedule											
227	9	3	1	2	5	6	4	12	10	8	7	11
234	9	1	2	4	12	3	5	6	8	7	11	10
237	1	2	12	6	4	7	5	3	8	9	10	11
231	9	1	2	3	4	5	6	12	7	8	10	11
234	3	10	5	12	8	7	1	2	6	4	9	11

- Randomly generating 20,000 work schedules gave similar results to the GA solutions (costs = 234, 232, 234, 230, 232).

Notice that

- If the series of operations is such that
  Time at stage 1 > Time at stage 2 > ⋯ > Time at stage n,
  the optimal schedule will be such that the final job is chosen so that (time at stage 2 + ⋯ + time at stage n) is minimised and no other calculation is required.

- If the series of operations is such that
  Time at stage 1 < Time at stage 2 < ⋯ < Time at stage n,
  the optimal schedule will be such that the first job is chosen so that (time at stage 1 + ⋯ + time at stage $n-1$) is minimised and no other calculation is required.

Consequently, in practice scheduling jobs through an n-stage flow shop may reduce to a very simple problem to be solved "by hand".

7.2 Transport Paradoxes and Traffic Planning

Transport paradoxes are based on the assumption that travellers are free to choose a route or transport mode that gives them, what they believe to be, the fastest possible journey time for themselves. The result of this freedom, or "anarchy", of choice is that relative travel times between routes and modes tend to a suboptimal equilibrium [1–3]. This equilibrium—Nash equilibrium as it is often known—is suboptimal insofar as had travellers chosen, or been allocated to, some other route or mode many would have had faster journeys, and none a slower journey. This selfish, if understandable, behaviour is based on the anarchy of choice and results in car travellers on average having longer journey times than need be—the "Price of Anarchy" [4–6].

Possibly the best-known transport paradox is due to Braess [7], who argued that on certain road networks, the addition of road capacity may, paradoxically, increase rather than decrease journey times due to drivers' suboptimal choice of route. Conversely, road capacity reductions can result in faster journey times. A considerable literature exists focusing on identifying the conditions under which Braess' Paradox could occur [8–12] (see Nagurney 1993). Braess' paradox, however, just considers road networks while the perhaps lesser known Downs-Thompson paradox focuses on combined road and rail networks and it is this paradox which this paper reconstructs and extends.

The basis of this paradox is that road improvements that cause a modal shift from rail to road transport and result in a consequent disinvestment in the rail network can ultimately aggravate congestion on the road and so nullify the road improvements [13] (see Arnott and Small [14]; Ding et al. 2009).

Case studies 1–4 are considered to demonstrate that different solution techniques can be the most appropriate to solve what seem, at first, to be similar problems.

7.2.1 *Case Study 1*

There are several independent routes from an origin point to a destination, a city centre, and the objective here is to determine the benefits to be gained from the imposition of a traffic management system, measuring the benefits through the total or average travel time.

The cost/time per travelling unit of travel along each route when x_i units access the route is given by

$$f_i(x_i) \quad i = 1, \ldots, n$$

and the cost/time for all travellers accessing route i will be given by

$$\sum x_i f_i(x_i) \quad i - 1, \ldots, n$$

Traffic flow without a traffic management system: Without a traffic management system, theoretically, the demand for each route will be such that

$$f_i(x_i) = f_j(x_j) \text{ all } i, j$$

giving the equilibrium solution.

Traffic flow with a traffic management system: If there could be a perfect traffic management system, then the demand for each route will be such that the total travel time/cost T is minimised, where

$$T = \sum x_i f_i(x_i), \quad \sum x_i = N$$

at the optimal solution, the demand for each of these routes will be such that

$$f_i(x_i) + x_i f_i'(x_i) = f_n(x_n) + x_n f_n'(x_n) \quad i = 1, \ldots, n-1$$

In both cases, there is a set of $n-1$ simultaneous nonlinear equations to be solved.

Illustrative example: There are 20 routes available to travellers from an origin to a destination point, and the total demand is for 120 units of travellers.

A tabu search methodology was first employed to obtain a good solution to this problem:

for the equilibrium solution minimising the variance of $f_i(x_i)$, and
for the optimal solution minimising the variance of $f_i(x_i) + x_i f_i'(x_i)$.

The same methodology was then employed to investigate the effects of investing to improve travel time along the slowest or along the fastest routes or closing routes.

The results from these investigations are shown in Table 7.3.

These results demonstrating the following:

- A tabu search methodology provides an effective means of solving this problem.
- Some savings could be made from the use of a traffic control system.
- Expensive road improvements may not lead to significant improvements in travel times.
- It is better to improve the faster roads rather than the slower roads.
- Closing routes does not always significantly change the overall cost/time.
- A "perfect" traffic management system would allow the closure of the slower, more local, roads with travel times still much less than the equilibrium time when all roads are available for use.

Table 7.3 Solutions to traffic distribution problems

	Optimal solution		Equilibrium solution	
	Cost	Percentage change (%)	Cost	Percentage change (%)
20 routes	404		473	
20 routes, improving slowest	403	−0.25	473	0.00
20 routes, improving fastest	400	−1.00	470	0.40
20 routes, slowing fastest	405	0.25	474	0.21
19 routes, closing slowest	411	1.75	476	0.20
18 routes, closing two	431	6.70	496	5.10
17 routes, closing three	615	50.2	680	44.0

The savings from the use of a "perfect" traffic management system compared with cost of the no control system were between 10 and 15%

7.2.2 *Case Study 2*

Here, there are several routes from source to destination, but now, some road segments are common to more than one route, for example:

Route	Road segments	Demand
1	1, 2	x_1
2	1, 7, 4	x_2
3	1, 7, 9, 6	x_3
4	3, 8, 2	
5	3, 4	
6	3, 9, 6	
7	5, 6	
8	5, 10, 4	
9	5, 10, 8, 2	x_9

The equilibrium flow can be determined using a tabu search, as before, aiming to minimise the variance of the travel times on these routes.

However, the optimal flow will be determined by minimising the total travel time summing the total times for each road segment.

Where, for example, the cost from road segment 1 is $(x_1 + x_2 + x_3)f_1(x_1 + x_2 + x_3)$.

Here, the determination of the optimal solution, a genetic algorithm-based approach, has been employed to obtain a good solution.

The genetic algorithm string would indicate the number accessing each of the available routes, as an example if the total demand were for 25 units of travellers, each string would indicate the number accessing each of the 9 available routes:

$$\{1.0,\ 2.0,\ 4.0,\ 6.0,\ 3.0,\ 2.0,\ 2.1,\ 3.0,\ 1.9\},\ \text{or}$$
$$\{5.0,\ 2.0,\ 1.0,\ 1.0,\ 3.0,\ 5.5,\ 3.9,\ 2.4,\ 1.2\}$$

Selection, for crossover, used tournament selection, and crossover was carried out in two stages: first crossover, to give

$$\{1.0,\ 2.0,\ 4.0,\ 1.0,\ 3.0,\ 5.5,\ 3.9,\ 2.4,\ 1.2\} \quad \text{Total} = 24.0$$
$$\{5.0,\ 2.0,\ 1.0,\ 6.0,\ 3.0,\ 2.0,\ 2.1,\ 3.0,\ 1.9\} \quad \text{Total} = 26.0$$

and then rescaling, to give the new strings

$$\{1.04, 2.08, 4.17, 1.04, 3.13, 5.73, 4.06, 2.50, 1.25\}$$
$$\{4.81, 1.92, 0.96, 5.77, 2.88, 1.92, 2.02, 2.88, 1.83\}$$

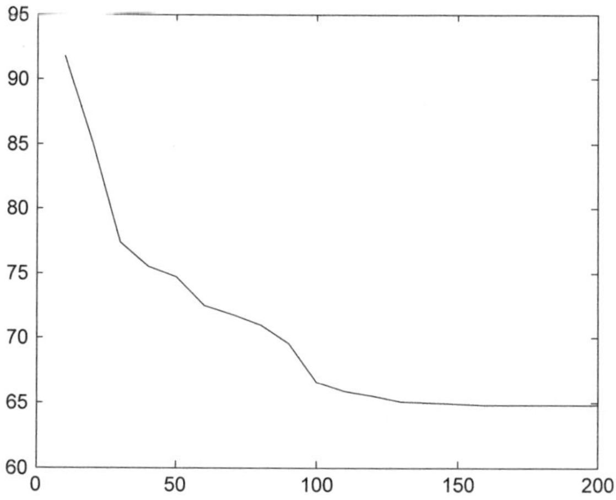

Fig. 7.1 Genetic algorithm results to convergence

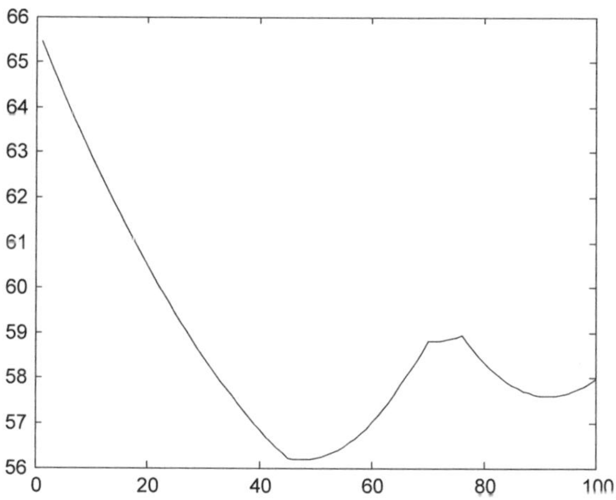

Fig. 7.2 Tabu search results

The genetic algorithm is used to convergence; then, the solution, from this solution as shown in Fig. 7.1, is improved using a tabu search starting from this best (found) solution. Figure 7.2 shows the current points in a tabu search (from the starting point determined by the genetic algorithm) to determine the best solution.

This class of problem indicates the advantages and disadvantages of both a genetic algorithm and tabu search approach to problem solving.

Advantage of genetic algorithms is as follows:

- It can quickly find a good solution.

Disadvantages:

- Can find it hard to improve a good solution.

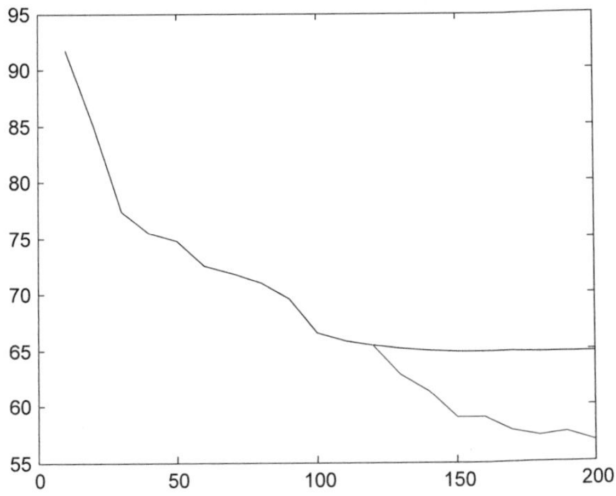

Fig. 7.3 Comparing genetic algorithm and tabu search solutions

Disadvantages of tabu searchare as follows:

- Efficiency of search is dependent on the starting point and
- The size of the tabu list or tabu duration.

Figure 7.3 shows the progression in the costs of the best solution costs obtained using

- Genetic algorithm (GA) only and
- Combining genetic algorithm and tabu search (GA + TS).

These plots show the benefits to be gained from the use of a hybrid search technique where the tabu search acts to improve the existing good solution when the genetic algorithm finds it hard to be able to move further towards a better solution, exploiting the advantages to be gained from both approaches to derive an effective solution technique.

7.2.3 *Case Study 3: To Demonstrate that a Seeming Similar Complex Problem Simplifies to a More Easily Solvable (Analytically or Using Numerical Methods) Problem*

Travellers from several, n, sources wish to reach a single destination point, and at each source, the travellers have a choice between two available routes (often road or rail) such that at source i, there are N_i travellers from which x_i choose to travel by $route_a$, and the remainder y_i by $route_b$.

Cost/time per traveller	Route$_a$	Route$_b$
From source 1 to source 2	$f_1(x_1)$	$g_1(y_1)$
From source 2 to source 3	$f_2(x_1+x_2)$	$g_2(y_1+y_2)$
From source n to source end	$f_n(\sum x_i)$	$g_n(\sum y_i)$

Thus, the unit cost for a traveller from source 1 to the end destination is

$$\begin{array}{ll} \text{Route}_a & f_1(x_1)+f_2(x_1+x_2)+\cdots+f_n(\sum x_i) \\ \text{Route}_b & g_1(y_1)+g_2(y_1+y_2)+\cdots+g_n(\sum y_i), \end{array}$$

and the total costs are given by

$$\begin{array}{ll} \text{Route}_a & x_1f_1(x_1)+x_1f_2(x_1+x_2)+\cdots+x_1f_n(\sum x_i) \\ \text{Route}_b & y_1g_1(y_1)+y_1g_2(y_1+y_2)+\cdots+y_1g_n(\sum y_i), \end{array}$$

Thus, the equilibrium solution is given as the solution to the simultaneous nonlinear equations:

$$\begin{array}{l} f_1(x_1)+\cdots+f_n(x_1+\cdots+x_n)=g_1(y_1)+\cdots+g_n(y_1+\cdots+y_n) \\ f_2(x_1+x_2)+\cdots+f_n(x_1+\cdots+x_n)=g_2(y_1+y_2)+\cdots+g_n(y_1+\cdots+y_n) \\ \\ f_n(x_1+\cdots+x_n)=g_n(y_1+\cdots+y_n) \end{array}$$

Subtracting gives the set of nonlinear equations to be solved sequentially

$$\begin{array}{l} f_1(x_1)=g_1(y_1) \\ f_2(x_1+x_2)=g_2(y_1+y_2) \\ \\ f_n(x_1+\cdots+x_n)=g_n(y_1+\cdots+y_n) \end{array}$$

Notice that if, for example, the alternative route is by rail where the travel time is independent of the number of travellers, these equations become

$$\begin{array}{l} f_1(x_1)=k_1 \\ f_2(x_1+x_2)=k_2 \\ \\ f_n(x_1+\cdots+x_n)=k_n \end{array}$$

Similarly, the optimal solution is obtained from

$$\begin{array}{l} T=x_1f_1(x_1)+(x_1+x_2)f_2(x_1+x_2)+\cdots+y_1f_1(y)+(y_1+y_2)f_2(y_1+y_2) \\ \quad +\cdots x_i+y_i=n_i \quad \text{all } i \end{array}$$

Differentiating leads to the equations, to be solved,

$$f_1(x_1) + x_1 f_1'(x_1) = g_1(y_1) + y_1 g_1'(y_1)$$
$$f_n(x_1 + \cdots + x_n) + (x_1 + \cdots + x_n) f_n'(x_1 + \cdots + x_n);$$
$$= g_n(y_1 + \cdots + y_n) + (y_1 + \cdots + y_n) g_n'(y_1 + \cdots + y_n)$$

Notice again that if, for example, the alternative route is by rail where the travel time is independent of the number of travellers, these equations become

$$f_1(x_1) + x_1 f_1'(x_1) = k_1$$
$$f_n(x_1 + \cdots + x_n) + (x_1 + \cdots + x_n) f_n'(x_1 + \cdots + x_n) = k_n$$

with solutions obtained analytically or using numerical methods.

7.2.4 *Case Study 4: Assessing the Green Benefits from a Traffic Management System*

Without a traffic management system, travellers will choose their mode of travel by road or by rail, so that the travel time from either mode is the same. This leads to a paradoxical situations where

- Road travellers choose to use a slow minor road rather than the faster improved major route and
- The green policy, "to open" more "halts" along an existing route, can lead more long-distance travellers to choose to travel by road a nongreen result.

7.2.4.1 Case Study 4a: Major Route or Minor Route

Road travel time is modelled by

$$T_{\text{road}} = a + b(x/c)^n,$$

where

x	number of travellers using the major, or improved, route
c	is the roads' planned capacity
a	the free flow time of travel, as c is large and $n > 1$ then $(1/c)^n \approx 0$
$a + b$	travel time when the road is at its planned capacity, $x = c$
n	a parameter defined by the road, this will have a value between 1 and 10 Jeong (2008) where $n = 10$ for an inner city road will and $n = 1, 2$ or 3 for a motorway

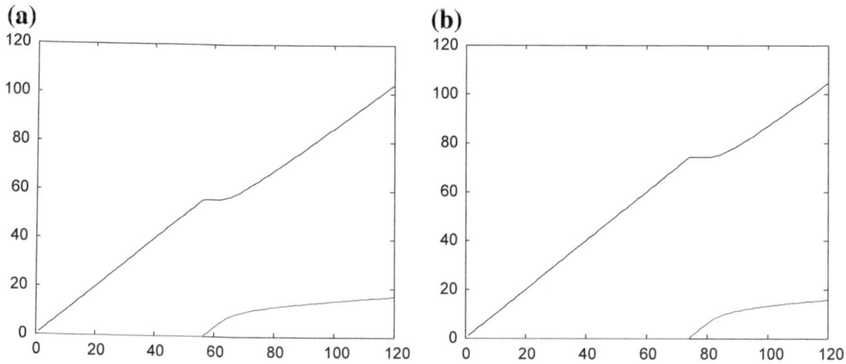

Fig. 7.4 Distribution of travellers between major and minor routes

For the major route, $T_{\mathrm{MA}} = 20 + 10\left(\frac{x}{100}\right)^4$ fast with a high capacity.

For the minor route, $T_{\mathrm{MI}} = 21 + 40\left(\frac{x}{20}\right)^7$ slow with a lower planned capacity.

The distribution of travellers between these two routes is shown in Fig. 7.4a, the usage on the minor route starts when the total demand exceeds 56.23 when the cost per traveller along the major route reaches 21 the base cost for the minor route.

Planners try to stop the use of minor roads through the application of traffic calming devices or speed limits; however, as shown in the next example, these merely postpone the use of these minor roads.

For example, replacing

$$T_{\mathrm{MI}} = 21 + 40\left(\frac{x}{20}\right)^7 \quad \text{with}$$

$$T_{\mathrm{MI}} = 23 + 50\left(\frac{x}{20}\right)^7$$

gave the traffic flow shown in Fig. 7.4b, showing that the minor route still experiences a significant demand from the travellers.

The optimal distribution of traffic (both cases) is given as the solution to the equation:

$$\text{First case} \quad \frac{\mathrm{d}}{\mathrm{d}x}\left(x\left(20 + 10\left(\tfrac{x}{100}\right)^4\right) + (N - x)\left(21 + 40\left(\tfrac{N-x}{20}\right)^7\right)\right) = 0$$

$$\text{Second case} \quad \frac{\mathrm{d}}{\mathrm{d}x}\left(x\left(20 + 10\left(\tfrac{x}{100}\right)^4\right) + (N - x)\left(23 + 50\left(\tfrac{N-x}{20}\right)^7\right)\right) = 0$$

A typical set of results as shown in Table 7.4.

The results demonstrating that there are no real benefits to be gained from a traffic control system when the alternative modes of travel are a major and a minor route.

Table 7.4 Changes in total travel costs

	Demand = 120		
	Optimal distribution of traffic cost	Equilibrium distribution of traffic cost	Percentage increase in costs (%)
Case 1	3765	3776	0.25
Case 2	3821	3853	0.8

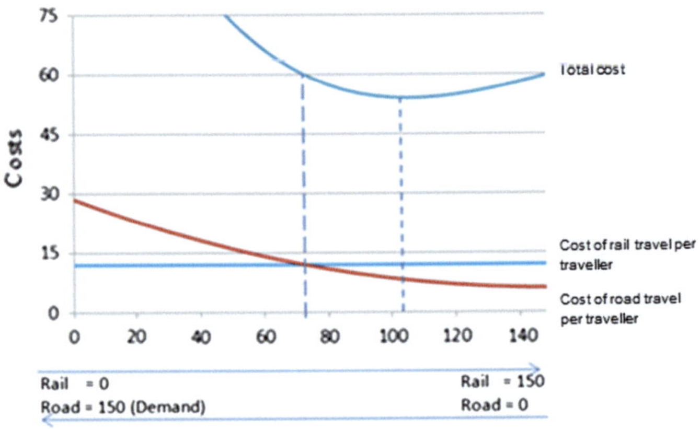

Fig. 7.5 Distribution of travellers between road and rail routes

7.2.4.2 Case Study 4b: Road or Rail Travel the "Green" Paradox

Without a traffic management system, travellers will choose their mode of travel by road or by rail, so that the travel time from either mode is the same.

As a first example, consider the case where travellers (150 in total) could use either road or rail transport to access a destination point. Figure 7.5 shows the cost per traveller for using each mode of transport.

This figure shows the alternative solutions:

Equilibrium solution, cost the same for both modes	road = 78, rail = 72 units
Optimal solution	road = 47, rail = 103 units

These results illustrate the first paradox that the optimal flow, minimising average travel time, can only occur when the rail travellers adopt the nonselfish policy of not transferring to road travel.

A second example concerns the current policy of opening, or reopening, more halts between an established origin point (start of a commuter line) and an established destination point (end of a commuter line). This leads to the paradoxical situation where a "perceived" green policy, open more rail halts, can lead more long-distance travellers choosing to travel by road, a "nongreen" result.

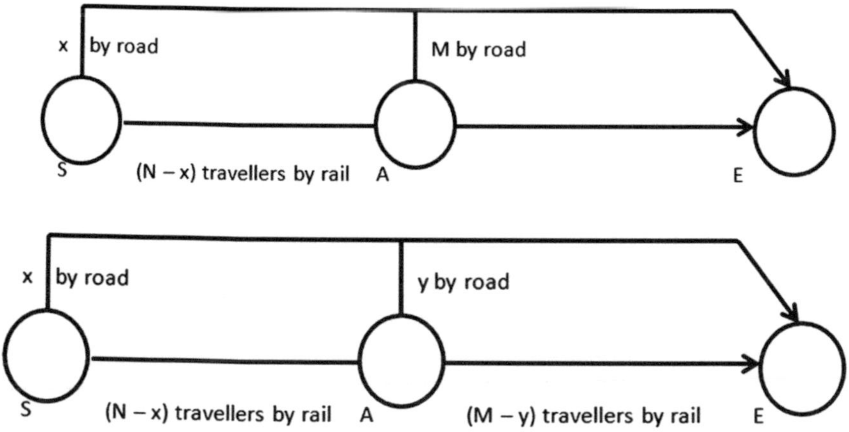

Fig. 7.6 Distribution of traffic

Illustrative example: Travel to work commuters from an origin point (S) to their final destination point (E) has the choice of travelling by road or by rail, see Fig. 7.6, N travellers in total of whom x choose to travel by road. All the M travellers from a second origin point A, no rail halt available, have to access the same road route as those travellers from origin point S.

In an attempt to encourage travel by rail, an additional halt is to be opened at point (A), y travellers from A will now choose to travel by road, see Fig. 7.6.

The travel times are as follows:

S to A by road	$f(n)$ when there are n travellers
A to E by road	$g(m)$ when there are m travellers
S to E by rail, no halt at A,	k
S to A by rail, halt at A,	k_1
A to E by rail, halt at A, k_2	$k_1 + k_2 > k$

Route costs per traveller before new halt:

S to E by road	$f(x) + g(x + M)$
S to E by rail	$k(N - x)$

Total cost all travellers before new halt:

$$xf(x) + (x + M)g(x + M)$$

Route costs per traveller after new halt:

S to E by road	$f(x) + g(x + y)$
A to E by road	$g(x + y)$
S to E by rail	$(k_1 + k_2)\ (N - x)$
A to E by rail	$k_2(M - y)$

Total cost all travellers after new halt:

$$xf(x) + (x + M)g(x + M)$$

Then, if routes from A are chosen so that road and rail times are equal:

before the new halt $f(x) + g(x + M) = k(N - x)$ solution $x = X$

after the new halt $f(x) + g(x + M) = (k_1 + k_2)(N - x)$

$g(x + y) = k_2(M - y)$

giving $f(x) + k_2(M - y) = (k_1 + k_2)(N - x)$

as $f(x) + g(X + M) = k(N - X)$ and as $y < M$

it then follows that $f(x) + g(X + y) < f(X) + g(X + M)$, and

$k(N - X) < (k_1 + k_2)(N - X)$, then

$$f(X) + g(X + y) - (k_1 + k_2)(N - X) < f(X) + g(X + M) - k(N - X) = 0$$

and thus, it follows that after the opening of the new halt, more travellers will choose to travel by road from the more distant destinations, an undesired "non-green" result.

7.3 Transportation Management: Methods and Algorithms for Solution

A dilemma often facing transport planners is to choose whether to leave motorists free to make their own route choices or to try and actively manage the traffic flows in order to minimise the journey times for all motorists travelling between origin and destination, i.e. whether "to plan" or "not to plan"? It is the resolving of this dilemma that provides the focus of this paper.

Assuming that journey time is the only criteria for route choice, car travellers may be seen as optimisers insofar as they usually want to minimise their own journey times. Consequently, in the absence of effective traffic control measures, there tends to be an approximate equilibrium travel time resulting between the routes available.

In what has become known as Braess' paradox, the difference between equilibrium and optimal travel times can lead to the decidedly counter-intuitive result that adds to road capacity, typically through more road construction, and lead to slower not faster car journey times [7].

7.3.1 General Cost of "Origin–Destination" Traffic Flow

A fundamental feature of road transportation is that car travel time is dependent on the number of cars accessing the route. If there are $m \geq 2$ routes between the origin and destination points, the time t_i for a car accessing route $i(i = 1, \ldots, m)$ is a monotonic increasing polynomial function of the traffic flow x_i as measured in "units of vehicles" per "unit of time" accessing route i, namely

$$t_i = f_i(x_i) = a_i + b_i x_i^{p_i}, \tag{7.1}$$

where $p_i \geq 1, a_i \geq 0$ and $b_i > 0$, as suggested by the model proposed by Youn et al. [5].

Denoting the cost of transporting x_i vehicles along route $i(i = 1, \ldots, m)$ by

$$g_i(x_i) = x_i f_i(x_i), \tag{7.2}$$

the total travel time for n vehicles distributed on m routes is given by formula

$$T(x) = \sum_{i=1}^{m} g_i(x_i) = \sum_{i=1}^{m} x_i f_i(x_i), \tag{7.3}$$

where $x = (x_1, \ldots, x_m) \in \mathbb{N}^m$ and $x_1 + \cdots + x_m = n$.

An example in this sense is where travellers can choose, or be directed to, one from several routes accessing a city centre. Assume that there are three types of routes available for a commuter

- main highway to city centre (direct and popular) and
- side routes (or rat runs, offering short distance but being easily congested), or ring roads/motorways (faster, but longer and indirect).

Travel time functions (7.1) for these routes are given in Table 7.5. Here, moving x_i cars along route i costs $g_i(x_i) = x_i f_i(x_i)$.

Therefore, the total travel time of $n = x_1 + \cdots + x_m$ cars along these routes is

$$T(x) = g_1(x_1) + g_2(x_2) + g_3(x_3), \text{from} \tag{7.4}$$

For example, when $x = (x_1, x_2, x_3) = (2, 1, 2)$, the total travelling cost given by (7.4) is $T(x) = 25.4$.

Table 7.5 Three routes traffic model

Route type	Travel time	1 "car unit" transit time
Main highways to city centre	$f_1(x_1) = 2 + 0.2x_1^3$	2.2
Side routes (rat runs)	$f_2(x_2) = 5 + 0.4x_2^5$	5.4
Ring roads/motorways	$f_3(x_3) = 6 + 0.1x_3^2$	6.1

7.3.1.1 Optimal Versus Equilibrium Solutions

In this section, two main types of traffic management approaches are considered.

First, one may try to find the number of cars to be directed along each route in such a way that the total travelling time is minimised. Therefore, we consider the following discrete optimisation problem

$$\begin{cases} \text{Minimise } T(x_1, \ldots, x_m) \\ \text{subject to} \\ x_1 + \cdots + x_m = n \\ x_1, \ldots, x_m \in \mathbb{N}. \end{cases} \tag{7.4}$$

Second, admitting that all drivers are allowed to seek to minimise their own individual travel times, an equilibrium would develop where all travel times are equal and the steady-state traffic flow along each route could be obtained as a solution of the following system of equations:

$$\begin{cases} f_1(x_1) = \cdots = f_m(x_m) \\ x_1 + \cdots + x_m = n \\ x_1, \ldots, x_m \in \mathbb{N}. \end{cases} \tag{7.5}$$

7.3.1.2 Price of Anarchy

The effect/cost of allowing travellers the free choice of route can be measured using the price of anarchy is defined by

$$P_{\mathrm{A}} = \frac{\text{Cost } T \text{ at equilibrium}}{\text{Cost } T \text{ at optimum}}. \tag{7.6}$$

When the optimal and equilibrium solutions are the same, P_{A} equals 1. Then, the greater the value of P_{A}, the greater the effect/cost of allowing travellers the free choice of route and the greater the benefits (travel time), to be gained through effective traffic management.

7.3.2 Introduction to Solution Methodologies

Considering the following polyhedral set in $\mathbb{R}^m$:

$$S = \{ x = (x_1, \ldots, x_m) \subset \mathbb{R}^m | x_1 + \cdots + x_m = n, x_1, \ldots, x_m > 0 \}, \tag{7.7}$$

we observe that all problems (7.4) and (7.5) have the same feasible domain, namely

$$S \cap \mathbb{N}^m,$$

whose cardinality is the number of multisets of length m on n, cf. Stanley [15]:

$$\mathrm{card}(S \cap \mathbb{N}^m) = \binom{n+m-1}{m-1} = \frac{(n+1)(n+2)\cdots(n+m-1)}{(m-1)!}. \tag{7.8}$$

Showing that, for a fixed value m of routes, the total number of feasible points, $\mathrm{card}(S \cap \mathbb{N}^m)$, is the output of n by a polynomial of degree $m-1$. For $m=2$, the feasible set $S \cap \mathbb{N}^m$ contains $n+1$ points.

For $m=3$, the number of feasible solutions is given by $(n+1)(n+2)/2$, representing the sequence A000217 of triangular numbers in OEIS [16] which is approximately 5×10^5 for $n = 10^3$ and 5×10^7 for $n = 10^4$.
For $m = 4$, the formula gives $(n+1)(n+2)(n+3)/6$, representing the sequence of tetrahedral numbers A000292 in OEIS which is approximately 2×10^8 for $n = 10^3$ and 2×10^{11} for $n = 10^4$.

Therefore, for a few links m one can solve the problems (7.4) and (7.5) by exhaustive search. However, as expected, this method is not suitable for a large number of routes m or cars n. For this reason, in the next section we present other efficient methods for the numerical solution of large problems.

7.3.2.1 Solution of the Optimisation Problem (7.4) by Dynamic Programming

The cost function T of the optimisation problem (7.4) has separable variables, being a sum of terms containing independent variables (7.3).

Moreover, as we have seen in the previous section, the feasible set $S \cap \mathbb{N}^m$ is finite.

Therefore, problem (7.4) can be solved using Bellman's algorithm of dynamic programming (see [17] or [18]).

This approach is very efficient even for a large number of links and vehicles.

We start by defining recursively the functions

$G_1, \ldots, G_m : [0, n] \cap \mathbb{N} \to \mathbb{R}$ for all $c \in [0, n] \cap \mathbb{N}$ by Bellman's functional equations

$$\begin{cases} G_1(c) = g_1(c), \\ G_k(c) = \min_{c \in [0,n] \cap \mathbb{N}} [g_k(x) + G_{k-1}(c-x)], \quad k = 2, 3, \ldots, m. \end{cases} \tag{7.9}$$

Then, the optimal value of problem (7.4) is given by

$$\min_{x \in S \cap \mathbb{N}^m} T(x) = G_m(n), \tag{7.10}$$

and an optimal solution

$$x^0 = (x_1^0, \ldots, x_m^0)$$

of problem (7.4) can be deduced by the following backward recursive procedure:

Let $c := n$ and choose $x_m^0 \in \text{argmin}_{x \in [0,c] \cap \mathbb{N}}[g_m(x) + G_{m-1}(c - x)]$,
Let $c := n - x_m^0$ and choose $x_{m-1}^0 \in \text{argmin}_{x \in [0,c] \cap \mathbb{N}}[g_{m-1}(x) + G_{m-2}(c - x)]$,
...
Let $c := n - x_m^0 - \cdots - x_3^0$ and choose $x_2^0 \in \text{argmin}_{x \in [0,c] \cap \mathbb{N}}[g_2(x) + G_1(c - x)]$,
Let $x_1^0 := n - x_m^0 - \cdots - x_3^0 - x_2^0$.

Example 7.1 Consider the particular case when $m = 3$ and $n = 5$ and the travelling time functions f_1,f_2 and f_3 for moving along routes 1, 2, 3 are given in Table 7.5.

According to (7.2), the values of g_1, g_2 and g_3 involved in Bellman's algorithm are listed in the table below:

x	$g_1(x)$	$g_2(x)$	$g_3(x)$
0	0	0	0
1	2.2	5.4	6.1
2	7.2	35.6	12.8
3	22.2	306.6	20.7
4	59.2	1658.4	30.4
5	135	6275	42.5

By (7.9), the values for $G_1(c) = g_1(c)$ are given by

c	0	1	2	3	4	5
$G_1(x)$	0	2.2	7.2	22.2	59.2	135

In order to compute $G_2(c)$, we build the following table, where the numbers $g_2(x) + G_1(c - x)$ corresponding to $x = 0, 1, \ldots, c$ are listed along antidiagonals.

	$c - x$	0	1	2	3	4	5
X	$g_2(x)$	$G_1(c - x)$					
		0	2.2	7.2	22.2	59.2	135
0	0	0	2.2	7.2	22.2	59.2	135

(continued)

(continued)

	$c-x$	0	1	2	3	4	5
X	$g_2(x)$	$G_1(c-x)$					
		0	2.2	7.2	22.2	59.2	135
1	5.4	5.4	7.6	12.6	27.6	64.6	
2	35.6	35.6	37.8	42.8	57.8		
3	306.6	306.6	308.8	313.8			
4	1658.4	1658.4	1660.6				
5	6275	6275.0					

Then, for every $c \in \{0, \ldots, 5\}$, the value $G_2(c) = \min_{0 \le x \le c}[g_2(x) + G_1(c-x)]$ corresponds to the least entry in the antidiagonal corresponding to c, giving

c	0	1	2	3	4	5
$G_2(x)$	0	2.2	7.2	12.6	27.6	57.8

Similarly, we can compute the values of $G_3(c)$ by means of the next table:

	$c-x$	0	1	2	3	4	5
X	$g_3(x)$	$G_2(c-x)$					
		0	2.2	7.2	12.6	27.6	57.8
0	0	0	2.2	7.2	12.6	27.6	57.8
1	6.1	6.1	8.3	13.3	18.7	33.7	
2	12.8	12.8	15.0	20.0	25.4		
3	20.7	20.7	22.9	27.9			
4	30.4	30.4	32.6				
5	42.5	42.5					

The values of $G_3(c) = \min_{0 \le x \le c}[g_3(x) + G_2(c-x)]$ are given by

c	0	1	2	3	4	5
$G_3(x)$	0	2.2	7.2	12.6	18.7	25.4

According to (7.10), the minimum of T is $G_3(x) = 25.4$.

Finally, by going backwards, we deduce an optimal solution $x^0 = (x_1^0, x_2^0, x_3^0)$ of problem (7.4) as follows:

$$x_3^0 \in \operatorname{argmin}_{x\in[0,5]\;\cap\mathbb{N}}[g_3(x)+G_2(5-x)] = \{2\},$$
$$x_2^0 \in \operatorname{argmin}_{x\in[0,3]\;\cap\mathbb{N}}[g_2(x)+G_1(3-x)] = \{1\},$$
$$x_1^0 = 5 - x_3^0 - x_2^0 = 2.$$

7.3.2.2 Transforming the Equilibrium System (7.5) into Optimisation Problems

Since the equilibrium system (7.5) may be inconsistent (in the absence of integer solutions), we propose the following approach. For any $x = (x_1, \ldots, x_m) \in \mathbb{N}^m$ such that $x_1 + \cdots + x_m = n$, consider the mean and normalised standard deviation of the vector $(f_1(x_1), \ldots, f_m(x_m))$ defined by

$$\mu(x) = \frac{1}{m}\sum_{i=1}^{m} f_i(x_i)$$

and

$$\sigma(x) = \sqrt{\frac{1}{m-1}\sum_{i=1}^{m}|f_i(x_i) - \mu(x)|^2}.$$

Therefore, a counterpart of the equilibrium system (7.5) is represented by the following discrete optimisation problem:

$$\begin{cases} \text{Minimise}\,\sigma(x_1, \ldots, x_m) \\ \text{subject to} \\ x_1 + \cdots + x_m = n \\ x_1, \ldots, x_m \in \mathbb{N}. \end{cases} \tag{7.11}$$

Notice that whenever the equilibrium system (7.5) is consistent, its solutions are optimal solutions of problem (7.11).

Conversely, if $x^0 = (x_1^0, \ldots, x_m^0)$ is an optimal solution of (7.11), then x^0 is a solution of (7.5) if and only if $\sigma(x^0) = 0$.

Also, in contrast to (7.4), the optimisation problem (7.11) cannot be solved by Bellman's algorithm, because its objective function σ does not have separable variables. For this reason, we propose two approaches for solving (7.11). One is to solve the problem with a heuristic method. Alternatively, one may consider the relaxed version of (7.11) over nonnegative real numbers:

$$\begin{cases} \text{Minimise } \sigma(x_1, \ldots, x_m) \\ \text{subject to} \\ x_1 + \cdots + x_m = n \\ x_1, \ldots, x_m \geq 0. \end{cases} \tag{7.12}$$

This is solved in the following section by the *fmincon* routine available in MATLAB®.

7.3.2.3 Counterparts of the Optimisation Problem (7.4)

We consider the following relaxation of the optimisation problem (7.4):

$$\begin{cases} \text{Minimise } T(x_1, \ldots, x_m) \\ \text{subject to} \\ x_1 + \cdots + x_m = n \\ x_1, \ldots, x_m \geq 0 \end{cases} \tag{7.13}$$

and its counterpart

$$\begin{cases} \text{Minimise } T(x_1, \ldots, x_m) \\ \text{subject to} \\ x_1 + \cdots + x_m = n \\ x_1, \ldots, x_m > 0. \end{cases} \tag{7.14}$$

Denote by cl A and ri A the closure and the relative interior of any set $A \subseteq \mathbb{R}^m$, in the sense of convex analysis [19].

Since the feasible domain of problem (7.13) is the set S introduced in (7.7), it follows that the feasible domain of problem (7.14) is ri$S$.

Taking into account that $S \cap \mathbb{N}^m \subseteq S = \text{cl}(\text{ri } S)$ and T is continuous, we infer that the optimal values of problems (7.4), (7.13) and (7.14) satisfy

$$\min_{x \in S \cap \mathbb{N}^m} T(x) \geq \min_{x \in S} T(x) = \inf_{x \in \text{ri}S} T(x).$$

Notice that problems (7.4) and (7.13) always have optimal solutions, while problem (7.14) possesses minimising sequences, but not necessarily minimal solutions.

The optimisation problem (7.13) will be solved numerically in the next section.

In what concerns the constrained optimisation problem (7.14), we can prove that it is equivalent to an equilibrium-type system. To this aim, we attach to (7.14) the Lagrangian function $L : \text{int } \mathbb{R}^m_+ \times \mathbb{R} \to \mathbb{R}$ defined by

$$L(x, \lambda) = T(x) + \lambda H(x), \tag{7.15}$$

where the objective function $T(x) = \sum_{i=1}^{m} g_i(x_i)$ is given by 7.3 and the constraint function is defined by

$$H(x) = x_1 + \cdots + x_m - n, \tag{7.16}$$

for all $x = (x_1, \ldots, x_m) \in \text{int } \mathbb{R}^m_+$.

By the classical necessary condition of optimality, it follows that whenever $x^0 = (x_1^0, \ldots, x_m^0) \in \text{int } \mathbb{R}^m_+$ is an optimal solution of (7.14) there exists $\lambda_0 \in \mathbb{R}$ such that (x^0, λ_0) is a stationary point of L, that is,

$$\begin{cases} \frac{\partial L}{\partial x_i}(x^0, \lambda_0) = g_i'(x_i^0) + \lambda_0 = 0, \quad i = 1, \ldots, m \\ \frac{\partial L}{\partial \lambda}(x^0, \lambda_0) = H(x^0) = x_1^0 + \cdots + x_m^0 - n = 0. \end{cases} \tag{7.17}$$

In particular, this shows that x^0 is a solution of the following system:

$$\begin{cases} g_1'(x_1) = \cdots = g_m'(x_m) \\ x_1 + \cdots + x_m = n \\ x_1, \ldots, x_m > 0. \end{cases} \tag{7.18}$$

On the other hand, let $L_0 : \text{int } \mathbb{R}^m_+ \to \mathbb{R}$ be defined for all $x \in \text{int } \mathbb{R}^m_+$ by

$$L_0(x) = L(x, \lambda_0).$$

By means of (7.1) and (7.2), we deduce that

$$\begin{aligned} d^2L_0(x^0)(h) &= \sum_{i=1}^{m}\sum_{j=1}^{m} \frac{\partial^2 L_0}{\partial h_i \partial h_j}(h)h_i h_j = \sum_{i=1}^{m} g_i''(x_i^0)h_i^2 \\ &= \sum_{i=1}^{m} b_i(1+p_i)p_i x_i^{p-1} h_i^2 > 0 \end{aligned}$$

for all $h = (h_1, \ldots, h_m) \in \mathbb{R}^m \backslash \{0_m\}$ such that $\mathrm{d}H(h) = h_1 + \cdots + h_m = 0$, and hence, the sufficient optimality condition is fulfilled.

Consequently, a point $x^0 = (x_1^0, \ldots, x_m^0) \in \mathbb{R}^m$ is an optimal solution of (7.14) if and only if it is a solution of the system (7.18).

Comparing (7.18) to (7.5), it is natural to consider the equilibrium-type system:

$$\begin{cases} g_1'(x_1) = \cdots = g_m'(x_m) \\ x_1 + \cdots + x_m = n \\ x_1, \ldots, x_m \in \mathbb{N}. \end{cases} \tag{7.19}$$

For any $x = (x_1, \ldots, x_m) \in \mathbb{N}^m$ such that $x_1 + \cdots + x_m = n$, we consider the mean and normalised standard deviation of the vector $(g'_1(x_1), \ldots, g'_m(x_m))$, defined by

$$\bar{\mu}(x) = \frac{1}{m}\sum_{i=1}^{m} g'_i(x_i)$$

and

$$\bar{\sigma}(x) = \sqrt{\frac{1}{m-1}\sum_{i=1}^{m}|g'_i(x_i) - \bar{\mu}(x)|^2}.$$

As an alternative of the equilibrium system (7.19), we consider the optimisation problem:

$$\begin{cases} \text{Minimise } \bar{\sigma}(x_1, \ldots, x_m) \\ \textit{subject to} \\ x_1 + \cdots + x_m = n \\ x_1, \ldots, x_m \geq 0. \end{cases} \tag{7.20}$$

Since $\bar{\sigma}$ does not have separable variables, we cannot use Bellman's algorithm. Hence, this problem will be solved using a heuristic algorithm.

7.3.2.4 Heuristic Algorithms

A tabu search was implemented to determine both the equilibrium and optimal solutions to this traffic planning problem, solving the discrete optimisation problems (7.11) and (7.20).

To locate the equilibrium solution, the search starting with an initial allocation, vehicles to routes, and the resultant cost per vehicle along each route

$$x_i \quad \text{and} \quad R_i = f(x_i)$$

The search starts by locating the following roads:

Route m where $f_m(x_m) \geq f_i(x_i)$ all i and $x_m > h$ quantity to be reduced

Route n where $f_n(x_n) \leq f_i(x_i)$ all i quantity to be increased

Reassigning the traffic so that

$$x_m - h \quad \text{and} \quad x_n + h$$

Updating the tabu list, reducing the tabu value for all routes and increasing the tabu values for routes m and n, so that:

x_m cannot be increased until TabuTime has elapsed and

x_n cannot be decreased until TabuTime has elapsed.

Updating routine

$$\begin{aligned} U(i) &= \max(U(i) - 1, 0), \\ D(i) &= \max(D(i) - 1, 0) \\ U(m) &= U(m) + \text{TabuTime} \\ D(n) &= D(n) + \text{TabuTime} \end{aligned}$$

At the subsequent stages, find

$$\begin{array}{llllll} \text{Route } m & \text{where} & f_m(x_m) \geq f_i(x_i) & \text{all} & i, x_m > h \text{ and } D(m) = 0 \\ \text{Route } n & \text{where} & f_n(x_n) \leq f_i(x_i) & \text{all} & i, \text{ and } U(n) = 0 \end{array}$$

At each stage, calculate the variance of the route costs, var(R), and record that solution with the lowest variance as the best found solution.

To locate the optimal solution, the search starts with an initial allocation,

$$x_i$$

and the "optimality measure"

$$R_i = f_i(x_i) + x_i f_i'(x_i)$$

Calculate the variance of the route costs, var(R), and record that solution with the lowest variance as the best found solution.

Note: the efficiency of the search is dependent upon the values of the:

TabuTime, and
The stepsize h.

7.3.3 Numerical Results for a Problem with 3 Routes

This example highlights some key features of the traffic problems formulated in the introduction. For simplicity, we denote $x = x_1, y = x_2$ and $z = x_3$.

The individual cost functions f_1,f_2 and f_3 for routes 1, 2 and 3 given in Table 7.5 are

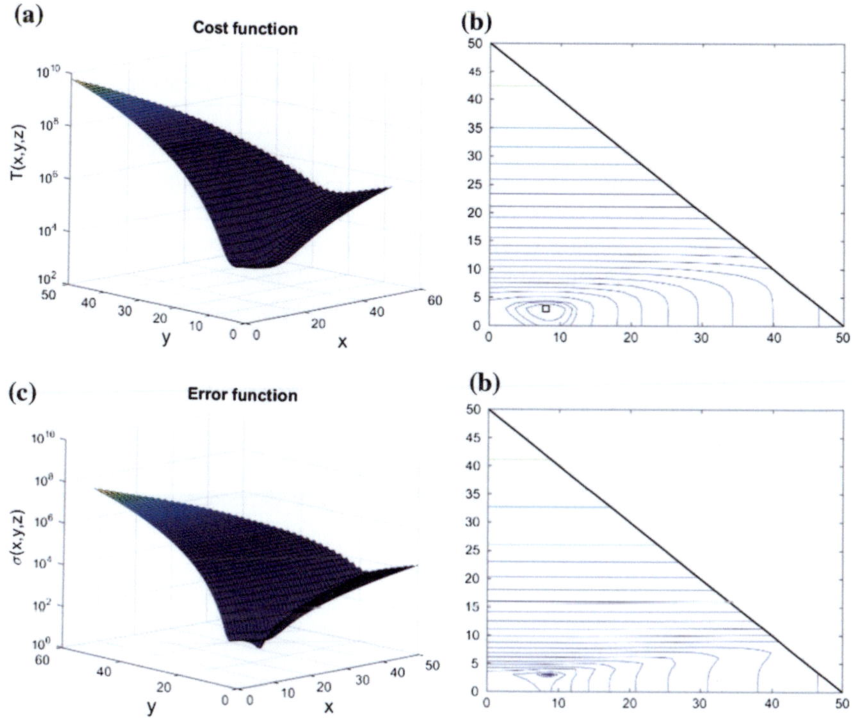

Fig. 7.7 Results for $x+y+z=n=50$. **a** Total cost $T(x;y;z)$ (surface); **b** $T(x;y;z)$ (contour plot); **c** $\sigma(x,y,z)$ (surface plot); **d** $\sigma(x,y,z)$ (contour plot)

$$f_1(x) = 2+0.2x^3, \quad f_2(y) = 5+0.4y^5, \quad f_3(z) = 6+0.1z^2.$$

In this case, the polyhedral set

$$S = \{(x,y,z) \in \mathbb{R}^3 \ |x+y+z=n, x,y,z \geq 0\} \tag{7.21}$$

depends upon two parameters x and y (as $z=n-x-y$), while the total number of feasible points in S is $(n+1)(n+2)/2$.

Some basic features of the solutions of problems (7.4) and (7.11) can be obtained by computing the values of the cost function $T(x,y,z)$ and the error function $\sigma(x,y,z)$ and by evaluating their optimum values by exhaustive search.

The results for $n=50$ links are depicted in Fig. 7.7. Numerically, the minimum cost is 7307.7 obtained for $(x,y,z)=(8,3,39)$, while the minimum error is 38.3378, obtained for $(x,y,z)=(9,3,38)$.

7.3.3.1 Optimal Cost Solutions

Solutions of the optimisation problem for $n = 1, \ldots, 100$ vehicles are shown in Fig. 7.8.

In Fig. 7.8a, the cost functions f_1, f_2 and f_3 for the three routes given in Table 7.5 suggest that vehicles follow the route of lower cost (here f_3), to minimise cost. Once the curves intersect, some vehicles are diverted along alternative routes. The integer solutions for the optimisation problem are shown in Fig. 7.8b.

Notice that there is an almost linear dependence of the solutions with respect to the traffic size. Also, most of the vehicles travel along the route of minimal power. For $n = 100$, the optimal solution is $(x, y, z) = (14, 4, 82)$.

Figure 7.8c depicts the cost obtained if all vehicles were directed along a single route. One may notice that the diversion of some vehicles to alternative routes has an important effect in bringing the cost down. For $n = 100$, the optimal cost is 64,998.4, while for $n = 1000$ this increases to 83,059,738.

Finally, Fig. 7.8d illustrates the computational time dependence on the problem size. As expected for this approach, the time seems to increase quadratically with the problem size, as suggested by the complexity analysis.

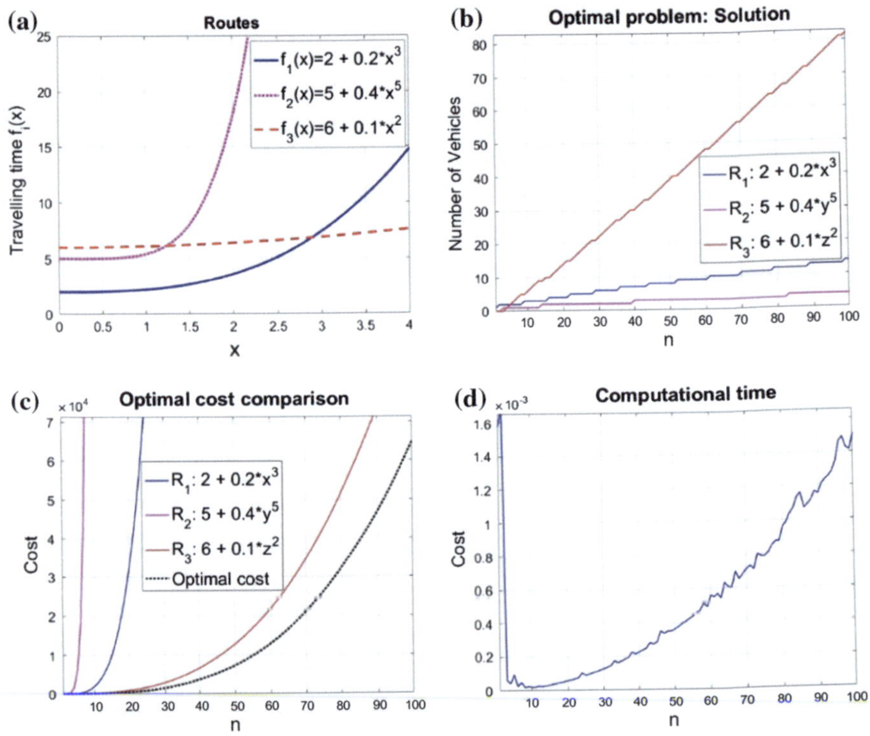

Fig. 7.8 Solution for $n = 1, \ldots, 100$. **a** Route equations; **b** optimal solution (x, y, z) for the three routes; **c** Total cost compared against $f_i(n)$; **d** Execution time

For $n = 1000$ vehicles, one can use Bellman's algorithm to obtain the solution

$$(x, y, z) = (68, 10, 922).$$

A numerical method using the *fmincon* MATLAB® routine is also possible, which generates the solution

$$(x, y, z) = (68.2920, 10.1203, 921.5875).$$

The cost difference is less than 0.001% (83,058, 316.9974 vs. 83,059,738).

7.3.3.2 Equilibrium Problem and Price of Anarchy

At equilibrium

$$2 + 0.2x^3 = 5 + 0.4y^5 = 6 + 0.1z^2, \tag{7.22}$$

or when this has no integer solutions, the aim could be to try to minimise the function

$$\sigma(x, y, z) = \sqrt{\frac{|2 + 0.2x^3 - \mu|^2 + |5 + 0.4y^5 - \mu|^2 + |6 + 0.1z^2 - \mu|^2}{2}}, \tag{7.23}$$

where

$$\mu = \mu(x, y, z) = \frac{(2 + 0.2x^3) + (5 + 0.4y^5) + (6 + 0.1z^2)}{3},$$

with the constraints $x + y + z = n$ and $x, y, z \in \mathbb{N}$.

The integer solutions for the equilibrium problem are shown in Fig. 7.9a.

The equilibrium and optimal solutions appear to be quite similar. At $n = 100$, both solutions yield $(14, 4, 82)$, while at $n = 1000$ the optimal solution is $(68, 10, 922)$ whereas the equilibrium one $(77, 12, 911)$.

Figure 7.9b depicts the individual vehicle cost/route, while the normalised standard error σ/n with σ given by (7.23) shown in Fig. 7.9c presents oscillations.

The price of anarchy shown in Fig. 7.9d is very close to 1, which suggests a strong coupling between the optimisation and equilibrium problems.

Also, the solution of the equilibrium problem (7.11) is

$$(74.73536,\ 11.58528,\ 913.6794),$$

which compares well against the equilibrium integer solution

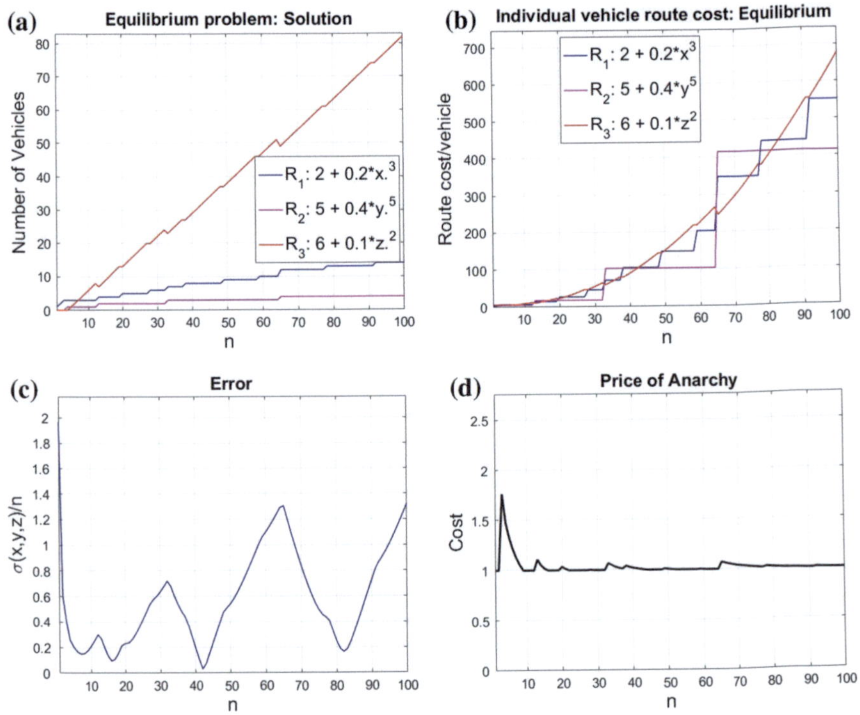

Fig. 7.9 Solution for $n = 1, \ldots, 100$. **a** Equilibrium solution (x, y, z) for the three routes; **b** single vehicle time/route at equilibrium; **c** $\sigma(x, y, z)/n$; **d** price of anarchy

$$(77, 12, 911),$$

This solution produces an standard deviation $\sigma(x, y, z)$ given by formula (7.23) of 0.0021486. The predicted price of anarchy is $P_A = 1.005$.

7.3.4 A Problem with 20 Routes

When a large number of routes is considered, the problem becomes very complex. The exhaustive search is no longer an option, and alternative approaches are required. In this section, we investigate a problem with 20 routes.

We first discuss the optimisation problem, which is solved in integers with Bellman's algorithm, then by a heuristic method, and finally, by nonlinear optimisation solvers which produces real solutions. The equilibrium problem is solved over integers with a heuristic method and also over real numbers using numerical methods. Here, the focus is the variance of $f_i(x_i)$, the travel time along route i when x_i vehicles are present.

7.3.4.1 Optimal Cost Solutions

In this section, we examine results obtained for the problem optimising the travel cost. The integer optimisation problem (7.4) is solved by Bellman's algorithm. Its discrete alternative (7.20) is solved by the heuristic algorithm, while the solution of the continuous counterpart (7.13) is obtained by numerical methods.

Solution of (7.4) by Bellman's Algorithm

Table 7.6 presents the route costs for a problem involving 20 routes, and the solution produced by Bellman's algorithm for $n = 2000, 3000, 5000$ or 10,000 cars.

The distribution of vehicles at the optimal solution is presented in Fig. 7.10. It can be seen that more than 50% of the vehicles follow link 14, while the links with power 1 (links $4, 6, 14, 16$) took the majority of the traffic.

Table 7.6 Optimal distribution of n cars along $m = 20$ links obtained by Bellman's algorithm

Link Nr	Link cost function	$n = 2000$	$n = 3000$	$n = 5000$	$n = 10{,}000$
1	$x(2 + 0.1x^3)$	6	7	9	11
2	$x(2.3 + 0.02x^5)$	374	6	5	5
3	$x(3.5 + 0.01x^2)$	58	72	95	135
4	$x(1.8 + 0.15x^1)$	346	530	901	1839
5	$x(0.01 + 0.04x^2)$	30	37	48	68
6	$x(0.8 + 0.25x^1)$	210	320	542	1105
7	$x(0.04 + 0.05x^2)$	27	33	43	61
8	$x(0.11 + 0.025x^3)$	10	12	14	18
9	$x(0.075 + 0.125x^2)$	17	21	27	38
10	$x(0.99 + 0.07x^4)$	4	5	5	6
11	$x(2.1 + 0.001x^5$	7	8	9	10
12	$x(0.01 + 0.5x^2)$	8	10	13	19
13	$x(1.5 + 0.1x^3)$	6	7	9	11
14	$x(2.8 + 0.05x^1)$	1027	1578	2692	5507
15	$x(0.1 + 0.02x^4)$	6	6	7	9
16	$x(3.8 + 0.25x^1)$	204	314	537	1099
17	$x(0.4 + 0.5x^3)$	4	4	5	7
18	$x(0.11 + 0.0025x^6)$	4	5	5	6
19	$x(1.75 + 0.125x^2)$	17	21	27	38
20	$x(0.099 + 0.03x^4)$	5	6	7	8
Cost $\times 10^5$		1.035	2.367	6.694	27.319

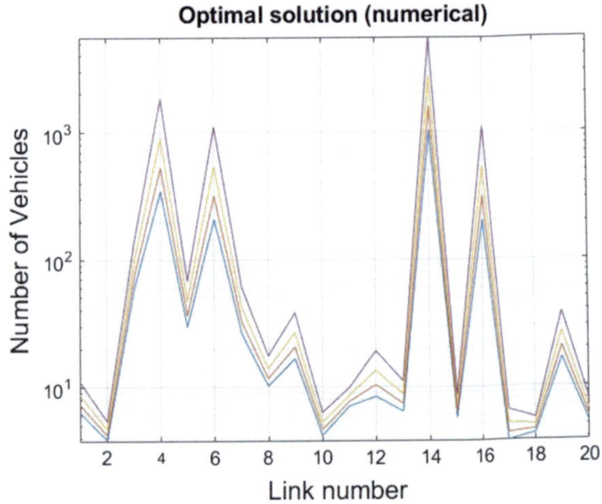

Fig. 7.10 Log plot of the solution of the optimisation problem (7.4), obtained by Bellman's algorithm for $m = 20$ links and $n = 2000, 3000, 5000, 10{,}000$ vehicles (*bottom* to *top*)

Bellman's Versus Heuristic Algorithm

Table 7.7 shows the small difference between the optimum solution obtained by Bellman's method and the heuristic method (maximum difference is 3 out of 2692 vehicles at link 14). As expected, the results produced by Belmann's algorithm give lower figures for the total cost, as this method is exact. However, the cost difference is negligible (less than 0.1%).

Bellman's Versus Nonlinear Optimisation Solution

Table 7.8 compares solutions of the optimisation problem (7.4) given by Bellman's algorithm, against numerical solutions of the alternative optimisation problem (7.13), solved numerically by the MATLAB® solver *fmincon*.

One can notice that the difference between the solutions is again negligible. Also, while the cost obtained for the relaxed problem over integers is slightly smaller, the corresponding cost functions only differ by less than 0.03%.

7.3.4.2 The Equilibrium Problem

As the variables are no longer separated, Bellman's algorithm can no longer be applied, while due to the problem's complexity (for $m = 20$ and $n = 1000$, one would have to evaluate approximately 10^{30} configurations) exhaustive search is not an option.

For this reason, the optimisation problem (7.11) is solved by a heuristic algorithm and its relaxed continuous counterpart (7.12) by numerical methods.

Table 7.7 Optimal solution x_i and total solution cost: Bellman (B) versus Heuristic (H)

Link Nr	$n = 2000$		$n = 3000$		$n = 5000$		$n = 10,000$	
	B	H	B	H	B	H	B	H
1	6	6	7	8	9	8	11	12
2	4	4	4	4	5	5	5	6
3	58	58	72	72	95	95	135	135
4	346	345	530	529	901	902	1839	1840
5	30	29	37	37	48	48	68	68
6	210	211	320	320	542	542	1105	1105
7	27	27	33	33	43	43	61	60
8	10	10	12	12	14	14	18	18
9	17	17	21	21	27	26	38	38
10	4	4	5	4	5	5	6	7
11	7	7	8	8	9	8	10	9
12	8	8	10	11	13	14	19	19
13	6	6	7	8	9	8	11	11
14	1027	1027	1578	1578	2692	2695	5507	5506
15	6	6	6	6	7	7	9	8
16	204	204	314	314	537	537	1099	1100
17	4	4	4	4	5	5	7	7
18	4	4	5	4	5	5	6	5
19	17	17	21	21	27	27	38	39
20	5	6	6	6	7	6	8	7
Cost $\times 10^5$	1.0355	1.0359	2.3670	2.3675	6.6941	6.6947	27.3241	27.3290

Table 7.8 Optimal solution x_i and cost: Bellman (B) versus real numerical method (R)

Link Nr	$n = 2000$		$n = 3000$		$n = 5000$		$n = 10,000$	
	B	R	B	R	B	R	B	R
1	6	6.37	7	7.35	9	8.77	11	11.13
2	4	3.86	4	4.21	5	4.68	5	5.40
3	58	58.33	72	72.39	95	94.63	135	135.40
4	346	345.93	530	529.77	901	901.16	1839	1838.85
5	30	29.66	37	36.60	48	47.62	68	67.91
6	210	209.56	320	319.86	542	542.70	1105	1105.31
7	27	26.53	33	32.73	43	42.59	61	60.74
8	10	10.18	12	11.71	14	13.96	18	17.69
9	17	16.77	21	20.70	27	26.94	38	38.41
10	4	4.16	5	4.62	5	5.28	6	6.30
11	7	7.04	8	7.66	9	8.52	10	9.83
12	8	8.39	10	10.35	13	13.47	19	19.21
13	6	6.38	7	7.36	9	8.78	11	11.13

(continued)

Table 7.8 (continued)

Link Nr	$n = 2000$		$n = 3000$		$n = 5000$		$n = 10{,}000$	
	B	R	B	R	B	R	B	R
14	1027	1027.78	1578	1579.30	2692	2693.46	5507	5506.46
15	6	5.70	6	6.33	7	7.22	9	8.62
16	204	203.56	314	313.86	537	536.70	1099	1099.31
17	4	3.75	4	4.31	5	5.14	7	6.51
18	4	4.27	5	4.58	5	5.00	6	5.62
19	17	16.64	21	20.59	27	26.85	38	38.36
20	5	5.15	6	5.72	7	6.53	8	7.79
Cost × 10^5	1.0355	1.0352	2.367	2.366	6.694	6.693	27.324	27.322

Table 7.9 Equilibrium solution x_i and cost: Heuristic (H) versus numerical method (R)

Link Nr	$n = 2000$		$n = 3000$		$n = 5000$		$n = 10{,}000$	
	H	R	H	R	H	R	H	R
1	7	7.97	10	9.21	11	11.01	14	14.00
2	5	4.79	5	5.22	6	5.82	7	6.72
3	71	70.13	88	87.47	115	114.88	165	165.22
4	340	339.31	519	521.71	888	892.13	1822	1837.98
5	36	36.29	45	44.72	58	58.19	83	83.13
6	207	207.58	315	317.00	537	539.17	1097	1106.11
7	32	32.45	39	39.99	53	52.04	74	74.35
8	13	12.81	15	14.73	18	17.56	22	22.27
9	21	20.51	26	25.29	33	32.91	47	47.02
10	5	5.21	5	5.80	7	6.62	8	7.92
11	9	8.73	9	9.51	11	10.59	12	12.24
12	11	10.26	13	12.65	16	16.46	23	23.51
13	11	8.00	9	9.23	11	11.02	14	14.01
14	996	996.59	1547	1541.25	2648	2641.86	5444	5408.75
15	7	7.16	8	7.95	9	9.07	11	10.84
16	196	195.58	305	305.00	525	527.17	1085	1094.12
17	5	4.71	5	5.42	6	6.46	8	8.20
18	5	5.25	6	5.63	6	6.15	7	6.93
19	20	20.19	26	25.02	33	32.70	47	46.88
20	6	6.47	7	7.18	8	8.20	10	19.80
Cost × 10^5	1.053	1.053	2.398	2.396	6.764	6.768	27.508	27.503

The solution of the integer optimisation problem (7.11) with 20 links obtained by the heuristic algorithm is compared against the numerical solution in Table 7.9. The solutions and their respective costs are very close to each other.

Table 7.10 Equilibrium solution single route cost $f_i(x_i)$ and variance: Heuristic (H) versus numerical method (R)

	$n = 2000$		$n = 3000$		$n = 5000$		$n = 10{,}000$	
Link Nr	H	R	H	R	H	R	H	R
1	36	52.68	102	80.01	135	135.44	276	276.41
2	64	52.68	64	80.01	157	135.44	338	276.41
3	53	52.68	80	80.01	135	135.46	272	276.48
4	52	52.70	79	80.04	135	135.62	274	276.50
5	51	52.68	81	80.01	134	135.44	268	276.45
6	52	52.69	79	80.05	135	135.59	275	276.33
7	52	52.68	76	80.01	140	135.45	273	276.44
8	55	52.68	84	80.01	145	135.44	266	276.42
9	55	52.68	84	80.01	136	135.45	276	276.43
10	44	52.68	44	80.01	169	135.44	287	276.41
11	61	52.68	61	80.01	163	135.44	250	276.41
12	60	52.68	84	80.01	128	135.44	264	276.42
13	52	52.68	74	80.01	134	135.44	275	276.41
14	52	52.63	80	79.86	135	135.89	275	276.24
15	48	52.68	82	80.01	131	135.44	200	276.41
16	52	52.70	79	80.05	135	135.59	275	276.33
17	62	52.68	62	80.01	171	135.44	364	276.41
18	39	52.68	39	80.01	116	135.44	294	276.41
19	51	52.68	86	80.01	137	135.45	277	276.43
20	38	52.68	72	80.01	122	135.44	300	276.41
Variance	58	0.00017	210	0.00143	211	0.01992	1015	0.68122

The integer part of the heuristic results is displayed, while the results of the nonlinear method are truncated after two decimal places

Table 7.10 presents the route cost $f_i(x_i)$. One can see that the variance of the continuous problem (7.12) is actually minimised to nearly zero, although the numerical solution stopped after the default number of 1500 iterations.

Solving the optimisation and equilibrium problems by the heuristic method when the number of links is reduced by removing inefficient links produces solutions satisfying $x_1, \ldots, x_m > 0$, if sufficiently many vehicles are travelling. This leads to the minimum traffic demand levels ensuring nonempty routes given in Table 7.11.

7.3.4.3 Effects of Road Closures and the Price of Anarchy

Here, we investigate the effects of route closures upon the overall travelling cost in the optimisation problem (7.4), and the impact on the price of anarchy.

Table 7.11 Number of vehicles required for strictly positive integer solutions

Nr of links	Optimal	Equilibrium
20	86	132
19	67	107
18	53	78
17	46	66
16	42	58
15	41	55
14	37	49
13	35	46
12	33	40
11	30	33
10	25	31
9	22	28
8	20	21
7	13	13
6	11	10
5	9	8
4	7	6
3	5	4

Table 7.12 Links sorted by the number of vehicles/route when $n = 10{,}000$

Link Nr	Link cost function	n = 2000	n = 3000	n = 5000	n = 10,000
14	$x(2.8 + 0.05x^1)$	1027	1578	2692	5507
4	$x(1.8 + 0.15x^1)$	346	530	901	1839
6	$x(0.8 + 0.25x^1)$	210	320	542	1105
16	$x(3.8 + 0.25x^1)$	204	314	537	1099
3	$x(3.5 + 0.01x^2)$	58	72	95	135
5	$x(0.01 + 0.04x^2)$	30	37	48	68
7	$x(0.04 + 0.05x^2)$	27	33	43	61
9	$x(0.075 + 0.125x^2)$	17	21	27	38
19	$x(1.75 + 0.125x^2)$	17	21	27	38
12	$x(0.01 + 0.5x^2)$	8	10	13	19
8	$x(0.11 + 0.025x^3)$	10	12	14	18
1	$x(2 + 0.1x^3)$	6	7	9	11
13	$x(1.5 + 0.1x^3)$	6	7	9	11
11	$x(2.1 + 0.001x^5$	7	8	9	10
15	$x(0.1 + 0.02x^4)$	6	6	7	9
20	$x(0.099 + 0.03x^4)$	5	6	7	8
17	$x(0.4 + 0.5x^3)$	4	4	5	7

(continued)

Table 7.12 (continued)

Link Nr	Link cost function	$n = 2000$	$n = 3000$	$n = 5000$	$n = 10{,}000$
10	$x(0.99 + 0.07x^4)$	4	5	5	6
18	$x(0.11 + 0.0025x^6)$	4	5	5	6
2	$x(2.3 + 0.02x^5)$	4	6	5	5

Table 7.13 Proportion of vehicles as a function of power, for the optimum solution

Power	$n = 2000$	$n = 3000$	$n = 5000$	$n = 100, 00$
1	0.8935	0.9140	0.9344	0.9550
2	0.0785	0.0647	0.0506	0.0359

Table 7.14 Optimal solution when the most productive $m = 10$ links are maintained

Link Nr	Link cost function	$n = 2000$	$n = 3000$	$n = 5000$	$n = 10{,}000$
14	$x(2.8 + 0.05 \cdot x^1)$	1058	1615	2735	5557
4	$x(1.8 + 0.15 \cdot x^1)$	356	542	915	1856
6	$x(0.8 + 0.25 \cdot x^1)$	216	327	551	1115
16	$x(3.8 + 0.25 \cdot x^1)$	210	321	545	1110
3	$x(3.5 + 0.01 \cdot x^2)$	59	73	95	136
5	$x(0.01 + 0.04 \cdot x^2)$	30	37	48	68
7	$x(0.04 + 0.05 \cdot x^2)$	27	33	43	61
9	$x(0.075 + 0.125 \cdot x^2)$	17	21	27	39
19	$x(1.75 + 0.125 \cdot x^2)$	17	21	27	39
12	$x(0.01 + 0.5 \cdot x^2)$	9	10	14	19
Cost $m = 10(\times 10^5)$		1.083	2.448	6.852	27.715
Cost increase		4.64%	3.42%	2.36%	1.45%

Links sorted by the number of vehicles/route when $n = 10{,}000$

Remark from Table 7.6 that link 14 attracts more than half of the vehicles, while links with power 1 (4, 6, 14, 16) take more than 95.5% of the traffic for $n = 10{,}000$. To confirm that the most loaded segments correspond indeed to the routes with smaller powers, we sort the routes in the decreasing order of traffic for $n = 10{,}000$ vehicles, as illustrated in Table 7.12.

Table 7.13 shows that the proportion of vehicles taking routes of power 1 increases from 89.35% at $n = 2000$ to 95.5% at $n = 10{,}000$, while the percentage of vehicles taking routes of power decreases from 7.85% at $n = 2000$ to 3.59% at $n = 10{,}000$.

A legitimate question is then whether it is worth investing in maintaining low-capacity routes, if the high-capacity roads can accommodate the traffic flow. For this reason, we shall check what happens if some of the lower capacity routes are deleted. Road segments are deleted in the decreasing order of the power.

First, roads of power greater than or equal to 3 (10 routes) are closed and the optimal solution and minimum cost are displayed in Table 7.14. The relative

Table 7.15 Optimal solution when the most productive $m = 4$ links are maintained

Link Nr	Link cost function	n = 2000	n = 3000	n = 5000	n = 10,000
14	$x(2.8 + 0.05 \cdot x^1)$	1151	1728	2882	5766
4	$x(1.8 + 0.15 \cdot x^1)$	387	579	964	1926
6	$x(0.8 + 0.25 \cdot x^1)$	234	350	580	1157
16	$x(3.8 + 0.25 \cdot x^1)$	228	343	574	1151
Cost $m = 4$ ($\times 10^5$)		1.204	2.671	7.336	29.095
Cost increase		16.33%	12.84%	9.59%	6.50%

Links sorted by the number of vehicles/route when n = 10,000

Table 7.16 Price of anarchy

	$n = 2000$	$n = 3000$	$n = 5000$	n = 10,000
Price of anarchy (m = 20)	1.0173	1.0136	1.0100	1.0068
Price of anarchy (m = 10)	1.008	1.007	1.005	1.004
Price of anarchy (m = 4)	1 + 485e−5	1 + 2.19e−5	1 + 7.97e−6	1 + 2.01e−6

increase in the total travel cost compared to the original problem with 20 links decreases from 4.64% at $n = 2000$ to 1.45% at $n = 10{,}000$ vehicles.

Further, the routes of power 2 are also closed (6 routes) and the results for the remaining 4 routes of power 1 are displayed in Table 7.15. The relative increase in the total travel cost compared to the original problem with 20 links decreases from 16.33% at $n = 2000$ to 6.50% at $n = 10{,}000$ vehicles.

The results indicate that the closure of lower capacity roads does increase the overall optimal cost, especially at low traffic. However, closing roads have a positive effect on the price of anarchy, which goes down, as shown in Table 7.16.

References

1. Fisk C (1980) Some developments in equilibrium traffic assignment. Transp Res Part B Methodol 14(3):243–256
2. Horowitz J (1984) The stability of stochastic equilibrium in a two link transportation network. Transp Res Part B Methodol 18(1):13–28
3. Hartman J (2009) Special issue on transport infrastructure: a route choice experiment with an efficient toll, networks and spatial economics. Accessible at: http://www.springerlink.com/content/372117228926871O/fulltext.pdf
4. Roughgarden T (2005) Selfish routing and the price of anarchy. MIT Press, London
5. Youn H, Jeong H, Gastner M (2009) The price of anarchy in transportation networks: efficiency and optimality control. Phys Rev Lett 101 (19 Sept 2009)
6. Skinner, Carlin (2013) The price of anarchy. Significance
7. Braess D, Nagurney A, Wakolbinger (2005) On a paradox of transport planning (a translation of the 1968 article). Transp Sci 39(4):446–450
8. Steinberg R, Zangwill W (1983) The prevalence of Braess' paradox. Transp Sci 17(3):301–318

9. Dafermos SC, Nagurney A (1984) On some traffic equilibrium theory paradoxes Transp Res Ser B 18(2):101–110
10. Catoni S, Pallottino S (1991) Traffic equilibrium paradoxes. Transp Sci 25(3):240–244
11. Pas E, Principio S (1997) Braess' paradox: some new insights. Transp Res B 31(3):265–276
12. Abrams R, Hagstrom J (2006) Improving traffic flows at no cost. In: Lawphongpanich S, Hearn D, Smith M (eds) Mathematical and computational models for congestion changing. Springer, New York
13. Mogoridge M (1997) The self-defeating nature of urban road capacity policy. Transp Policy 4 (1):5–23
14. Arnott R, Small K (1994) The economics of traffic congestion. Am Sci 82:446–455
15. Stanley RP (2012) Enumerative combinatorics, vol 1 (2nd ed.). Cambridge Studies in Advanced Mathematics, 49, Cambridge University Press, Cambridge
16. The On-Line Encyclopedia of Integer Sequences (2011) https://oeis.org. OEIS Foundation Inc
17. Bellman R (1957) Dynamic programming. Princeton University Press, Princeton
18. Bazaraa MS, Sherali HD & Shetti CM (2006) Non linear programming. Wiley, New York
19. Rockafellar RT (1970) Convex analysis. Princeton Mathematical Series 28, Princeton University Press, Princeton

Chapter 8
Air Traffic Controllers Planning: A Rostering Problem

Richard Conniss

The air traffic controllers (ATC) rostering problem shares some features with standard rostering problems reported in the literature, and at the same time has some unique features that required special attention. All controllers start their careers by attending a specialised training college to learn the basic skills of ATC. Once a new controller arrives at their unit, they must undertake a period of on-the-job training. At any given unit, there will be a set of controlling tasks, or positions, for which the new controller must become proficient.

Most ATC units have multiple control positions, each with unique demands and training requirements. Ideally, all controllers will eventually become endorsed (qualified) in all positions and this is where the first main difference with other scheduling problems occur. If a controller holds an endorsement in a position, they are expected to be able to staff that task as required.

This is quite different to the use of qualifications in other rostering problems. As an example, in many nurse rostering problems qualifications denote the seniority of an employee. If a senior or more qualified nurse is assigned to a task that would more normally be undertaken by a more junior colleague, this assignment is penalised in some fashion by the solution method. The senior nurse is qualified to undertake the task, but it is seen as an inefficient use of resources as salary is linked to seniority.

For controllers, the need to maintain familiarity with all tasks for which they are qualified is safety critical. Skill fade is a significant problem and can induce potentially catastrophic effects on the safe movement of aircraft. Controller's salaries are excluded from rostering decisions as their remuneration has no effect on their ability to execute a task. As such, controllers should be regularly assigned to each of the positions for which they hold endorsements. One way of measuring this

---

R. Conniss (✉)
University of Derby, Derby, UK
e-mail: r.conniss@derby.ac.uk

S. Berry et al. (eds.), *Guide to Computational Modelling for Decision Processes*,
Simulation Foundations, Methods and Applications,
DOI 10.1007/978-3-319-55417-4_8

familiarity is known as currency, which is used as a measure of an individual's competence for a task.

Currency is defined as the number of days since a controller worked productively in a given position. What is meant by productive work is that a controller has completed a reasonable amount of work in a position as opposite to just awaiting traffic. Merely scheduling someone to a task is not sufficient to maintain their skill set, and as such the values for currency are difficult to predict. They are updated on a daily basis. The current limit is set at 30 days, and if a controller were to violate this restriction they would have to undergo a period of retraining and re-qualify for that position. Clearly, this situation will add to the training burden of a unit and should be avoided wherever possible; achieving this can be a difficult task for most units.

Like any employee, controllers require rest breaks throughout the working day. In the civilian ATC world, there are very strict and legally binding working rules and conditions to ensure the safety of aviation operations. The rules include maximum durations for a controller to work in a position (usually 2 h) and frequency of rest breaks and meal breaks. For UK ATC operations these rules are defined in the Scheme for Regulation of Air Traffic Controllers Hours (SRATCOH) which is published by the Civil Aviation Authority.

To give a controller a break in a particular position, another qualified controller must replace them. This transfer of responsibility requires a formal handover procedure to ensure that the incoming controller is aware of the location and intentions of all aircraft receiving a service, the local weather conditions, unusual variations to normal procedures, temporary airspace restrictions and any other information deemed necessary for safe operations. This requirement prevents controllers from switching tasks instantaneously, as this hand-over process will always require at least a few minutes to complete. Usually, this is accomplished by separating controller assignments with a break. The problem is exacerbated when multiple controllers require breaks over several time periods. In this scenario, some chain of moves must be found that simultaneously gives all controllers a suitable set of breaks and maintains the required staffing for tasks throughout the day. The goal of rostering is to produce a single day roster that ensures that qualified controllers are appropriately assigned to positions, given breaks and whilst maintaining currency.

One of the most difficult aspects of the process for the watch supervisors, whom are the senior controllers in charge of daily operations, is the initial creation of the daily schedule. With so many permutations of controllers and qualifications, it can be extremely difficult to construct a roster that is feasible. An inordinate amount of a supervisor's time is spent managing the roster to meet the goals of the day. This distracts from their core responsibilities to monitor staff and maintain safe ATC operations, therefore any automated approach that could achieve this part of their daily responsibility would not only simplify their working day, but could also have positive effects on flight safety in general. Breaks are monitored continuously, which places additional pressure onto the supervisor's workload, and occasionally controllers can be left in position for an unsuitable length of time. Often late notice changes to staffing can cause problems, last minute medical appointments, meetings off site and even the rare controlling incident can all cause disruption. Planning for these events is almost impossible and as such supervisors are constantly dealing with new inputs of information, throughout a shift.

An effective algorithm to produce valid rosters has to consider all of the above restrictions placed on controlling staff. During conversations with senior ATC staff at RAF Cranwell, a number of key requirements for a daily roster have been identified.

These are as follows:

- All operational demand for the flying program must be met by qualified controllers.
- Controllers must have suitable rest breaks.
- The system must be able to adapt to sudden changes in staffing or the operational flying task.

8.1 Mathematical Model

An appropriate model for a single day roster is required to understand the complexities involved. Given a set of controllers C with $c \in C$, a set of positions P with $p \in P$ and a set of qualifications Q containing tuples c, p, f where f denotes a currency (familiarity) value in the range $\{0, \ldots, 30\}$ for controller c on position p. The shift is divided into a set of fixed interval time slots T with $t \in T$, and the task is to find a set of assignments A, containing tuples c, p, t which represent a roster for an entire shift.

In the model defined, the day is divided into T time slots of 30 min duration for reasons of simplification. With n controllers $i = 1, \ldots, n$, m positions $j = 1 \ldots m$ and $t \in \{1, \ldots T\}$, the following matrices are defined.

$R_{i,j,t} = 1$	controller i is in position j at time t, 0 otherwise
$Q_{i,j} = 1$	controller i is qualified in position j, 0 otherwise
$D_{j,t} = 1$	position j is to be staffed at time t, 0 otherwise
$A_{i,t} = 1$	controller i is available to work at time t, 0 otherwise
$C_{i,j} = \{0, \ldots, 30\}$	currency of controller i in position j, measured in days

This leads to the following set of hard constraints:

A controller must be qualified to work in a position:

$$R_{i,j,t} \leq Q_{i,j}, \quad \forall i, j, t$$

A controller must be available to work in a position:

$$R_{i,j,t} \leq A_{i,t}, \quad \forall i, j, t$$

If there is a demand for a position, that position must be staffed:

$$\sum_{j} R_{i,j,t} = D_{j,t}, \quad \forall j, t$$

Each controller can only be assigned to a single position at a given time slot:

$$\sum_{j} R_{i,j,t} \leq 1, \quad \forall i, t$$

Controllers cannot instantaneously switch tasks, and must have a break before being assigned to a position. This allows for a formal handover of responsibility as new controllers take on a task.

$$R_{i,j,t} - \sum_{k=1}^{j} R_{i,k,t} + 1 \geq R_{i,j,(t+1)}, \quad \forall i, j, t$$

A controller must be current in a position to work:

$$R_{i,j,t} \leq C_{i,j}, \quad \forall i, j, t$$

Additionally, controllers will require sufficient rest breaks during their shift. As it stands, the RAF has no formal system for defining controller worker hours and as such rules from civilian ATC have been incorporated into the model. Essentially, the main rule to consider is that a controller cannot work for longer than 2 h in a position without a break.

$$\sum_{t=s}^{s+4} \sum_{j=1}^{m} R_{i,j,t} \leq 4, \quad \forall i \in \{1, \ldots, n\}, s \in \{1, \ldots, T-4\}$$

This problem is formulated as constraint satisfaction problem and therefore no formal objective function is required. However, an evaluation function is defined to compare solutions quality. Once a valid roster is produced, the currency value for each (controller, position, time) assignment is retrieved, and the sum of all these assignments is used as a measure of roster quality. Larger values for this function are better, implying that controllers with high currency values in a position at the beginning of the rostering period are assigned to that task in the roster.

$$\sum_{i} \sum_{j} \sum_{t} (R_{i,j,t} \cdot C_{i,j})$$

8.2 Methodology

Given the above restrictions and requirements for successfully scheduling controllers, our proposed algorithm constructs a roster and satisfies the requirements of a particular day. It ensures controllers are able to take reasonable breaks. It is also capable of re-rostering due to short notice events.

The rostering problem can be treated as a tree of fixed size. Therefore, the structure of the developed algorithm is similar to that of the depth first search (DFS) algorithm. DFS is a procedure for traversing every node in a given graph or tree, via their connecting vertices. It does this by first selecting a starting node and then travelling along vertices, always trying to get as far from the start as possible.

Each level of the tree represents a combination of a position and time slot value. The first layer is the first position to staff in the first time slot of the shift, the second is the second position and first time slot etc. Each node is the assignment of a controller at a particular position and time. Figure 8.1 shows the structure of this approach. Only those controllers who are qualified for a position will appear as nodes at each level, and the depth of the tree is equal to the product of the number of positions and the total number of time slots determined by the required length of planning horizon. A valid roster is any path that stretches from the root node to the lowest level.

The process to select each new node begins with the creation of a list of all available controllers and is then divided into three distinct phases. Each phase filters this list, based on the particular requirements at each stage.

The first phase focusses on ensuring that only suitably qualified controllers are assigned to a given position. The node under consideration has a position parameter attached to it. The list of controllers is filtered such that all unqualified controllers are removed and the resultant list is passed to the next phase.

The second phase considers the temporal restrictions on an assignment and considers the following three rules as follows:

1. No controller can be assigned to more than a single position per time slot.
2. No controller can change position in consecutive time slots.
3. Maximum work time limits must be enforced, to allow for controller rest breaks.

Any controller which will violate any of the above restrictions will be excluded from the list.

Fig. 8.1 Example of a roster tree

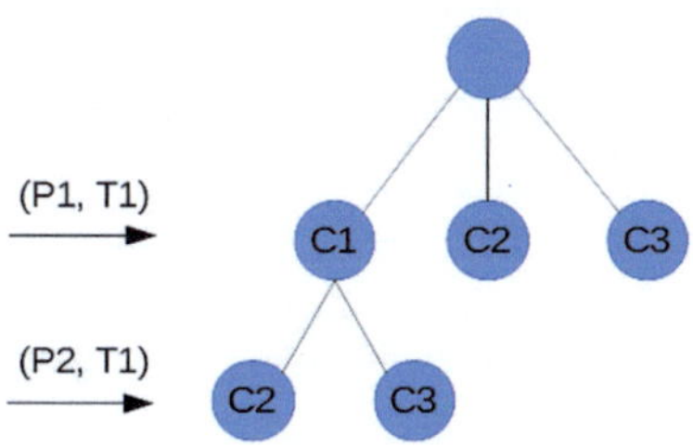

The final phase sorts the remaining controllers in the list into a specified order. In this instance, controllers are ranked according to their currency values, with the least current controller being assigned to the first element of the list, with the remaining controllers assigned to subsequent elements in decreasing order of number of day's currency. The resultant list is then stored with the node for use by the search algorithm.

The search is controlled by the state of the list of controllers. If after the filtering process a controller remains in the list, then this controller will be assigned to the current node and the search will move on to the next node in the tree. If at any point the list is found to be empty, this signals to the search that no feasible solution can be obtained with the current set of assignments. The search then backtracks to the previous node, removes the assigned controller and inserts the next controller in the list. The search then continues forward until another empty list of controllers is found. If after the removal of a controller from a node the search finds that the list is empty, the search immediately backtracks once more, removes the controller from the earlier node and continues this process until another controller can be assigned to a node. The pseudocode representation of the algorithm is shown in Fig. 8.2.

```
Input:
timeslots = A list of Timeslots [0,…,T]
positions = A list of Positions [0,…,m]
controllers = A list of all controllers to be considered
solution = A list of type assignment(c,p,t), where each assignment is initialised
as a position and a timeslot only. The values for the controller are left empty as
this is what will be found in the search.
solutionPointer = 0, an integer which tracks the current          assignment for
consideration in the solution. The value must be between 0 and
length(positions)*length(timeslots) i.e. the total number of assignments in the
roster.
stack = Empty stack of type assignment.
Output:
// Set up the stack for first use
Stack.push(Null)
// Null is used to show when the assignment under consideration // changes and
triggers a backtrack.
Stack.push(controllers)
While 0≤ solutionPointer ≤ total number of assignments
      If the first element in the stack is Null
            Then decrement the solutionPointer by 1
            And discard the Null value

      If the first element in the stack is a valid assignment
            Add the controller to the assignment in the solution
            Increment the solutionPointer by 1
            Push a Null and the controller list to the stack

      Else
            Discard the top element in the stack

End
Return solution
```

Fig. 8.2 Pseudocode for depth first search algorithm

8.3 Results

The algorithm as proposed has been implemented in C# and suit of experiments designed to validate the algorithm.

The aim of the experiments is as follows:

1. To test if the algorithm can produce feasible rosters which satisfy all the problem constraints.
2. To determine if a heuristic ordering on the list of controllers can improve performance.
3. To assess the effect of ordering the nodes in the search by the qualification requirements for each position.

To examine the effect of different heuristic sorting methods on performance, four variants of the search process were created. They are defined as follows:

- dfs: The basic version of the search. No ordering is applied to controller's currency or to the order of assignment of controllers to positions. This variant is useful for finding rosters that are at minimum feasible.
- dfsQ: The set of assignments required are ordered by the number of controllers qualified for each task. The algorithm attempts to first assign tasks with the fewest number of qualified controllers, for each time slot. The goal here is to try and force the search to backtrack as early as possible and leave the most flexibility and choice for assignment to positions with the most number of qualified controllers.
- dfsC: The set of controllers that are qualified for each position are ordered by descending currency value and presented for assignment in turn. The intention is for controllers with the most need to become current. In a particular position to be the first selected for assignment to any given task. Using currency to order the newly generated nodes is equivalent to expressing a preference to assign controllers to positions that they have not worked in for some time. It does not guarantee that the least current controller in a position is always assigned.
- dfsQC: The final variant is a mix of dfsQ and dfsC. The set of assignments are first ordered as in dfsQ and then controllers are presented in order of currency as in dfsC.

To compare the behaviour of each variant, a shared set of controller and task data was produced for each experiment. Initially, 20 controllers each with their own set of qualifications and 10 positions were considered. The planning horizon was a single 9 h day shift consisting of 18 time slots of 30 min duration. Currency values were randomised at the start of each of the four search process and each of the variants produced a roster subject to its respective heuristic ordering method. A total of 30 comparisons were produced, with each comparison comprising a single solution attempt using each of the 4 algorithms. For each new comparison, a common set of randomised currency values was used as input values for each algorithm.

Figure 8.3 shows an example roster produced by the dfs variant of the algorithm, which shows the assignment of controllers (A–Z) to positions for a particular time slot (Controller H's work schedule highlighted as an example).

In the real world, it is unlikely that all staff will be available for every shift, so the algorithm needs to be able to successfully produce rosters with reduced controller numbers. Fewer controllers imply a different distribution of qualifications and an increase in the difficulty of creating a valid schedule. Figure 8.4a shows the distribution of qualifications for the set of controllers and indicates some of the difficulties associated with finding feasible solutions, Fig. 8.4b shows the qualifications of the controllers used to construct the roster shown in Figs. 8.3 and 8.4c, d re-formulate the earlier tables to highlight the controllers needing retraining. Figure 8.4f drawn from Fig. 8.3 indicates where the controllers have gained or refreshed endorsements.

The experiment was then extended to include a reduced numbers of controllers. The aim here is to constrain the search as much as possible to check for changes in solution time and to validate the algorithms ability to deal with more realistic staffing levels. As before, each set of comparisons attempts to roster a number of controllers to 10 tasks and ensures that each controller in the roster receives an adequate number of rest breaks.

The number of controllers considered in each experiment was gradually reduced, starting with the most qualified controllers, the full titles for the positions are defined in Table 8.3.

Table 8.1 shows the average time required to generate a solution given in milliseconds for each set of comparisons, for different numbers of controllers. The basic dfs and dfsC variants have the largest average time to find a solution, although they are both relatively quick to find a feasible solution.

The dfsQC and dfsQ variants produce rosters in the shortest time. One interesting feature to note is the trend in solution time decreases from 20 to 17 controllers and then spikes at 16.

Position/Time Slot	1	2	3	4	5	6	7	8	9	10	11	12	13	14	15	16	17	18
APP	S	S	S	X	X	X	K	K	K	K	Y	Y	Y	Y	S	S	S	S
CWL DIR	T	T	T	O	O	O	O	X	X	X	X	K	K	K	K	Y	Y	Y
BKN DIR	O	O	Y	Y	Y	**H**	**H**	**H**	**H**	S	S	S	S	O	O	O	O	X
CWL DEPS	**H**	**H**	**H**	**H**	S	S	S	S	O	O	O	O	X	X	X	X	Z	Z
BKH DEPS	X	X	Z	Z	Z	Z	Y	Y	Y	T	T	Z	Z	Z	Z	**H**	**H**	**H**
ADC	M	M	B	D	D	D	E	E	F	F	**H**	**H**	**H**	**H**	E	E	E	E
GND	Y	E	E	E	E	W	W	W	W	Q	Q	Q	Q	M	M	M	M	T
PAR	W	W	W	W	Q	Q	Q	Q	M	M	M	M	B	B	B	B	K	K
CWL SRA	Q	Q	Q	M	M	M	M	B	B	B	B	E	E	V	V	W	W	W
BKN SRA	B	K	K	K	K	V	V	Z	Z	Z	W	W	W	W	Q	Q	Q	Q

Fig. 8.3 Example roster produced by depth first search

(a)

Position/Controller	B	C	D	E	F	G	H	J	K	L	M	N	O	P	Q	R	S	T	V	W	X	Y	Z
APP		q	q		q	q		q	q	q		q		q		q	q	q	q		q	q	
CWL DIR		q	q		q	q		q	q	q		q	q	q		q	q	q	q		q	q	
BKN DIR		q	q		q	q	q	q	q	q		q	q	q		q	q	q	q		q	q	
CWL DEPS		q	q		q	q	q	q	q	q		q	q	q		q	q	q	q		q	q	q
BKH DEPS		q	q		q	q	q	q	q	q		q	q	q		q	q	q	q		q	q	q
ADC	q	q	q	q	q	q	q	q	q	q	q	q	q	q		q	q	q	q		q	q	
GND	q	q	q	q	q	q	q	q	q	q	q	q	q	q	q	q	q	q	q	q	q	q	q
PAR	q	q	q	q	q	q	q	q	q	q	q	q	q	q	q	q	q	q	q	q	q	q	q
CWL SRA	q	q	q	q	q	q	q	q	q	q	q	q	q	q	q	q	q	q	q	q	q	q	q
BKN SRA	q	q	q	q	q	q	q	q	q	q	q	q	q	q	q	q	q	q	q	q	q	q	q

(b)

Position/Controller	K	S	T	V	X	Y	O	H	Z	B	E	M	Q	W
APP	q	q	q	q	q	q								
CWL DIR	q	q	q	q	q	q	q							
BKN DIR	q	q	q	q	q	q	q	q						
CWL DEPS	q	q	q	q	q	q	q	q	q					
BKH DEPS	q	q	q	q	q	q	q	q	q					
ADC	q	q	q	q	q	q	q	q		q	q	q		
GND	q	q	q	q	q	q	q	q	q	q	q	q	q	q
PAR	q	q	q	q	q	q	q	q	q	q	q	q	q	q
CWL SRA	q	q	q	q	q	q	q	q	q	q	q	q	q	q
BKN SRA	q	q	q	q	q	q	q	q	q	q	q	q	q	q

(c)

Position/ Controller	C	D	F	G	J	K	L	N	P	R	S	T	V	X	Y	O	H	Z	B	E	M	Q	W
APP	q	q	q	q	q	q	q	q	q	q	q	q	q	q	q								
CWL DIR	q	q	q	q	q	q	q	q	q	q	q	q	q	q	q	q							
BKN DIR	q	q	q	q	q	q	q	q	q	q	q	q	q	q	q	q	q						
CWL	q	q	q	q	q	q	q	q	q	q	q	q	q	q	q	q	q	q					
BKH DEPS	q	q	q	q	q	q	q	q	q	q	q	q	q	q	q	q	q	q					
ADC	q	q	q	q	q	q	q	q	q	q	q	q	q	q	q	q	q		q	q	q		
GND	q	q	q	q	q	q	q	q	q	q	q	q	q	q	q	q	q	q	q	q	q	q	q
PAR	q	q	q	q	q	q	q	q	q	q	q	q	q	q	q	q	q	q	q	q	q	q	q
CWL SRA	q	q	q	q	q	q	q	q	q	q	q	q	q	q	q	q	q	q	q	q	q	q	q
BKN SRA	q	q	q	q	q	q	q	q	q	q	q	q	q	q	q	q	q	q	q	q	q	q	q

Fig. 8.4 **a** Qualifications of controllers by position (*q* indicates qualified). **b** Controllers used to construct roster. **c** Grouping staff by qualifications. **d** Grouping staff by qualifications and currency, indicating training needs. **e** Controllers used to construct roster maintaining currency. **f** Endorsements completed or renewed

(d)

Position/ Controller	C	F	J	P	S	V	Y	Q	D	G	K	L	N	R	T	X	O	H	Z	B	E	M	W
APP	3	6	3	4	4	3	6		5	4	3	4	2	6	3	2							
CWL DIR	1	6	5	3	3	2	7		5	3	6	2	2	5	6	7	5						
BKN DIR	7	1	3	1	1	6	5		3	3	7	6	4	5	6	6	2	7					
CWL	7	6	6	2	3	7	6		2	7	7	6	6	7	5	2	7	6	4				
BKH DEPS	7	5	1	4	4	6	6		1	3	6	6	5	5	2	5	7	2	3				
ADC	6	7	7	6	6	7	5		7	2	4	4	2	6	5	5	6	2		7	5	6	
GND	3	3	3	2	3	7	7	1	6	4	2	5	4	3	2	3	5	2	7	6	5	6	3
PAR	3	3	7	6	5	2	1	4	7	6	6	2	3	4	2	7	6	2	2	7	2	4	4
CWL SRA	6	7	6	6	3	7	7	4	7	4	4	7	4	5	2	5	5	3	3	2	6	5	3
BKN SRA	7	3	6	2	4	1	2	5	7	4	3	3	4	3	7	2	4	4	3	3	3	2	6

(e)

Training Allocated	C	S	F	V	J	Y	P						Q	
Position/Controller	C	S	F	V	J	Y	P	H	Z	B	E	M	Q	W
APP	q	q	q	q	q	q	q							
CWL DIR	q	q	q	q	q	q	q							
BKN DIR	q	q	q	q	q	q	q	q						
CWL DEPS	q	q	q	q	q	q	q	q	q					
BKH DEPS	q	q	q	q	q	q	q	q	q					
ADC	q	q	q	q	q	q	q	q		q	q	q		
GND	q	q	q	q	q	q	q	q	q	q	q	q	q	q
PAR	q	q	q	q	q	q	q	q	q	q	q	q	q	q
CWL SRA	q	q	q	q	q	q	q	q	q	q	q	q	q	q
BKN SRA	q	q	q	q	q	q	q	q	q	q	q	q	q	q

(f)

S	3	APP;	BKN DIR;	CWL DEPS		
X	5	APP;	CWL DIR;	BKN DIR;	CWL DEPS;	BKH DEPS
K	4	APP;	CWL DIR;	PAR;	BKN SRA	
Y	5	APP;	CWL DIR;	BKN DIR;	BKH DEPS;	GND
T	3	CWL DIR;	BKH DEPS;	GND		
O	4	CWL DIR;	BKN DIR;	BKN DIR;	CWL DEPS	
H	4	BKN DIR;	CWL DEPS;	BKH DEPS;	ADC	
Z	3	CWL DEPS;	BKH DEPS;	BKN SRA		
M	4	ADC;	GND;	PAR;	CWL SRA	
B	4	ADC;	PAR;	CWL SRA;	BKN SRA	
E	3	ADC;	GND;	CWL SRA		
W	4	GND;	PAR;	CWL SRA;	BKN SRA	
Q	4	GND;	PAR;	CWL SRA;	BKN SRA	
V	2	CWL SRA;	BKN SRA			
Total	52					

Fig. 8.4 (continued)

Table 8.1 Average time required to generate a solution given in milliseconds

Search/number of controllers	20	19	18	17	16	15	14	13
dfs	1.13	19,515	9441	5627	17,322	11,050	32,078	187,795
dfsC	2.43	19,495	9447	5632	17,276	11,047	32,074	187,969
dfsQ	1.30	80	35	71	1076	593	884	7527
dfsQC	1.30	80	34	71	1080	593	884	7495

Table 8.2 Average currency value for solution

Controllers	20	19	18	17	16	15	14	13
dfs	1968	1989	1922	1887	1917	1898	1851	1860
dfsC	2408	2375	2364	2303	2069	2147	2122	2019
dfsQ	1968	1490	1922	1887	1917	1898	1851	1860
dfsQC	2408	2375	2364	2303	2069	2147	2122	2019

Table 8.3 Positions within ATC planning

Position	
APP	Approach
CWL DIR	Cranwell director
BKN DIR	Barkston heath director
CWL DEPS	Cranwell departures
BKH DEPS	Barkston heath departures
ADC	Aerodrome control
GND	Ground control
PAR	Precision approach radar
CWL SRA	Cranwell search radar approach
BKN SRA	Barkston heath search radar approach

The trend then re-emerges and continues to reduce, which implies some form of phase change is occurring in the algorithm. One possible explanation for this is that the search space size is reduced as the number of controllers is lowered. Early on in this reduction, the change in size offers a performance boost. At some point, the effect of making the problem more constrained begins to overcome the gains from the reduction in search space size, to increase the solution times.

Clearly, the heuristic sorting tends to speed up the search. Table 8.2 shows the average solution quality for each set of comparisons. The solution quality is the sum of the currencies of controllers for each assignment. Larger values suggest more high currency controllers have been assigned and therefore will have the opportunity to reset their currency. This has the effect of preventing all controllers from going out of currency over time.

These results imply that dfs is the worst performing variant, with dfsQC producing better quality rosters. Intuitively, these results make sense. The dfsQ search tries to assign the most difficult (in terms of number of qualified controllers) first, which means the search is more likely to backtrack earlier. Each time the search backtracks, it removes large parts of the search space and reduces the number of possible nodes to be evaluated. This should have the effect of decreasing the solution time.

The dfsC presents the least current controllers for assignment first. This should lead to an upward pressure on the values of currency for the solution. The result being that on average, higher total currency values are found as the first solution. These values are not optimal, but they are larger than both the dfs and dfsQ algorithms.

Due to the structure of the implementation of the algorithm, it is relatively simple to combine both the above sorting methods simultaneously which as demonstrated, leads to an improvement in both speed and quality of generated rosters.

8.4 Discussions and Future Research

This case study has presented a new class of rostering problem, which is focussed on a real world application and genuine need for a solution method. The ability to algorithmically generate daily rosters is of great use to operational controllers and watch supervisors, alike. The above results show that feasible and useful daily rosters can be created automatically to satisfy ATC needs, in reasonable time.

The next objective is being able to plan over a longer planning horizon. ATC units tend to fix staff onto rotating shift patterns to simplify this process. At Cranwell, staff tends to work a week of the same shift type at a time. These shifts are usually designated as early, day and late shifts and their durations tend to overlap. Using the current planning system, this eliminates the need for any consideration of shift patterns. However, there remains the problem of deciding how to allocate controllers to shift patterns to ensure suitable coverage so that normal operations can occur.

One possible approach would be when a valid roster is generated for a given planning horizon, it can be used as a template to create a more general version for future use. Some controllers share equal sets of qualifications and each such group can be classified as a controller type. Due to the structure of the training program at most units, there will only ever be a few of these groups and they will be relatively stable over time. When planning shift staffing, it then becomes possible to set limits on which type of controller can be assigned to which shift and still produce a valid roster. If instead of assigning specific controllers to positions, types of controllers were assigned then this would allow the system to create general rosters and experiment with different configurations of staff. As the algorithm is deterministic and can be restarted, there is no reason for the algorithm to ever truly terminate. As each new type categorised roster is generated it can be stored and used as the basis of a new roster, by replacing each type with a controller. These general rosters can also be rated for specific attributes, e.g. the average number of positions worked by a controller in a day, as a more varied work day can assist in preventing boredom and dissatisfaction.

Other preferences can be easily added to the system. This is useful for temporary situations like annual competency checks. Controllers are regularly checked by a colleague to ensure they are capable to continue controlling in a particular position. Usually, the scheduling of these checks is complicated by the need to maintain ATC operations and currency. Using this algorithm, check preferences can be added to a roster before it is generated thereby removing the difficulty and because the algorithm is deterministic it is guaranteed that if a valid solution exists it will be found.

The proposed extension of the algorithm will prioritise training, by using the preference ordering rules explained above. Training starts with a ground school which lays down the basic rules for a position and includes local peculiarities in procedures and processes. On completion, the student controller is placed into the live training environment and for the first time begins to work with real aircraft. Clearly, this is a critical and potentially dangerous situation and as such an experienced instructor will be given the responsibility of guiding the student through their training. What this entails is having both the instructor and student work on the same position until such point as the student is prepared for examination. After a successful exam, the student becomes endorsed and can now control independently for that single position. At RAF air traffic control units, controllers are usually only posted to that unit for 3–5 years. This causes a constant turnover of staff and creates a training burden that must be satisfied to ensure effective operations. The current method employed by the RAF is to define suitable time periods in a day that would afford the best opportunity to train and to then attempt to roster student/instructor pairs to those positions. This could be achieved by adding exceptions to the algorithm that attempt to satisfy these requests, but if such rosters are infeasible the algorithm would default to producing feasible rosters.

Finally, ensuring a fair allocation of tasks to controllers would be an obvious next step for the research. Initially, this type of investigation was hampered by the lack of any method to generate feasible rosters. Using fairness as measure of roster quality is a recent addition to the literature, but one which requires further investigation. One possible approach would use a multi-phase approach, beginning with the current algorithm to generate feasible rosters and then use a secondary heuristic method to maintain a fair task allocation over time.

Chapter 9
Solving Multiple Objective Problems: Modelling Diet Problems

Val Lowndes and Stuart Berry

The first objective in solving a diet problem is to select the best set of foods, from a given list, so that the resultant diet will both satisfy a set of nutrient restrictions and minimise the total cost of the diet; this cost could be expressed in monetary terms, for example, as the fat content of the diet.

Optimal solutions to diet problems have been obtained through the use of linear programming techniques [1–3]. However, although this approach produces the optimal solution to the problem, (the cheapest diet), the resultant diet tends to be both unpalatable (typically implying the consumption of the same foods every day) and unworkable (specifying unrealistic quantities of the chosen food types).

Subsequent investigations have, therefore, concentrated on reformulating the problem to be able to incorporate a consideration of "taste" into the problem. Where "taste" could be considered to have been incorporated either through the provision of a range of diets close to the optimal solution, or by the production of a diet which is close to a "patients" chosen diet, [2], either requirement tending to produce a multi-objective linear programming problem or a goal programming problem.

The models and approaches described here are based around this "diet construction problem"; this case study shows how models and solution methodologies have been developed to enable the construction of a satisfactory diet (both economical and palatable) and the effect of this development on the total cost of the implied daily food intake.

V. Lowndes
University of Derby, Kedleston Road, Derby DE22 1GB, UK
e-mail: V.P.Lowndes@derby.ac.uk

S. Berry (✉)
College of Engineering and Technology, University of Derby,
Kedleston Road, Derby DE22 1GB, UK
e-mail: s.berry@derby.ac.uk

S. Berry et al. (eds.), *Guide to Computational Modelling for Decision Processes*,
Simulation Foundations, Methods and Applications,
DOI 10.1007/978-3-319-55417-4_9

More acceptable solutions, palatable and varied, were initially obtained using an "iterative" linear programming approach where additional constraints were added to produce an acceptable (tasty/palatable) diet (see Table 9.1); this analysis demonstrates the shortcomings of the linear programming approach, suggesting that the model needs to have multiple objectives and that these may be satisfied using an approach based around genetic algorithm allied to fuzzy logic to synthesise an alternative (solution) methodology.

9.1 Diet Modelling Development

The basic requirement, from this model, is the derivation of an acceptable (healthy) diet often minimising costs while supplying attractive diets.

The traditional cost-minimising linear programming formulation of the diet problem aims to select a set, and quantity, of foods, x, which satisfy the dietary restrictions, typically

$$Ax \geq b$$

while at the same time minimises the value of a cost function where

$$\text{cost} = cx$$

Note:

1. Matrix $\boldsymbol{A}$ defines the nutrient content of each food; $\boldsymbol{a}_{ij}$ indicating the quantity of nutrient i in one unit of food j;
2. $\boldsymbol{b}_i$ defines the restriction on the quantity of nutrient i in the chosen diet, here assumed to be the minimum intake per day;
3. $\boldsymbol{c}_j$ gives the cost of a unit of food j; and
4. $\boldsymbol{x}_j$ gives the quantity of food j in the diet.

The formulated problem will have m variables (foods or food types) and n constraints (nutrient restrictions), where in general m will be much greater than n; consequently the optimal solution will consist of at most n nonzero variables (foods).

The deficiencies of a linear programming, cost-minimising approach can be demonstrated through the models and modelling of George Stigler in 1938 [4], solution obtained by "Inspection" and then George Dantzig in 1947 [2] solution obtained through the use of a "computer program solving linear programming problems".

9.2 Evaluating the Stigler Diet

An early attempt to solve this (diet) problem was by Stigler who utilised heuristic methods in order to find a solution. The original formulation was concerned with the construction of an economic, and healthy, diet for a 154-lb male from a list of 77 different foods, in order to fulfil the recommended intake of 9 different nutrients while keeping expense at a minimum but with consideration of taste.

Stigler's heuristic method used "trial and error, mathematical insight and agility", to eliminate 62 of the foods from the original 77 and from the reduced list calculating the required amounts of each of the remaining **15 foods** to arrive at a "cost-minimising" solution.

The annual cost of this solution was $39.93 (1939 dollars). When corrected for inflation using the consumer price index, the cost of the diet in 2005 dollars is $561.43.

The specific combination of foods and quantities is as follows (Fig. 9.1).

The 9 nutrients that Stigler's diet took into consideration and their respective recommended daily amounts were as follows (Fig. 9.2).

Seven years after Stigler made these estimates, the availability of computers and Dantzig's simplex algorithm made it possible to solve the problem without relying on heuristic methods. The exact value was determined to be $39.69 (using the

Stigler's 1939 Diet

Food	Annual Quantities	Annual Cost
Wheat Flour	370 lb.	$13.33
Evaporated Milk	57 cans	3.84
Cabbage	111 lb.	4.11
Spinach	23 lb.	1.85
Dried Navy Beans	285 lb.	16.80
Total Annual Cost		**$39.93**

Approximate quantities per day:	
Wheat flour	1 lb
Evaporated milk	0.14 can
Cabbage	0.33 lb
Spinach	0.06 lb
Dried Navy Beans	0.8 lb

Fig. 9.1 Stigler solution

Fig. 9.2 Recommended daily consumption of vitamin and nutrients

Calories	3,000 Calories
Protein	70 grams
Calcium	.8 grams
Iron	12 milligrams
Vitamin A	5,000 IU
Thiamine (Vitamin B_1)	1.8 milligrams
Riboflavin (Vitamin B_2)	2.7 milligrams
Niacin	18 milligrams
Ascorbic Acid (Vitamin C)	75 milligrams

original 1939 data), a saving of less than 1% from the cost of Stigler's diet, an outline of Dantzig's approach and commentary is available at, and his subsequent (iterative) attempt to incorporate taste into the diet through the addition of extra constraints.

These diets seem very unacceptable, no variety, consuming the same foods each day. Notice that these do have a slight resemblance to the diet in a prison camp suggested in "One Day in the Life of Ivan Denisovich" Solzhenitsyn [5].

The unpalatability and repetitiveness of the resultant diets follows from the fact that the formulated problem will have m variables (foods or food types) and n constraints (nutrient restrictions), where in general m will be very much greater than n,; consequently the optimal solution will consist of at most n nonzero variables (foods).

> *While Stiglers solution cost* ***$39.93 a year and Dantzigs*** linear programming *solution cost* ***$39.69*** *a year, a dietician working with the same data at the same time produced a more palatable diet costing* ***$115*** *a year.*

These results highlighted the facts that

- these problems are easily solved by inspection, giving a near-optimal solution, and
- a palatable diet could be considered to be acceptable if its cost does not exceed 3 times the optimal cost.
- A linear programming approach cannot produce a varied diet; here the optimal solution would consist of, at most, 9 foods. To increase this number, extra constraints would have needed to be included, see Table 9.1 for example of this approach.

These evaluations not only suggest that this base model is easily solvable but also suggest that the derived solution is unacceptable, in most applications, and the model needs to be developed to incorporate elements of taste.

A first attempt to overcome this problem, lack of taste, is described in Dantzig where the model was extended through the iterative addition of extra constraints.

The deficiencies of this (iterative addition of constraints) approach are supported by the results from an investigation by S. Chung who constructed diets from the database and problem definition:

> There were 76 foods available, and a planner wishes to produce a diet that **minimises costs** while containing sufficient **calories and vitamin C.** An example containing 76 variables and two constraints.

Progression to a Solution: A series of 8 models/solutions were constructed during this investigation each model aiming to overcome the deficiencies, in the solutions, of the earlier models.

Model 1: Linear programming model 76 variables and 2 constraints
Model 2: Integer solution required

Table 9.1 Progressive diets obtained through the addition of extra constraints

Model	1	2	3	4	5	6	7	8
Variables	76	76	76	76	76	76	76	76
Constraints	2	2	3	3	9	9	10	11
Variables		Integers		Integers		Integers	Integers	
	Foods and units of foods chosen from each model							
Price	£0.48	£0.52	£0.76	£0.79	£0.74	£0.95	£1.11	£1.26
Bread	18.32	17	5	5	14.1	10	5	5
Cabbage	1.22	1			1.06	1	1	1
Banana		1	3.88	5			2	
Almonds			1.45	1	0.11	1	1	
Cereal					0.001			
Carrots					0.48	1	1	1
Kidney beans					1.82	1	3	
Milk					0.78	1	1	1
Chickpeas								2

Models 3–8: Adding extra constraints to limit the quantities of specific foods, model 8 with 11 constraints and an integer solution only requiring 5 foodstuffs.

With costs increasing from

Model 1: cost £0.48 to
Model 8: cost £1.26 cost increased by 250%.

The unsuitability of this approach can be seen from the fact that no reasonable meals could be constructed from the foods and quantities of foods selected by these models.

These results are similar, and probably as palatable as those presented by Stigler and Dantzig. The major problem with these solutions produced from these models is the lack of variety in the diets produced, the same small set of foods.

The results generated by these models were:

The next sections show how the use of genetic algorithms is allied to fuzzy logic can produce many varied (reasonably) economical diets.

9.3 Incorporating Patient Choice into Diets Using Genetic Algorithms and Fuzzy Logic

This case study discusses the ways in which multi-objective or goal programming problems can be solved using a combination of genetic algorithms and fuzzy logic.

To illustrate this approach, it simulates the ways in which patients, or dieticians, select the food to be included within a diet with the aim of producing varied palatable diets.

- The first part validates the procedure by considering the ways in which genetic algorithms together with linear programming can be used to incorporate taste into the construction of diets to satisfy a patient's nutritional needs.
- The second part discusses how an approach using genetic algorithms can be used when there is a nonlinear objective, typically optimising taste per cost.
- The final part indicates how fuzzy logic can be allied with the genetic algorithm approach illustrated in the first part to provide a better mechanism to enable the production of many efficient diets. This approach enables the quick solution of nonlinear multi-goal programming problems.

9.3.1 Introduction

Here, genetic algorithms, allied to the traditional linear programming approach, can provide an alternative approach to the incorporation of taste into diet problems. The effectiveness of the genetic algorithm approach is demonstrated firstly through its application to the basic cost-minimising formulation and then to the "taste" formulation [2].

- Firstly, genetic algorithms are shown to provide a range of solutions close to the optimal solution (diet), thus satisfying the initial criterion defining the incorporation of taste into the solution.
- Secondly, how nonlinear objectives can be represented using GAs.
- Finally, they will show how fuzzy logic can be allied with genetic algorithms to provide an alternative approach, solving the resultant multiple objective goal programming problem.

These approaches act by simulating the behaviour of the patients, or dieticians, in continually refining the content of a diet until the resultant diet satisfies all the constraints and is acceptable to the dietician, or patient.

9.3.2 Genetic Algorithm Implementation

Genetic algorithms can be employed to obtain a set of near-optimal solutions to a "diet-type" problem, thus incorporating taste into the provision of many diets.

The structure of the LP model of the problem, i.e., many variables, few constraints, and the consequent limited number of foods in the optimal solution suggested that the use of a genetic algorithm approach to the problem might prove to be a viable alternative to the normal linear programming approach. In this approach, each GA string represents a series of foods which might be included with the diet.

For example, the GA string $[3, 5, \ldots, 78]$ contains m^L values. Each value corresponds to a food to be considered for inclusion in the diet; here foods 3, 5, etc.

A set of "m^L variable" sub-problems are chosen because it is known that the optimal solution will consist of no more than n nonzero values, and so $m^L \geq n$. The "included" foods are chosen randomly in the initial set of strings.

The kth sub-problem, n constraints, m^L variables at the first iteration can be represented by

$$\begin{array}{ll} \text{Minimise cost} & \mathbf{C}_{k1} = \mathbf{c}_k \mathbf{x}_{k1} \\ \text{subject to} & \mathbf{A}_k \mathbf{x}_{k1} \geq \mathbf{b} \end{array}$$

where $\mathbf{x}_{k1}$ indicates the quantity of each food in diet k at iteration 1.

The optimal solutions to each of these sub-problems are obtained using linear programming techniques, giving an optimal set of foods and quantities and a cost for the diet, $\mathbf{C}_{k1}$.

These costs are used to define the fitness function within the genetic algorithm and are used to generate the next set of sub-problems $\mathbf{x}_{i2}$.

- Tournament selection is used to select the GA strings from the first iteration.
- Crossover is performed as described below, and mutation is implemented by replacing a food from the string with a randomly chosen other food.

For example, given the two strings and the indicated crossover point,

3, 5, 11, 35, 43, 89
9, 17, 24, 37, 77, 99

gives the new strings

3, 5, 11, 37, 77, 99
9, 17, 24, 35, 43, 89

This process is repeated, that is using the $\mathbf{x}_{i,t}$ (iteration t results) to generate the $x_{i,(t+1)}$ (the strings for iteration $t + 1$), continuing until the solution set converges on a solution or until a set number of iterations have been completed.

This approach was employed to analyse a 280-food 5-nutrient constraint problem. The results showed that this (GA) process converges on a solution close to the optimal solution. Thus, a set of near-optimal solutions are obtained, producing a set of alternative near-optimal diets, hence incorporating taste through the provision of a set of alternative diets.

9.3.3 Results

The sample set of data, 280 foods and 5 nutrient constraints, was used to evaluate this approach. The nutrient requirements to provide an acceptable diet were obtained from "McCanse and Widdowson's The Composition of Food's",

Table 9.2 Number of alternative diets

Specified problem	Optimal solution (0 cost)	Number of solutions with in 2 × 0 cost
A	116	49
B	90	41
C	108	57
D	109	46

and the food costs/units were generated randomly to produce a large set of trial problems.

In each case, the GA procedure performed 10 iterations, and every diet with a cost less than twice that of the optimal cost (obtained from the Linear Programming formulation) was recorded (a factor of 2 was chosen, so that the results could be consistent with the cost differential reported from the investigation into the Stigler diet); the number of such diets is given in Table 9.2.

On average, 48 alternative diets were generated using this approach. Thus, the provision of a variety of solutions close to the optimal solution validates the use of genetic algorithms as a technique to introduce taste (by variety) into the diet problem.

9.4 Taste Formulation

This section investigates the application of the genetic algorithm approach to the alternative "taste formulation" for a diet problem, and this can be compared with the approach adopted by Fletcher [2].

This approach starts with an existing diet, many foods, and investigates the necessary changes to this diet, so that a feasible diet, obeying the nutrient restrictions, could be generated. The original diet can be considered to represent the taste of the "patient", and the result of the analysis produces a set of diets as close as possible to this original diet.

9.4.1 GA Problem Formulation and Notation

In addition to the defined set of nutrient constraints

$$\boldsymbol{A}\boldsymbol{x} \geq \boldsymbol{b}$$

$$\text{and cost function} \quad \boldsymbol{cost} = \boldsymbol{c}\boldsymbol{x},$$

there now exists a patient-chosen diet, given by the vector of foods $\boldsymbol{x}_c$, where $\boldsymbol{x}_{ci}$ indicates the quantity of food i in the chosen diet.

This diet can be assumed to be infeasible; in that the nutrient content does not satisfy all of the constraints.

If t is the vector defining the quantity of nutrients in this diet from the pre-selected foods ($\boldsymbol{t} = \boldsymbol{A}\,\boldsymbol{x}_c$), then the problem is now reformulated as:

Choose $\boldsymbol{y}$ to satisfy the constraints

$$\boldsymbol{A}\,\boldsymbol{y} \geq \boldsymbol{b} - \boldsymbol{t}$$

and minimise the cost function

$$\boldsymbol{cost} = \boldsymbol{c}\,\boldsymbol{y}.$$

The final diet will be given by

$$\boldsymbol{x}_c + \boldsymbol{y}$$

with a cost given by

$$\boldsymbol{c}(\boldsymbol{x}_c + \boldsymbol{y})$$

The genetic algorithm approach described above (see Sect. 9.3.2) was applied to this problem, producing a range of diets close to the chosen diet.

9.4.2 Results

The same data set was used to evaluate each approach and to simulate the effect of allowing the customer to choose "favourite foods" 4 foods were randomly chosen to be included in the diet.

The genetic algorithm procedure employed in Sect. 9.3 was then used to determine "second" set of foods, such that these together with the pre-chosen foods would produce a diet that would satisfy the nutrient requirements as cheaply as possible. Because of the higher cost of the resultant diet (compared with those obtained in Sect. 9.2), only those diets with a cost less than (1.5 × optimal cost) were recorded. On average, 48 alternative diets were generated using this approach.

Table 9.3 displays a typical set of results; it must be noted, however, that the initial customer selection could lead to an unsolvable problem; for example, the total fat content from the chosen foods may be greater than the maximum allowed quantity.

These results show that this approach, using genetic algorithms but allowing the customer to choose a number of foods, produces more varied but more expensive

Table 9.3 Number of alternative diets Ga formulation

Specified problem	Optimal solution (0 cost)	Number of solutions with in 2 × 0 cost
E	222	45
F	212	37
G	184	24
H	195	49

diets than those produced in Sect. 9.2. Here, a diet will contain 4 preferred foods and 5 others, whereas in Sect. 9.2 a solution contained up to 5 foods (not necessarily favoured by the customer). A sample solution and the associated diet are given in Table 9.3.

9.5 Nonlinear Costs Can Be Employed to Incorporate Multiple Objectives

The problem

$$\begin{aligned} &Ax \geq b && \text{nutrient and cost constraints}\\ &\text{Cost} = cx \\ &\text{Taste} = tx \\ &\text{Maximise}(\text{Taste}/\text{Cost}) \end{aligned}$$

Here, the genetic algorithm string has the form

$$x = [1, 0, 0, 0, 1, 2, 0, 0, 1, \ldots, 0]$$

indicating that this diet contains

1 unit of foods 1 and 5
2 units of foods 6 and so on

Evaluation of solution given in x:

$$\begin{array}{llll l} \text{If} & Ax \geq b & \text{then} & \text{Taste}/\text{Cost} = tx/cx & \text{a valid diet} \\ \text{otherwise} & Ax < b & \text{then} & \text{Taste}/\text{Cost} = 0 & \text{an invalid diet} \end{array}$$

Then proceed by crossover and mutation to generate a new population with the aim of maximising the objective function.

9.6 Diet Problem Solution Using a Fuzzy Approach

The objective in this approach is still the selection of a set of quantities of foods, $\boldsymbol{x}$, which satisfy the dietary restrictions

$$\boldsymbol{Ax} \geq \boldsymbol{b},$$

and at the same time minimise the value of a cost function where

$$\boldsymbol{cost} = \boldsymbol{c}\,\boldsymbol{x}.$$

However, here the solution makes use of standard sized portions (for example, 200 g portion of minced beef) and starts by generating a set of "genetic algorithm strings" each representing a possible solution. Each string contains a randomly chosen set of food portions to be included within the diet, thus $x_i = 0$, or 1, or 2, or…. representing 0, or 1, or 2 portions of food x_i.

This procedure can be implemented on either the basic model (Sect. 9.2) or the patient choice model (Sect. 9.3).

The nutrient content, for each nutrient, V_i, and cost C of each diet is then calculated

for all i

$$V_i - \sum_j A_{ij} x_j$$

and

$$C = \sum_j c_{ij} x_j$$

Nutritional value and cost have been identified as linguistic variables.

The nutritional constraints and desirable cost are now represented by simple trapezoidal membership functions, Fig. 9.3. Here, for example, if the fat content of the diet is between a and c units, then this "diet" would be given a value of 1 (on target) with respect to this constraint.

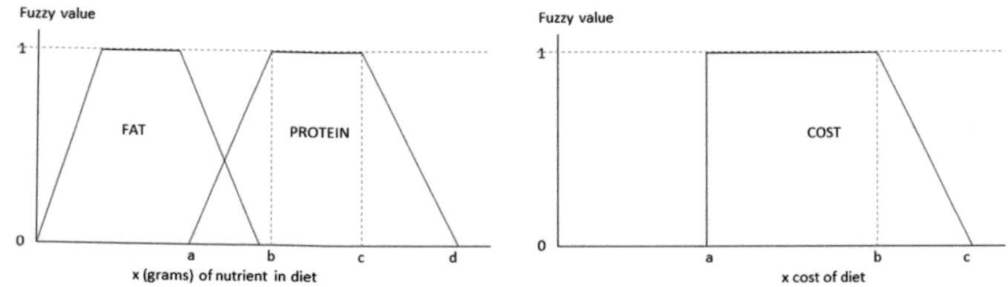

Fig. 9.3 Fuzzy membership for fat and protein and costs

The fat and protein constraints were obtained from "McCanse and Widdowson's The Composition of Food's", and the cost target from the optimal solution to the linear programming problem.

Here, "point a" is at the cost of the minimum cost diet, C_{OPT}, "point b" at $k_1 C_{\mathrm{OPT}}$ and 'point c' at $k_2 C_{\mathrm{OPT}}$.

Membership of cost, similar statements can be made for protein and fat, is given by

$$\begin{array}{ll} \mu(x) = 0 & x < C_{\mathrm{OPT}} \\ \mu(x) = 1 & C_{\mathrm{OPT}} \leq x \leq k_1 C_{\mathrm{OPT}} \\ \mu(x) = 1 - \frac{(x - k_1 C_{\mathrm{OPT}})}{(k_2 C_{\mathrm{OPT}} - k_1 C_{\mathrm{OPT}})} & k_1 C_{\mathrm{OPT}} \leq x \leq k_2 C_{\mathrm{OPT}} \\ \mu(x) = 0 & x \geq k_2 C_{\mathrm{OPT}} \end{array}$$

The value of each diet is then determined by combining the fuzzy membership values for V_i and C using the fuzzy operation of intersection.

$$\mathrm{N} := \bigcap_{j=1,J} V_j : \mu_N(x) = \inf_{j=1,J} \mu_{V_j}(x) \quad \text{for all} \quad x \in X$$

The value N now represents the fitness of each string within the current population of the genetic algorithm. This value is being used as the fitness function within the genetic algorithm, and new strings are being generated using crossover and mutation. After a number of iterations, many alternative good diets were produced.

9.6.1 *Example*

A problem containing 200 food types and 6 constraints was solved using 40 genetic algorithm strings. The initial string population was chosen so that on average 5% of the food stock was included within each string as a single portion (starting the process with a set of infeasible diets).

The genetic algorithm procedure outlined in 9.3 was then employed until 10 efficient diets had been obtained, where an efficient diet would have a fuzzy value of 1.

9.6.2 *Conclusion*

It can be seen, from the results given in Sect. 9.3, that the use of genetic algorithms does allow the production of a range of nearly optimal diets. Thus, this approach can be considered to provide a viable solution to the problem of constructing a set of alternative diets given a single set of constraints.

The successful results are due to the implicit parallelism within the genetic algorithm approach that allows an efficient search through the feasible solution space. This approach has the effect of investigating many solution paths simultaneously, thus demonstrating its ability to produce a range of acceptable near-optimal solutions and hence incorporating taste into diet problems.

The use of fuzzy logic has been shown to produce an equally good variety of solutions without the need for any integer programming packages.

References

1. Dantzig GB (1990) The diet problem: interfaces. Pract Math Program 20(4):43–47, Jul–Aug
2. Fletcher LR (1994) LP techniques for the construction of palatable human diets. J Oper Res Soc 68:489–496
3. Smith VE (1964) Linear programming models for determination of palatable diet. J Farm Econ 41(2) 272–283
4. Stigler G (1945) Cost of subsistence. J Farm Econ 25:303–314
5. Solzhenitsyn A (1962) One day in the life of Ivan Denisovich

Chapter 10
Fuzzy Scheduling Applied to Small Manufacturing Firms

Val Lowndes

10.1 Scheduling and Small Manufacturing Firms

A scheduling problem can be considered to be an exercise in finding an appropriate timetable for the processing of jobs, by machines, such that some performance measure achieves its optimal value. Within this definition, it can be seen that there are two aspects to be considered concurrently, the satisfaction of constraints (e.g. availability of resources) and the optimisation of objectives (e.g. flow-times).

In general, such problems are known to be NP hard and probably as a consequence of this, scheduling has been an active area of research for many years.

Pinedo lists a number of important requirements of real manufacturing that are not normally met by Operational Research-based models. An example of this is the existence of multiple objectives, i.e. there is not a single objective but multiple objectives to be optimised.

Illustrative example, in a job shop, with random job arrivals, where all jobs are processed on a single machine, the scheduler may need to consider the following goals:

> Satisfy all due dates, however, certain jobs are for particularly important customers and it is a major priority to ensure that these jobs are completed on time.

These observations are particularly relevant when considering small manufacturing firms. These firms experience the same problems as larger firms (or even worse problems) without the same resources to solve the problem (i.e. create an effective schedule) Berry [1], typically within a small firm only the owner-manager is allowed to set up (and change) a production schedule.

V. Lowndes (✉)
University of Derby, Kedleston Road, Derby DE22 1GB, UK
e-mail: V.P.Lowndes@derby.ac.uk

S. Berry et al. (eds.), *Guide to Computational Modelling for Decision Processes*,
Simulation Foundations, Methods and Applications,
DOI 10.1007/978-3-319-55417-4_10

10.2 Small Manufacturing Firms

The definition employed here has been synthesised from Berry [1] which established a model for a small manufacturing firm as follows:

- The manager is dominant in a small firm imposing the planning and control system onto the firm,
- Information about the state of the production system and demand is low in small firms,
- Small firms do not forecast (but respond to demand),
- There are only a few product types in a small firm,
- There are only a few production stages in a small firm, typically 3 with one dominant stage and
- Effective planning can be achieved by scheduling the dominant stage in the production system.

These factors indicate that a small manufacturing firm needs a mechanism by which it can improve its performance through more effective scheduling. The next section shows how fuzzy logic can be used to schedule production in a typical small manufacturing firm.

10.3 Fuzzy Modelling

10.3.1 Fuzzification

The first stage in producing a model is to identify those linguistic variables to be included. It was decided that *due date* and *customer priority* were the most significant factors, with *processing time* being of lesser importance for the small firm model presented above (Sect. 10.3.1).

- Due date Z
- Customer Priority X
- Processing Time

10.3.2 Rule Evaluation

For example:

$$\begin{aligned}
&\text{IF customer priority is Bad AND due} - \text{date is Close} \quad \text{THEN} \quad \text{Reject.}\\
&\text{IF customer priority is Low AND due} - \text{date is Close} \quad \text{THEN} \quad \text{Sequence quite high.}\\
&\text{IF customer priority is High AND due} - \text{date is Distant THEN} \quad \text{Sequence quite low.}
\end{aligned}$$

10.3.3 Rule Matrix (R)

If more than one job has the same priority at the head of the sequence, then a job with 'shortest' processing time will be selected for processing (Table 10.1).

10.3.4 Sequencing Priority (P)

The general model can be described as a composition or relational product (Table 10.2).

Suppose $T = S \circ R$, where

$$\mathrm{R} \in \mathrm{F}(\mathrm{CLOSE} \times \mathrm{CUST} - \mathrm{PRI}), \mathrm{S} \in \mathrm{F}(\mathrm{CUST} - \mathrm{PRI} \times \mathrm{DISTANT}).$$
$$\forall(\mathrm{CLOSE}, \mathrm{DISTANT}) \in \mathrm{CLOSE} \times \mathrm{DISTANT}$$

$$\mu_T(\mathrm{CLOSE}, \mathrm{DISTANT}) = \sup_{z \in Z} \min\{\mu_R(\mathrm{CLOSE} \times \mathrm{CUST} - \mathrm{PRI}), \mu_S(\mathrm{CUST} - \mathrm{PRI} \times \mathrm{DISTANT})\} \tag{10.1}$$

Union

$$X = A \cup B \Leftrightarrow \forall x \in U$$
$$[\mu_X(x) = \mu_A(x) \vee \mu_B(x)] = \forall x \in U[\mu_X(x) = \max\{\mu_A(x), \mu_B(x)] \tag{10.2}$$

Table 10.1 Summary of sequencing priorities

Due date customer priority	Close	Distant
B bad	Reject	Reject
L low	Sequence quite high	Sequence very low
M medium	Sequence high	Sequence low
H high	Sequence very high	Sequence quite low
VI very important	Sequence extremely high	Sequence medium

Table 10.2 Ordering of sequence priorities

Sequence	
Extremely high	EH
Very high	VH
High	H
Quite high	QH
Medium	M
Quite low	QL
Low	L
Very low	VL
Reject	R

The fuzzy relation SP representing the sequencing priorities, is derived from an application of Eq. (10.1)

$$\mu_{SP}(\text{CLOSE}, \text{DISTANT}) = \sup \min\{\mu_R(\text{CLOSE} \times \text{CUST} - \text{PRI}), \mu_S(\text{CUST} - \text{PRI} \times \text{DISTANT})\} \quad (10.3)$$

10.4 Application

An example, based around a typical small manufacturing firm, will illustrate how the rule base enables a job to improve its sequencing priority as the due date gets closer. Note, however, that a job for a Bad customer will be rejected and not included in the sequencing schedule. The following example will illustrate the mechanics of the fuzzy algorithm.

There are six jobs waiting to be processed, one of which is for a customer considered to be of 'medium' importance and two for 'very important' customers. The example has been deliberately chosen to create problems for the scheduler in the light of conflicting priorities, i.e. of fulfilling all promised due dates whilst ensuring the satisfaction of the most significant customers.

The due dates range from 0 days for Job 1 (medium) to 28 days for Job 4 (very important).

The fuzzy values for 'customer priority', 'close' and 'distant' have been derived according to the definitions given in Appendix A4.

The sequencing priority is then determined by applying Eq. (10.3) in the form:

$$\mu_{SP}(c, d) = \min\{\mu_c, \mu_{cp}\} \vee \min\{\mu_{cp}, \mu_d\},$$

according to the rule matrix in Table 10.1.

Step 1 Consider Job 1:
Comparing the fuzzy value of 'customer priority' with 'close' and 'distant'—

$$\min\{\mu_c, \mu_{cp}) \vee \min\{\mu_{cp}, \mu_d\}$$

μ_{cp} = 0.5 (customer priority is medium)
μ_c = 1.0 (membership of 'close')
μ_d = 0.0 (membership of 'distant')

(min{1.0, 0.5} = 0.5) ∨ (min{0.5, 0.0} = 0.0)
max {0.5, 0.0} = 0.5 **'close'** (Application of equation 10.2)

Thus: Medium and close => Sequence high; μ_{SP} (according to Tables 10.1, 10.3).

Table 10.3 Test example—six jobs waiting to be processed

Job	1	2	3	4	5	6
Due date	0	10	6	28	26	14
Process time	5	1	8	6	2	4
Customer priority	M	L	V.I	V.I	H	L
Fuzzy customer priority	0.5	0.2	1.0	1.0	0.75	0.2
Fuzzy due date close	1.0	0.0	0.4	0.0	0.0	0.0
Fuzzy due date distance	0.0	0.21	0.0	1.0	1.0	0.5
Max–min	Close	Dist	Close	Dist	Dist	Dist
Sequence	H	VL	EH	M	QL	VL

The same procedure is followed for the remaining jobs resulting in the sequencing priority:

$$<3, 1, 4, 5, 2, 6>$$

Job 3 (the head of the sequence) is processed first with a duration of 8 days. *Note: This schedule can be compared with the "earliest due date" schedule*:

$$<1, 3, 2, 6, 5, 4>$$

Step 2 Implemented after Job 3 has been completed
This will repeat all the tasks in Step 1, for the remaining five jobs (Table 10.4).
The (now) current sequencing priority is now given by: $<1, 2, 6, 4, 5>$ thus, Job 1 is processed next—duration 5 days.

Step 3 Implemented after Job 1 has been completed (Table 10.5).
The sequencing priority for the current jobs is now: $<2, 6, 4, 5>$
This process continued until all jobs in the list of waiting jobs have been scheduled. The dynamic nature of this process can be demonstrated by considering the sequence priority each time the machine becomes available:

Table 10.4 Test example—five jobs in queue

Job	1	2	4	5	6
Due date	−8	2	20	18	6
Process time	5	1	6	2	4
Customer priority	M	L	V. I	H	L
Fuzzy customer priority	0.5	0.2	1.0	0.75	0.2
Fuzzy due date close	1.0	0.8	0.0	0.0	0.4
Fuzzy due date distant	0.0	0.0	0.93	0.79	0.0
Max–min	Close	Close	Distant	Distant	Close
Sequence	H	QH	M	QL	QL

Table 10.5 Test example—four jobs in queue

Job	2	4	5	6
Due date	−3	15	13	1
Process time	1	6	2	4
Customer priority	Low	V. Imp	High	Low
Fuzzy customer priority	0.2	1.0	0.75	0.2
Fuzzy due date close	1.0	0.0	0.0	0.9
Fuzzy due date distant	0.0	0.57	0.43	0.0
Max–min	Close	Distant	Distant	Close
Sequence	Quite high	Medium	Quite low	Quite high

Table 10.6 Comparing results fuzzy and EDD schedules

Schedule	Delay very important	Total all jobs delay
Fuzzy	2	23
EDD	7	20

Step 1 $<3, 1, 4, 5, 2, 6>$ Job 3 processed
Step 2 $<x, 1, 2, 6, 4, 5>$ Job 1 processed
Step 3 $<x, x, 2, 6, 4, 5>$ Job 2 processed
Step 4 $<x, x, x, 6, 4, 5>$ Job 6 processed
Step 5 $<x, x, x, 6, 4, 5>$ Job 5 processed

giving the full implemented job sequence $<3, 1, 2, 6, 5, 4>$
for comparison, the earliest due date (EDD) schedule would have been given by

$$<1, 3, 2, 6, 5, 4>$$

Notice that the status of Job 4 is continually updated, this demonstrates the dynamic nature of the scheduling system. The fuzzy schedule can now be compared with the alternative EDD schedule where it can be seen that the Fuzzy schedule minimises the delay to very important customers (as required in small manufacturing firms) but not the total delay (Table 10.6). This example only considers a static case (no new orders) to illustrate the methodology; however, in the full dynamic simulation when new orders are received they are immediately added to the list of waiting jobs and assigned places in the sequence of jobs at that time when the next job to be processed is chosen.

10.5 Computational Efficiency

An exhaustive search methodology in this example would investigate *6*! possible production schedules; here, *35* calculations to determine the fuzzy values for each job are carried out. In an n job problem, the comparative values are $n!$ and $n(n + 1)$.

Table 10.7 Simulation results summary

Planning system	n_T	T_{max}	VTJ	VIJ
Fuzzy logic	4	10.1	5	1.6
EDD	5.5	22	6	4.7

10.6 Dynamic Scheduling

The performance of the fuzzy scheduling approach in a small manufacturing firm was simulated (15 replications) to evaluate its suitability in a dynamic environment where jobs arrive randomly. The demand level was chosen to be close to the production capacity so that this methodology could be tested on a commonly occurring problem in SMF's, when growth in demand occurs and the planning system comes under more stress.

Table 10.7 contains the summarised results from all the simulations as percentages of jobs completed.

This table displays the information:

- Number of very important late jobs [n_T]
- Maximum lateness of a very important job [T_{max}]
- Number of very late jobs (>28 days) [VTJ]
- Number of very important jobs late (>7 days) [VIJ]

The fuzzy scheduling system is performing better with all these criteria, in particular acting to satisfy the demands of the very important customers.

10.7 Conclusions

Fuzzy set theory allows the complexity of real-life issues to be included within the confines and rigours of the mathematical model. In this paper, a theoretical model has been presented which demonstrates how fuzzy decision-making can support the dynamic scheduling process, enabling the conflicting priorities of multi-objectives to be managed effectively in polynomial time providing a mechanism for efficient planning, thus supporting the role of the manager, in a small manufacturing firm by acting as an automated scheduler.

Reference

1. Berry S (1999) Production planning in small manufacturing firms small manufacturing firms

Chapter 11
The Design and Optimisation of Surround Sound Decoders Using Heuristic Methods

Bruce Wiggins, Stuart Berry and Val Lowndes

11.1 Introduction

Since the introduction of the DVD (both video and audio) surround sound has become an affordable luxury, surround sound mixing and reproduction equipment is also in widespread use. The standard speaker configuration, as specified by the ITU, is a five speaker layout, as shown in Fig. 11.1 However, this is likely to be expanded upon in the near future, and other, larger, venues are likely to have more speakers in order to adequately cover a larger listening area.

Due to the likelihood of ever changing reproduction layouts, a more portable approach should be used in the creation of multichannel material, and such a system has been available since the 1960s [1].

Ambisonic systems are based on a spherical decomposition of the sound field to a set order (typically 1st or 2nd order Malham [2], and Leese [3]). The main benefit of the Ambisonic system is that it is a hierarchical system, i.e. once the sound field is encoded in this way (into four channels for 1st order and 9 channels for 2nd order), it is the decoder that decides how the sound field is reconstructed using the Ambisonic decoding equations [4]. The Ambisonic system was largely researched and developed by Gerzon, and in 1992, papers were published proposing a method for the optimisation of Ambisonic decoders for irregular speaker arrays [5]. This was necessary because the original decoding equations were difficult to solve for irregular speaker arrays in the conventional way (inverting a matrix of spherical harmonic coefficients).

B. Wiggins · S. Berry (✉)
College of Engineering and Technology, University of Derby,
Kedleston Road, DE22 1GB Derby, UK
e-mail: s.berry@derby.ac.uk

V. Lowndes
University of Derby, Kedleston Road, DE22 1GB Derby, UK
e-mail: V.P.Lowndes@derby.ac.uk

S. Berry et al. (eds.), *Guide to Computational Modelling for Decision Processes*,
Simulation Foundations, Methods and Applications,
DOI 10.1007/978-3-319-55417-4_11

11.2 Irregular Ambisonic Decoding

In order to quantify decoder designs, Gerzon chose two main criteria for designing and evaluating multi-speaker surround sound systems in terms of their localisation performance. The two criteria represent the energy and velocity vector components of the sound field [6]. The vector lengths represent a measure of the 'quality' of localisation, with the vector angle representing the direction that the sound is perceived to originate from. A vector length of one indicates a good localisation effect. These are evaluated using the equations shown in Eq. 11.1.

For regular speaker arrays, designing an optimised Ambisonics decoder is simply a case of using one virtual microphone response for low frequencies and a slightly different virtual microphone response for the mid- and high frequencies by the use of shelving filters [7] as shown in Figs. 11.2 and 11.3.

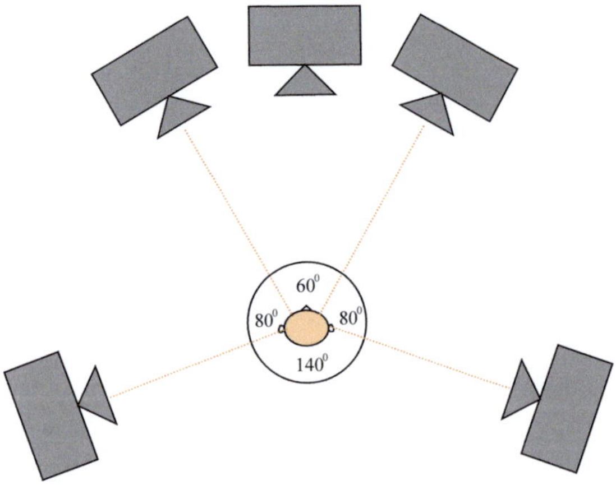

Fig. 11.1 Recommended loudspeaker layout, as specified by the ITU

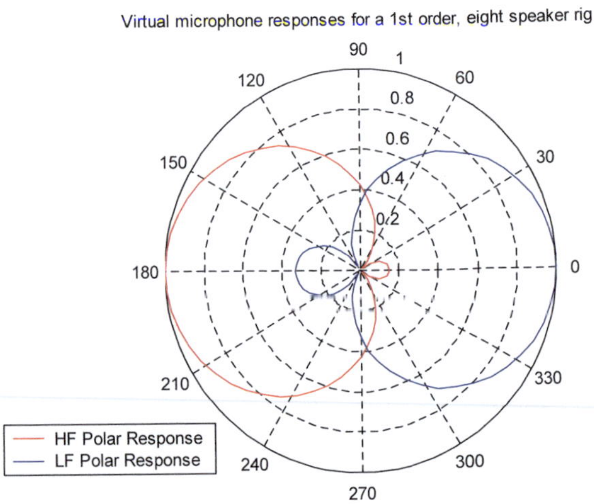

Fig. 11.2 Virtual microphone polar diagrams that satisfy Eq. 11.1 for a 1st order, eight speaker rig

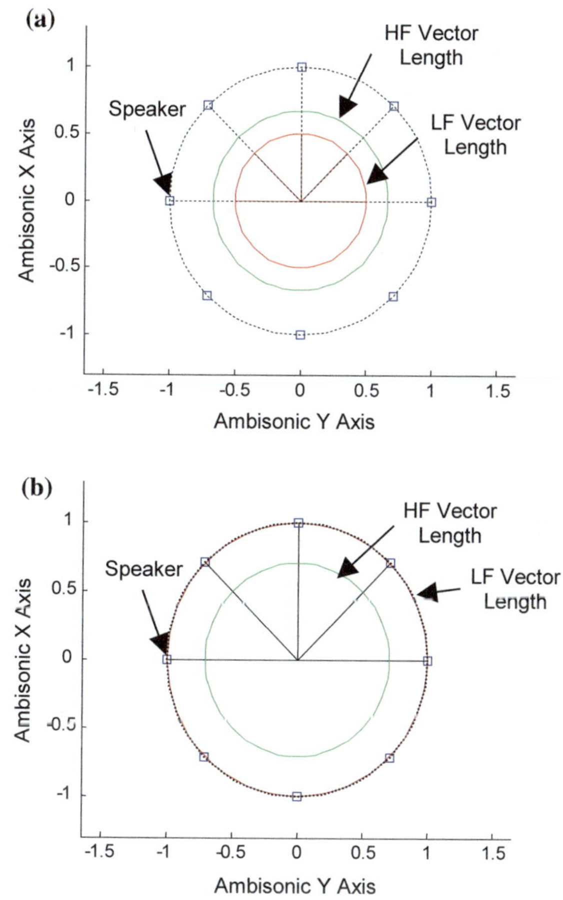

Fig. 11.3 **a** Velocity and energy localisation vectors. Magnitude plotted over 360° and angle plotted at five discrete values. *Inner circle* represents energy vector, and *outer circle* represents velocity vector. Using virtual cardioids. **b** Velocity and energy localisation vectors. Magnitude plotted over 360° and angle plotted at five discrete values. *Inner circle* represents energy vector, and *outer circle* represents velocity vector. Using virtual patterns from Fig. 11.2

As long as the virtual microphone patterns are the same for each speaker, the estimated localisation angle is always the same as the encoded source angle, with only the localisation quality (length of the vector) affected by changing the polar patterns.

Velocity and energy vector equations

$$
\begin{aligned}
P &= \sum_{i=1}^{n} g_i & E &= \sum_{i=1}^{n} g_i^2 \\
Vx &= \sum_{i=0}^{n} g_i \cos(\theta_i)/P & Ex &= \sum_{i=0}^{n} g_i^2 \cos(\theta_i)/E \\
Vy &= \sum_{i=0}^{n} g_i \sin(\theta_i)/P & Ey &= \sum_{i=0}^{n} g_i^2 \sin(\theta_i)/E
\end{aligned}
\tag{11.1}
$$

where

g_i represents the gain of a speaker (assumed real for simplicity)
n is the number of speakers
θ_i is the angular position of the ith speaker

When irregular speaker arrays are used, the vector magnitudes, reproduction angles and overall volume of the decoded sound field all require optimisation simultaneously, otherwise excessive decoding artefacts will be observed.

For example, consider the non-uniform speaker configuration of the ITU five speaker layout. If all speakers are fed by virtual microphones oriented in the direction of the loudspeakers, and with the same, cardioid, polar pattern, then a sound encoded to the front of a listener will be louder than a sound emanating from the rear. Also, the perceived direction of the reproduced sound will be distorted, as shown in Fig. 11.4.

These decoding artefacts are not a problem when the audio is produced for a fixed set-up (for example, amplitude panned 5.1) since the material is mixed to sound correct on the chosen speaker layout. This is in contrast to a truly hierarchical system in which, ideally, it would be possible to reproduce the audio material accurately regardless of the configuration of the output system. Such a hierarchical system requires corrections to be made at the decoding stage where the speaker layout is known.

Due to the added complexity of the speaker arrays response to an Ambisonic-type decode (see the reproduction angle discrepancies and vector lengths in Fig. 11.4), Gerzon and Barton [5] proposed that two separate decoders be used, one for low frequency ($<\sim 700$ Hz) and another for high frequencies ($>\sim 700$ Hz).

This can be achieved using a simple crossover network (preferably using linear phase, FIR, filters) feeding low and high passed versions of the Ambisonic,

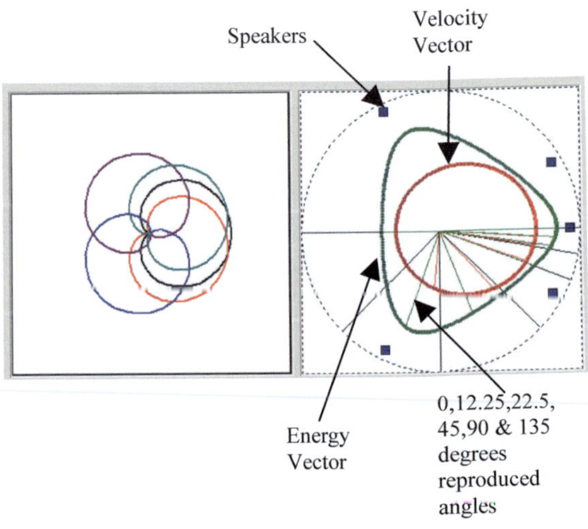

Fig. 11.4 Energy and velocity vector response of an ITU 5 speaker system, using virtual cardioids

B-format, signals to the two decoders where totally separate decoding can be achieved, not just a microphone polar pattern adjustment as in a regular speaker array decode.

11.3 Decoder System

1st order Ambisonics is based on four different signals, as shown in Fig. 11.5, an omnidirectional pressure signal (W), a front–back figure of eight (X), a left–right figure of eight (Y) and an up–down figure of eight (Z).

The 5 speaker system shown in Fig. 11.1 is a horizontal only system, and hence, only three of the four available B-format signals are required at the input of the decoder (W, X and Y). Also, the speaker array in Fig. 11.1 is left/right symmetric such that the decoder coefficients are arranged to work in mid- and side pairs (i.e. sum and difference). The Ambisonic encoding equations are given in Eq. 11.2.

The incorporation of a 'frontal dominance' control in the decoding system can also be considered; the definition for this is given in Eq. 11.3. Although this form of the frontal dominance equation exhibits a nonlinear response to the dominance parameter, it is used in this investigation to keep compatibility with Gerzon's previous paper on this subject [5].

Ambisonic encoding coefficients.

$$\begin{aligned} W &= 1/\sqrt{2} \\ X &= \cos(\theta) \\ Y &= \sin(\theta) \end{aligned} \tag{11.2}$$

where θ is the encoded angle, taken anticlockwise from straight ahead.

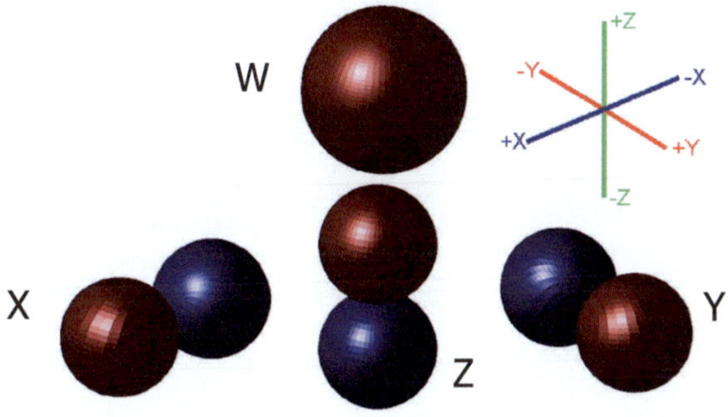

Fig. 11.5 Polar patterns of the four B-format signals used in 1st order Ambisonics. *Red* shows an in-phase response, and *blue* shows an out-of-phase response

Forward dominance equation.

$$
\begin{aligned}
W' &= 0.5\left(\lambda+\lambda^{-1}\right)W + 8^{-\frac{1}{2}}\left(\lambda-\lambda^{-1}\right)X \\
X' &= 0.5\left(\lambda+\lambda^{-1}\right)X + 2^{-\frac{1}{2}}\left(\lambda-\lambda^{-1}\right)W \\
Y' &= Y
\end{aligned}
\tag{11.3}
$$

where λ is the forward dominance parameter ($2 > \lambda > 1$ for front and $1 > \lambda > 0$ for rear dominance).

The frontal dominance terms are then substituted into the decoding equations to give a numerical value for each speaker output. Equation 11.4 shows the substitutions used for a 5 channel system.

Decoding equations for each of the five speakers.

$$
\begin{aligned}
C_F &= (kW_C \times W') + (kX_C \times X') \\
L_F &= (kW_F \times W') + (kX_F \times X') + (kY_F \times Y') \\
R_F &= (kW_F \times W') + (kX_F \times X') - (kY_F \times Y') \\
L_B &= (kW_B \times W') + (kX_B \times X') + (kY_B \times Y') \\
R_B &= (kW_B \times W') + (kX_B \times X') - (kY_B \times Y')
\end{aligned}
\tag{11.4}
$$

where k denotes a decoding coefficient.

The λ and 'k' values are chosen so as to optimise the decoded output, with λ having possible values between 0 and 2, and 'k' values having a nominal range between 0 and 1.

Equations used to measure the performance of a decoder design.

$$
\begin{aligned}
Vx &= \sum_{i=1}^{N} g_i \times \cos(SPos_i)/P_V \\
Vy &= \sum_{i=1}^{N} g_i \times \sin(SPos_i)/P_V \\
Ex &= \sum_{i=1}^{N} g_i^2 \times \cos(SPos_i)/P_E \\
Ey &= \sum_{i=1}^{N} g_i^2 \times \sin(SPos_i)/P_E \\
R_E &= \sqrt{E_x^2 + E_y^2} \quad \theta_E = \tan^{-1}\left(E_y/E_x\right) \\
R_V &= \sqrt{V_x^2 + V_y^2} \quad \theta_V = \tan^{-1}\left(V_y/V_x\right) \\
P_V &= \sum_{i=1}^{n} g_i \qquad P_E = \sum_{i=1}^{n} g_i^2
\end{aligned}
\tag{11.5}
$$

where

g_i = Gain of the ith speaker

$SPos_i$ = Angular position of the Ith speaker.

In order to optimise the solution, the equations used to measure the performance of the design are given in Eq. 11.5. The conditions that need to be met for the decoder to be deemed 'Ambisonic' are:

- Radius of the localisation vector lengths (R_V and R_E) should be as close to 1 as possible for all values of θ.
- $\theta = \theta_V = \theta_E$ for all values of θ (where θ is the encoded source angle).
- $P_V = P_E$ and must be constant for all values of θ.

In practice, the conditions defined in Eq. 11.5 are difficult to solve because the best result must be found over the whole 360° of encoded source positions. It is known that these equations are laborious to solve for five speaker systems Gerzon [5]. Furthermore, an increase in the number of speakers will result in a disproportionate increase in the complexity of the decoding optimisation problem. Also, more than one valid solution for each decoder design exists at low and high frequencies. This means that a group of solutions need to be found, followed by subjective listening tests in order to find the best performing coefficient set.

Due to the laborious and time-consuming nature of decoder optimisation, a method is needed that can automate this process so that the onus in designing Ambisonic decoders is shifted from the calculating of the decoding coefficients to listening to the different decoders, so the optimal system can be decided upon.

11.4 The Heuristic Search Methods

The word heuristic can be used to mean 'using trial and error', and mathematical searches using this technique can lead to the solutions of complex numerical problems. Heuristic search methods work on the simple principle that any result that is found (using, say, random starting values) can be tested on its 'correctness', with this then being used to decide on values to try next following some rule depending on the type of search method used.

As a result of the fact that each parameter has a value from a well-defined range, 0–1 or 0–2, a search method seemed to be a very viable solution. However, if we wish to determine the settings to two decimal places, there are 2×10^{18} possible solutions (given that there are 9 search parameters) and an exhaustive search is not feasible. The first avenue of research taken was that of a genetic algorithm approach. However, genetic algorithms are particularly well suited to problems that have large search spaces (i.e. large parameter ranges), and this is not the case here. Also, a genetic algorithm approach is very good at getting reasonably close to an accurate solution but will then need optimising further.

This was seen to be overly complicated for our needs, and a method based on the tabu search (memory-based search) was developed, which is a method that can achieve accurate results and is a viable option as long as the parameters that are to be altered have defined limits.

This, slightly adapted, form of a tabu search works by having the decoder coefficients initialised at random values (or values of a previous decoder, if these values are to be optimised further). Then, the tabu search program changes each of the tweakable values in turn, plus or minus the step size. The result that is deemed to be most correct is then kept, and the parameter changed is then restricted to only move in the successful direction for a set number of iterations (which, of course, will only happen if this parameter, again, is the best one to move). It must be noted that the random start position is of great importance, as it helps in the search for a wide range of solutions as, it the tabu search starts in exactly the same place each time, exactly the same results will be found (as there is no randomness in the search process itself, unlike a Genetic Algorithm). The most important part of the tabu search algorithm is the equations used to measure the fitness (or correctness) of the coefficients used as it is this one figure that will determine the course that the tabu search takes. As mentioned above, three parameters must be used in an equation that represents the overall fitness of the decoder coefficients presented. These are as follows:

- localisation measure (vector lengths, R_V and R_E).
- localisation angle (vector angles, θ_V and θ_E).
- volume (sound pressure gain, P_V and energy gain, P_E) of each encoded direction.

As each of these results must be as good a fit as possible for the whole 360° sound stage, the three parameters must be evaluated for a number of different encoded source positions. Gerzon evaluated these parameters at 14 points around the unit circle (7 around a semicircle assuming left/right symmetry), but as computers can calculate these results extremely quickly, it was decided that encoded sources at 4° intervals would be used (90 points around the unit circle). Due to the large number of results for each of the fitness values, an average was taken for each fitness parameter using a route mean square approach. If we take the example of the fitness of the vector lengths (localisation quality parameter), and if a mean average is taken, then a less than one vector length in one part of the circle could be compensated for by a greater than one vector length elsewhere. However, if we take a good fit to be zero and use a route mean square approach, then a non-perfect fit around the circle will always give a positive error value, meaning it is a true measure of the fitness. The equations used for each of the fitness parameters are shown in Eq. 11.6.

Equations of fitness used to evaluate the decoder coefficients.

$$
\begin{aligned}
VFit &= \sqrt{\sum_{i=0}^{n} \frac{(1-P_0/P_i)^2}{n}} && \text{where : } P_0 \text{ is the pressure at an encoded direction of } 0^\circ. \\
MFit &= \sqrt{\sum_{i=0}^{n} \frac{(1-R_i)^2}{n}} && n \text{ is the number of points taken around the unit circle.} \\
AFit &= \sqrt{\sum_{i=0}^{n} \frac{\left(\overset{Enc}{\underset{i}{\theta}} - \theta_i\right)^2}{n}} && \theta \text{ Enc is the encoded source angle and } \theta \text{ is the localisation angle.}
\end{aligned}
\tag{11.6}
$$

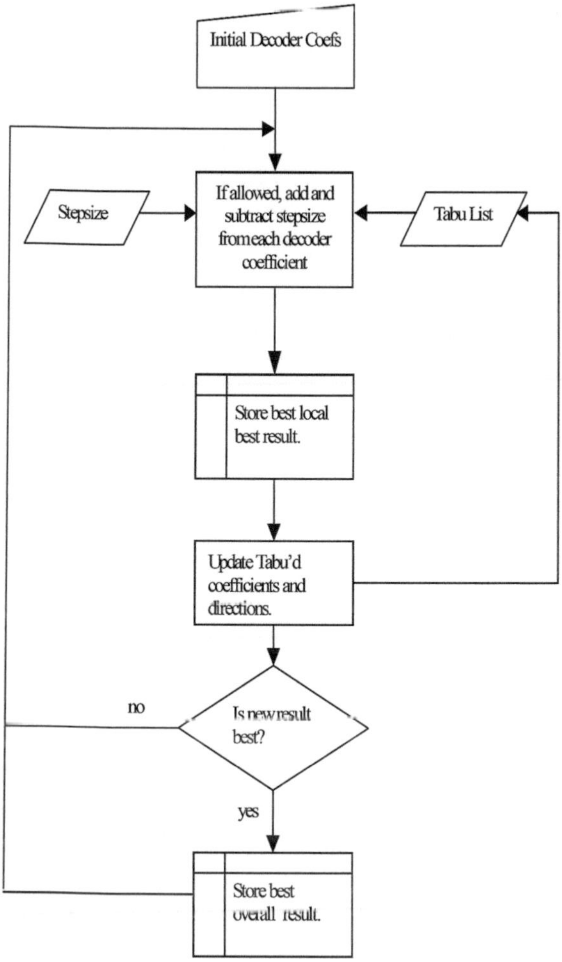

Fig. 11.6 Tabu search algorithm

Given the three measures of fitness in Eq. 11.6, the overall fitness for the high- and low-frequency versions of the decoder is actually calculated slightly differently. The low-frequency decoder can achieve a near perfect fit, but the best fit that the high-frequency decoder can expect to achieve is shown in Fig. 11.3. The best results were obtained from the tabu search algorithm if the overall fitness was weighted more towards the angle fitness (*AFit* from Eq. 11.6.) as shown in Eq. 11.7.

Fitness equations for low- and high-frequency models.

$$\begin{aligned} LFFitness &= AFit + MFit + VFit \\ HFFitness &= AFit + (MFit + VFit)/2 \end{aligned} \tag{11.7}$$

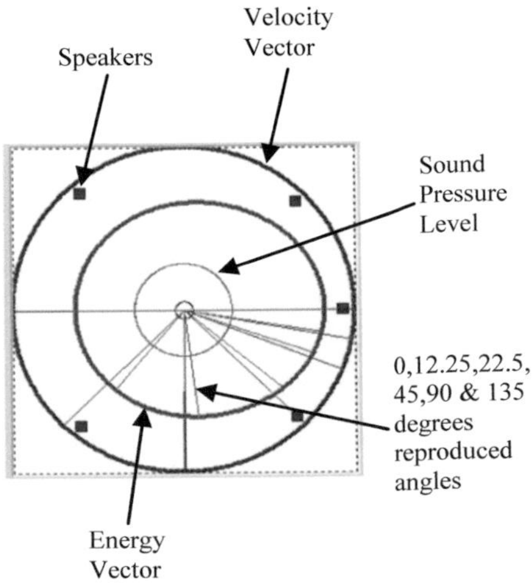

Fig. 11.7 Graphical plot of the Gerzon/Barton coefficients published in the Vienna paper. Encoded/decoded directions angles shown are 0°, 12.25°, 22.5°, 45°, 90°, 135° and 180°

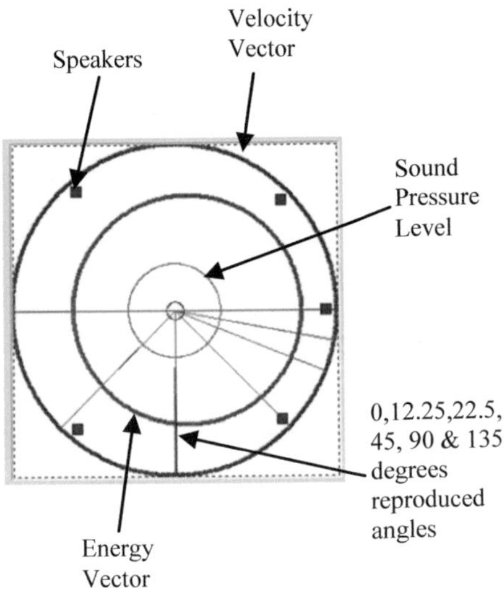

Fig. 11.8 Graphical plot of the coefficients generated using a tabu search algorithm. Encoded/decoded directions angles shown are 0°, 12.25°, 22.5°, 45°, 90°, 135° and 180°

The main benefit of the tabu search method is that all three of the conditions to be met can be optimised for simultaneously, which had not been accomplished in Gerzon's paper [5]. For example, if we take the speaker layout used in the Vienna paper, which is not the ITU standard but is very similar, then the coefficients derived by Gerzon and Barton [5] would give an energy and velocity vector response as shown in Fig. 11.7. Several observations can be made from this figure. There is a high/low localisation angle mismatch due the forward dominance being applied to the high-frequency decoders input *after* the localisation parameters were used to calculate the values of the coefficients. Or, if the frontal dominance is applied to both the high- and low-frequency decoders, a perceived volume mismatch occurs with the low-frequency decoder replaying sounds that are louder in

Fig. 11.9 Graphical representation of the decoder virtual microphone patterns obtained from the three optimum solutions indicated by the *squares* in Fig. 11.10

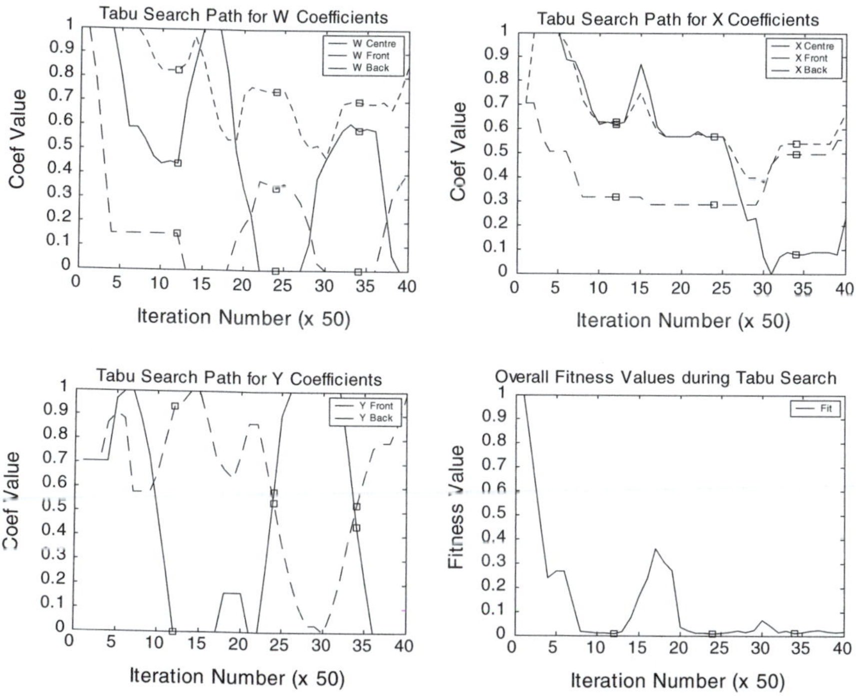

Fig. 11.10 A graph showing the transition of the eight coefficients in a typical low-frequency tabu search run (2000 iterations). The *square markers* indicate the three most accurate sets of decoder coefficients (low fitness)

the frontal hemisphere than in the rear. Also, even if these mismatches were not present, every set of results presented in the Vienna paper showed a distortion of the decoders reproduced angles. Figure 11.8 shows a set of coefficients calculated using the tabu search algorithm described in Fig. 11.6 and shows that if all three criteria are optimised simultaneously, a decoder can be designed that has no angle or volume mismatches and should reproduce a recording more faithfully than previously possible. Figures 11.9 and 11.10 show the results and fitness coefficients of a typical run of the Tabu search algorithm.

11.5 Conclusions

The tabu search algorithm has provided an efficient and effective methodology to optimise surround sound decoders. This methodology providing an improvement over the alternative approach [1], allowing for the Vienna equations [11.1] to be easily solved for virtually any arrangement of speakers and thus simplifying the design process for Ambisonic decoders. Although the software used to generate the results presented here concentrates on a typical five speaker, horizontal arrangement, the methodology is applicable to any configuration. This approach has the advantage of generating multiple sets of good solutions (alternative decoders) in a single execution of the tabu search program; the existing method generates a single solution, thus greatly increasing the number of decoders that can be realised and tested in a very short time.

Reference

1. Borwick J (1981) Could 'Surround sound' bounce back. Gramophone, 1125–1126.
2. Malham D, Second and third order ambisonics. http://www.york.ac.uk/inst/mustech/3d_audio/secondor.html
3. Leese M (2012) Ambisonic Surround Sound. http://members.tripod.com/martin_leese/Ambisonic/
4. Gerzon MA (1977) Multi-system ambisonic decoder, parts 1 & 2—Wireless World July & August 1977
5. Gerzon MA, Barton GJ (1992) Ambisonic decoders for HDTV—92nd AES convention, Vienna. Preprint 3345
6. Gerzon MA (1992) General methatheory of auditory localisation—92nd AES convention, Vienna. Preprint 3306
7. Farina A, Ugolotti E (1998) Software implementation of B-format encoding and decoding—104th AES convention, Amsterdam. Preprint 4691

Chapter 12
System Dynamics Case Studies

Chris Parkes, Stuart Berry and John Stubbs

12.1 Transport Planning and Transport Planning Paradoxes

When going green is not a green policy, this case study is used to demonstrate how a modelling can develop an understanding of a problem (planners continually make the same mistakes!).

Given that the current concern is the need for a "green" transport policy, where green is currently seen as a shift towards public/rail transport and away from a private/car based transport system, to be able to evaluate the "greenness" of this policy, it is informative to investigate the historic growth and subsequent decline of the dominant form of mass transportation of people in the UK. It can be suggested that the dominant transport system is continually changing in response to customer demand although it will be shown here that it can also be considered that customer demand is changing in response to transport changes. This case study investigates these claims by constructing models to describe and explain changes to the dominant mode of transport.

Over the period of 250 years, mass transport systems have continually developed with the dominant mode changing from canal to rail to road. The relationship between alternative modes of transport can be established and illustrated with influence diagrams, Coyle [1]; these diagrams showing that transport planning/developments have always encountered the same dilemma, that the newer form of transport continually expands at the cost of the older form of transport until

C. Parkes · S. Berry (✉)
College of Engineering and Technology, University of Derby,
Kedleston Road, Derby DE22 1GB, UK
e-mail: s.berry@derby.ac.uk

J. Stubbs
School of Sciences, University of Derby, Derby, UK

S. Berry et al. (eds.), *Guide to Computational Modelling for Decision Processes*,
Simulation Foundations, Methods and Applications,
DOI 10.1007/978-3-319-55417-4_12

it either reaches (effective) saturation or an alternative (better) transport system becomes available. However, while a transport system is experiencing its expansion phase, the net effect is to enable its customers to travel further increasing "urban sprawl" and hence acting against a move towards a green solution.

12.1.1 *Trains and Barges and Motor Vehicles*

The changes to the dominant modes of transport in the UK are given in Table 12.1 as

A caveat to this table follows from the development of urban rail systems (for example London Underground) which were still developing after 1920 with small extensions to the existing network during the 1920s and 1930s, and the Jubilee line in the 1970s.

UK Canal length in 1830 was 4000 miles, this length remaining almost constant until about 1950. As a result of the development of the national rail system, canals experienced a loss of 2/3 of their trade by 1850 but they retained profitability until the early twentieth century when the increasing availability of road transport caused the closure of many canals, most falling into disuse by 1940.

However, note that now increasing canal leisure usage has acted to enable the restoration of some canals, for example the restoration of the Rochdale canal through Sowerby Bridge, (T1).

Now, however, road improvements/developments to the road network are no longer seen to provide a feasible solution to the problem of transporting workers into the centre of large cities, and as a consequence, investments are being made into expanding the existing rail network and constructing new urban railways, for example the Crossrail project in London (T6); these developments being suggested to provide green solutions to the transport problem.

Section 20.1 contains a summary of the changing transport demand within London.

Table 12.2 shows the effect of road developments on rail usage, with the decline in rail usage occurring after the 1920s and the increasing investment in road transport occurring after the 1930s.

Table 12.1 Eras for dominant modes of transport

Existing major transport system		New replacement system
Rudimentary road system	To 1770	Canals replace roads
Golden age of canals	1770–1830	Rail replaces canals
Rail network growing	1830–1910	Roads replace rail
Road travel growing	1920 onwards	Urban railways and trams
Urban rail/tram systems developments	1980 onwards	

Table 12.2 Road and rail changes T2, T3

Road travel			Rail travel		
Road developments			Rail network developments and demand data		
Road construction development	Date	Biographical details	Year	Network length miles	Passenger journeys (millions)
North circular and east Lancashire	1930s		1840 1860	500 10,000	
First motorway	1958	"Preston By-pass"	1900	18,614	1100
M1	1959	"61.5 miles"	1923	20,289	1772
M62	1970s		1950	19,585	1010
M25	1986	started 1975 London orbital	1960	18,476	1037
M40	1991		1980	10,500	760
M60	2000	Manchester orbital	2014	10,000	1600

12.1.2 Models for New Transport System Developments

The models to be presented here indicate that the "new" mode of transport (currently suburban tram/rail systems) will enter a period of unconstrained (over) growth (Model 12.1a) causing its own problems, for example urban sprawl and increasing living costs, the exact opposite of the intended green policy, for example the urban sprawl resulting from the Metropolitan line extension in the 1920s T4.

Model 12.1a is a generic model illustrating the (over) development of the new transport system.

This network has a positive feedback loop indicating unconstrained growth in the new system giving the expected result

{System demand increased	leading to	System Growth
System growth	leading to	Reduced Travel Times
travel times reduced	leading to	Increased System demand
System Demand increased	leading to	Network Growth and so on}.

12.1.3 Model Validation: Changes in the Transport System, from Canals to Railways

As a first example of such developments, consider Models 12.1b and 12.2 describing the growth of railways and the parallel decline in canal usage, first a model for journeys between two points at the time when the a new transport system

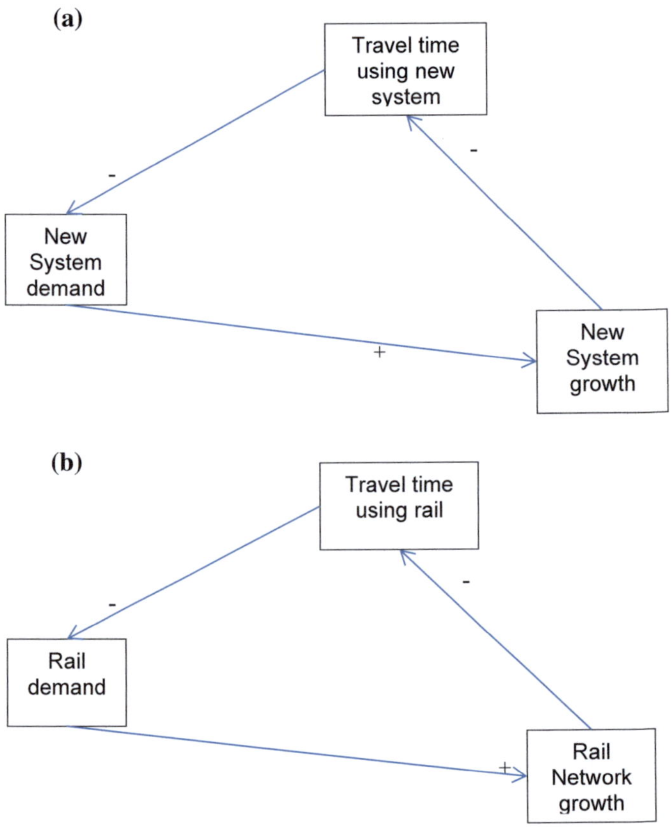

Model 12.1 **a** Modelling new system growth. **b** Modelling rail network growth

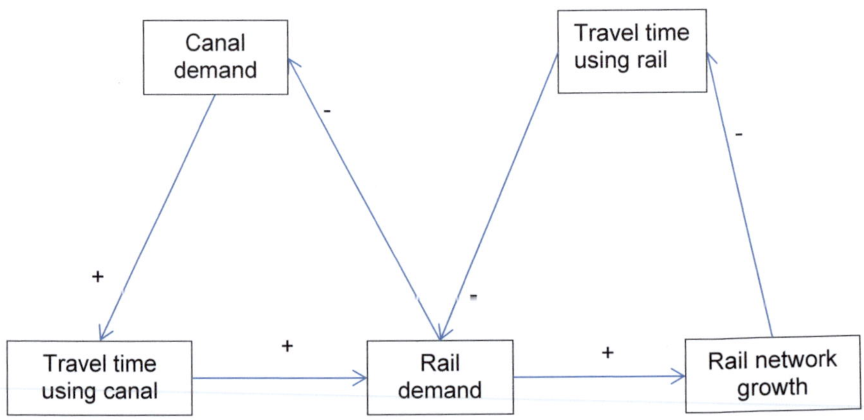

Model 12.2 Modelling the effect of rail growth on the canals

is expanding followed by a model demonstrating how this newer form of transport acts to cause a reduction in the demand for the alternative transport system.

This positive feedback loop indicating unconstrained growth in the new railways giving the expected result of unconstrained growth.

12.1.3.1 Model Validation

Consider the railway network growth, Railway Mania as occurred in the UK in the 1840s, the context of this model where:

{Rail demand increased	leading to	Rail network growth
network growth	leading to	Reduced Travel Times
travel times reduced	leading to	Increased Rail demand
Rail Demand increased	leading to	Rail Network Growth and so on}.

Now to consider the effect of this rail growth on the then existing (alternative) canal network, the next model shows how the development of rail travel caused the decline in commercial travel by canal and halted the further development of the canal system.

Here, there is a negative, equilibrium seeking, loop implying that alongside the growth in the rail network there will remain some commercial canal systems/usage,

{Increase in Rail Demand	leads to	Decline in Canal Demand
Decline in Canal Demand	leads to	Decline in Canal Travel Time
Decline in Canal Travel Time	leads to	Decline in Rail Demand
Decline in Rail Demand	leads to	Increase in Canal Demand
Increase in Canal Demand}	until a steady state has been achieved,	

Model validation: in the UK, canals were an important part of the transport system until the mid-twentieth century, T1.

12.1.4 Modelling the Effect of Changes to the Transport System, Railways and Roads

Now consider the more current road rail system developments, firstly the changes in road travel; here, Model 12.1a becomes Model 12.3.

This model contains a positive feedback loop implying continuing road developments, as occurred, major road developments with the motorway system being constructed after 1958. The first motorway was the Preston By-pass, opened in 1958. Currently the length of motorway is approximately 2300 miles, 1% of total road length but carrying over 20% of all road traffic.

Model 12.3 Modelling changes in road growth

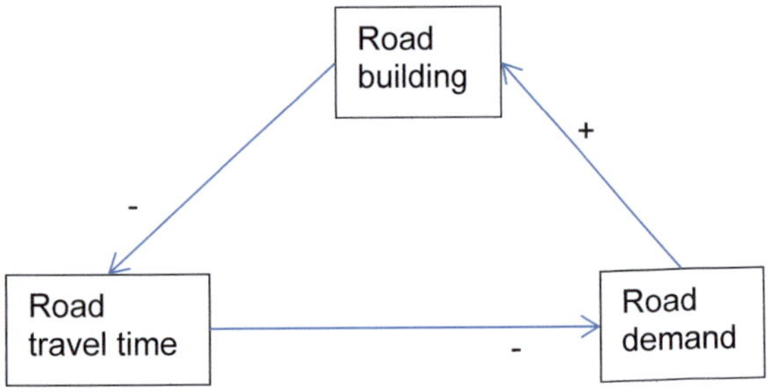

Model 12.4 Alternative mode choice model

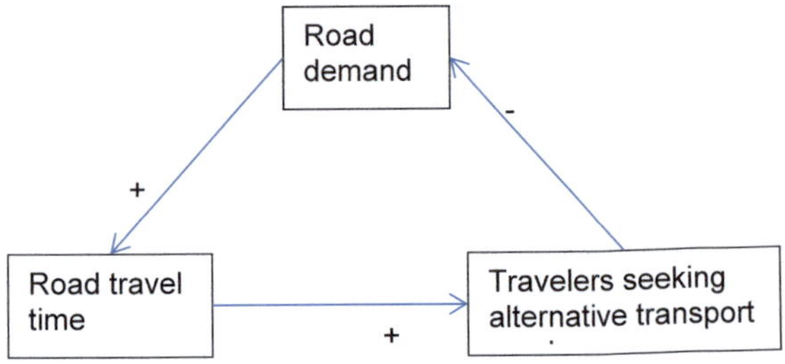

Note, however, were there no road development, no road building, Model 12.3 is replaced with Model 12.4 showing that if there is no road development travellers would tend to use the alternative system (now rail), a negative feedback loop.

For example, tram/light railways developments in Greater Manchester and Tyne and Wear.

Now consider the effect of road developments on the existing rail network/transportation system, shown in Model 12.5; this contains only positive feedback loops.

Model validation: What happened in the UK

	{road building went up,	then	road travel time went down
	road travel time went down,	then	road demand went up
	road demand went up,	then	road building went up
	road building went up,	then	rail demand went down}
Summary	[rail demand went down,	then	road demand went up, and so on]

Describing transport system changes with the shift from rail to road transport and the continuing development of road travel in the period from (about) 1920, see Table 12.1 for data.

The eventual result from this shift (from rail to road) resulted in the closure of almost 30% of Britain's railway route network between 1960 and 1970 and the majority of the commercial canals. This shift for rail to road transport can be described by a "Beeching" model describing this time when there were widespread rail closures resulting from this shift from rail to road transportation (Model 12.6).

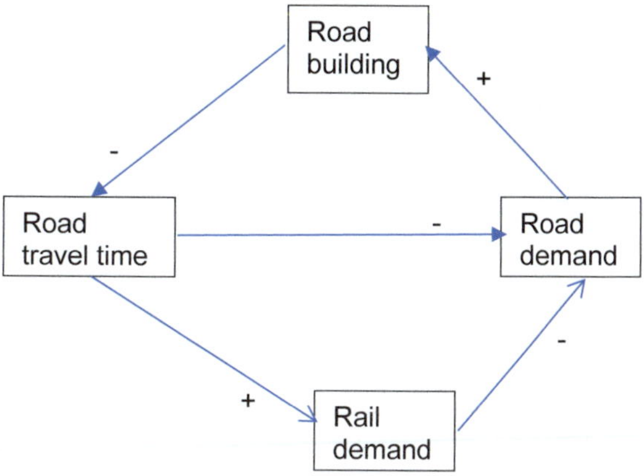

Model 12.5 Modelling interactions between demand for road and rail travel

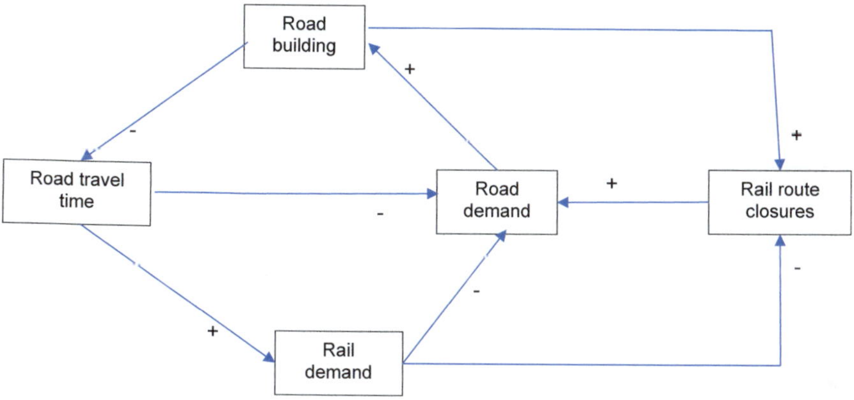

Model 12.6 The Beeching model. *Note* The "Beeching" report HM Treasury [2], HMSO [3]; this report/investigation recommended the closure of many railway stations and routes

12.1.5 A Non-green Result from a Green Policy

A consequence of the 1960s transport planning policies, road before rail, was that road development continued until, as now, in large conurbations road improvements are no longer feasible. As a consequence investments are being made into expanding the rail network, for example the Crossrail project in London, but Model 12.1 indicates that this "new" mode of transport (new rail) will enter a period of unconstrained (over) growth causing its own problems, for example urban sprawl and increasing living costs. Consider for example the rail growth/investment in and around London allowing commuters to travel greater distances to work with the subsequent ever increasing urban sprawl (Model 12.7).

This network has positive feedback loop indicating unconstrained growth in urban sprawl, non-green result.

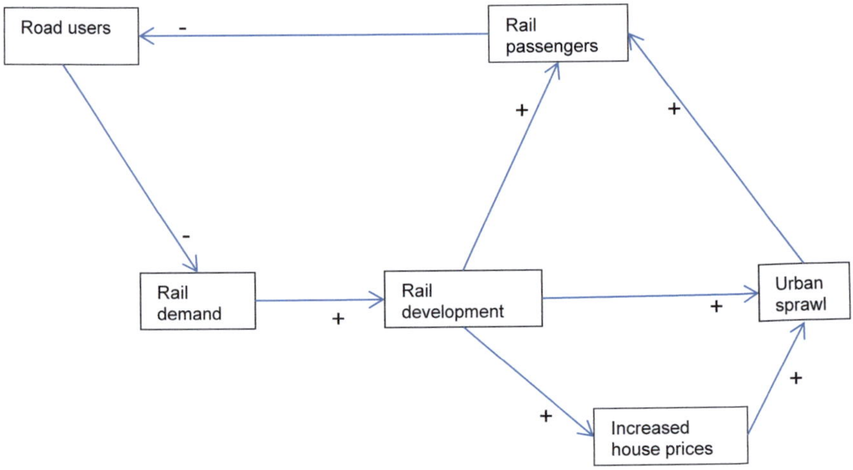

Model 12.7 Consequence of rail developments

{rail development increases,	then	House prices and Urban sprawl increases
House prices increase,	then	Urban sprawl increases
Urban sprawl increases,	then	Rail Passengers increase
Rail Passengers increase,	then	Road users decline
Road users decline,	then	Rail demand increases,
Rail demand increases	then	Rail development increases
Rail development increases	and so on}	

12.1.5.1 Model Validation

Historically, this model has been validated by the "metro land" developments in the 1920s (T4) caused by the metropolitan line (of the London Underground system).

Current validations demonstrating the "not green" results from green developments have occurred with the rationales for new rail developments. The Crossrail project in London and the Waverley line reopening into Edinburgh have presented the non-green consequences from this model as positives supporting these developments, for example:

quoted benefits from the Crossrail Project (T6) include:

> Residential capital values are projected to increase immediately around Crossrail stations in central London by 25 per cent, and by 20 per cent in the suburbs

- The impact on residential property market will also extend out to Berkshire and Essex.

quoted benefits Johnston and Causley [4] from the used to justify the Waverley route reopening in Scotland

- It is also expected that rail links will widen economic and housing opportunities

In both cases, the extended lines into the "countryside" producing non-green solutions and similar rationales have been given for both Crossrail2 and the Cambridge–Oxford link.

12.1.6 Conclusion

Thus, these models have demonstrated, paradoxically, that intended green transport policies do not have "green" results as witnessed by the current rail developments in and around London and Edinburgh, the models demonstrating (the positive feedback loops) that such investment results in increasing urban sprawl leading to commuters being prepared to travel greater distances to work, and the subsequent increase in housing prices in the newly reached areas, resulting in the demand for more rail developments and so on.

So "What is the best green policy?" is it to plan for slow rail journeys and congested roads? In particular, no extended planned mass (rail or road) transit systems in and accessing large conurbations? Is the Crossrail project counterproductive and a big expensive mistake? Should there be investment in better pedestrian provision and not in inner city cycleways? and should investment have been concentrated more on the provision of intercity rail links, to reduce long distance car travel into major centres, a genuine green policy, Chap. 20 gives relevant travel to work in London data.

12.2 Further Analysis of the Model for the Dow Jones Index

Figure 12.1 shows the changes in the Dow Jones index between 1924 and 1940 covering the periods before and after the "Wall Street Crash" in 1929.

The following analysis and models aim to describe the effect of investor confidence on the behaviour of this index as shown in Fig. 12.1 (Fig. 12.2).

Fig. 12.1 Plot of Dow Jones index 1924–1940

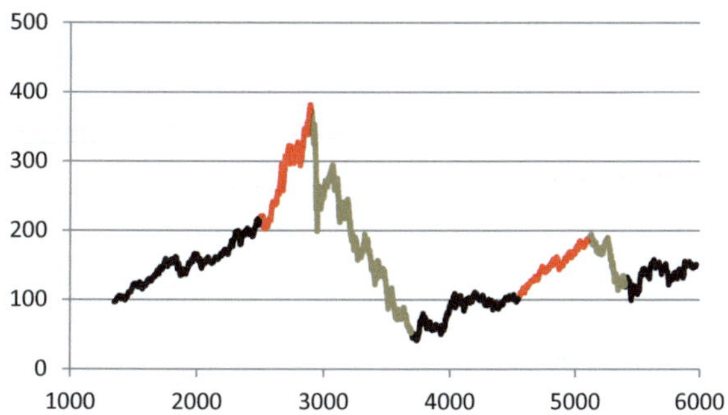

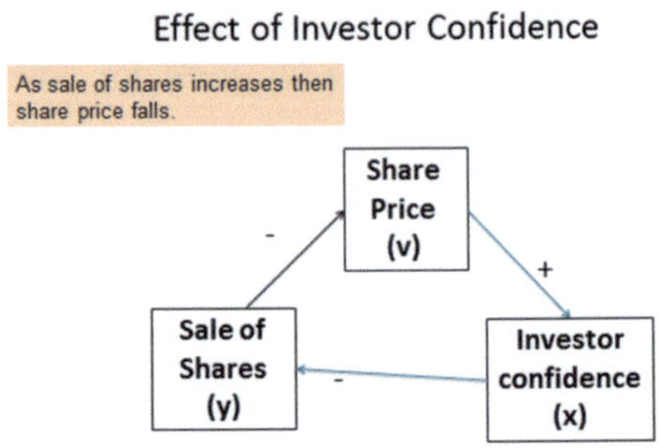

Fig. 12.2 Influence diagram investor confidence and share price

This model exhibits a positive feedback loop {+, −, −}; hence, either an uncontrolled increase or decrease in share values will result from this model:

Increase in share prices gives

Share Price up	leads to	Investor Confidence	up
Investor Confidence up	leads to	Sale of Shares	down
Sale of Shares down	leads to	Share Price	up

Decrease in share prices gives

Share Price down	leads to	Investor Confidence	down
Investor Confidence down	leads to	Sale of Shares	up
Sale of Shares up	leads to	Share Price	down

The Dow Jones index Fig. 12.1 (1924–1940, data points 1350–6000) acts to validate this model. A plot of the index for this period is shown in Fig. 12.4 with the "phases" within this date indicated by:

Normal growth	black	linear growth per day	0.085
Abnormal growth	red	linear growth per day	0.410
Fall	brown	linear fall per day	0.323
Normal growth	black	linear growth per day	0.087
Fall	brown	linear fall per day	0.251
New pattern	blue	linear growth per day	0.046

The "Break points", change of model, indicated by this plot approximately correspond to:

Normal growth period 1	Starts	1924	Calvin Coolidge elected president
	Ends	1928	end of Coolidge presidency
Abnormal growth	Starts	1928	Herbert Hoover elected president
	Ends	1929	
Abnormal fall	Starts	1929	Index doubled in 15 months, previously over 48
	Ends	1932	end of Hoover presidency
Normal growth period 2	Starts	1932	Franklin D Roosevelt elected president
	Ends	1937	
	Starts	1937	
	Ends	1938	Anschluss in Austria and Sudeten Crisis

The abnormal growth and subsequent dramatic fall as observed in this data can be modelled by Fig. 12.3.

The changes in direction, of the index, being caused by an external event that affected the confidence of the investors.

The Dow Jones index from 1994 to 2016 exhibits a similar pattern, growth then greater growth followed by a rapid fall, further validating this model (Fig. 12.4).

Abnormal Growth	Oil price rise started in mid 2006 (point 8000) to a peak price in late 2007 (point 8350)
Abnormal Fall	then falling to a low price in late 2008 (point 8500).

However, plotting the index from 2006(start) to 2009(end) indicates the possibility of an underlying long-term trend in the data (Fig. 12.5).

The bounds shown in Fig. 12.6 indicate the (approximate) times when the short-term trend in the data changes (from rising to fall, or from falling to rise).

The plots of crude oil prices and Dow Jones index, Fig. 12.7, shows the relationship between these two indexes.

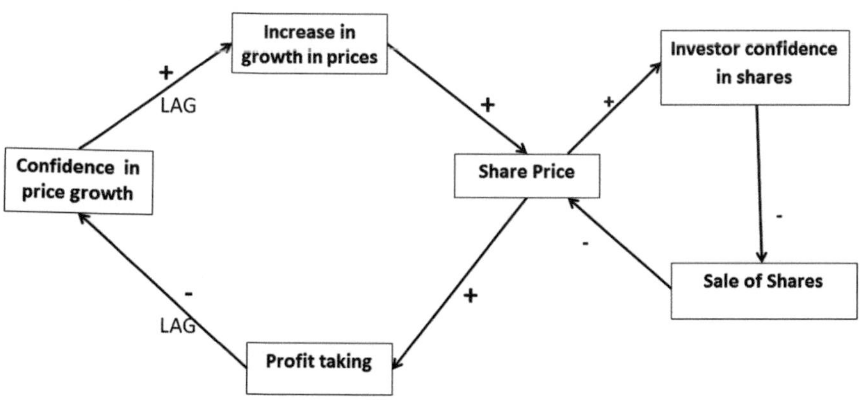

Fig. 12.3 Extended model

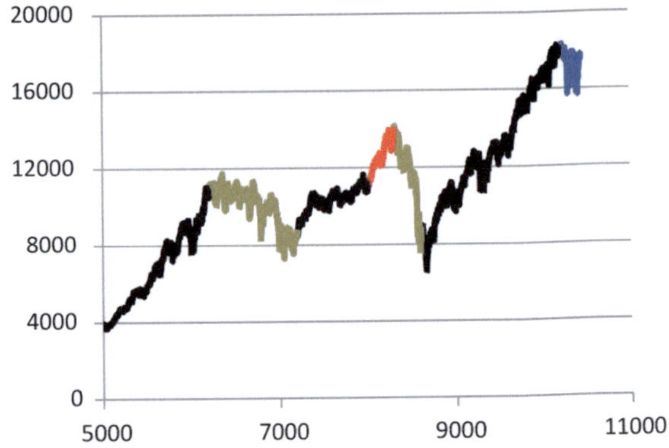

Fig. 12.4 Dow Jones index 1994–2016

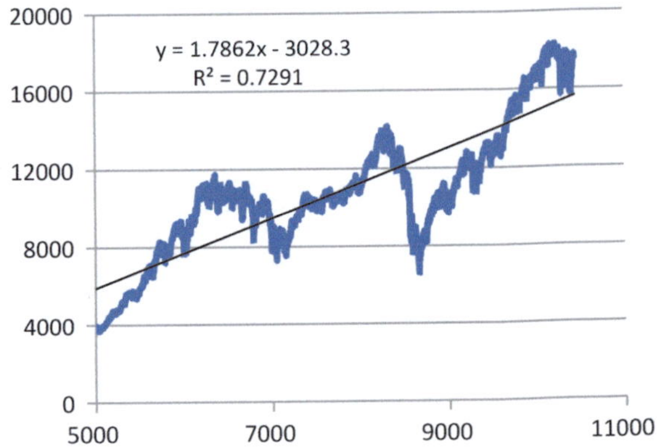

Fig. 12.5 Linear model for Dow Jones price changes

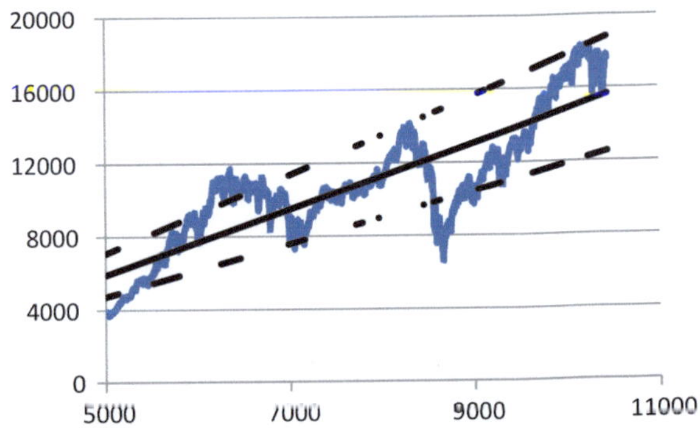

Fig. 12.6 Illustrating possible buy sell bounds for the Dow Jones Index

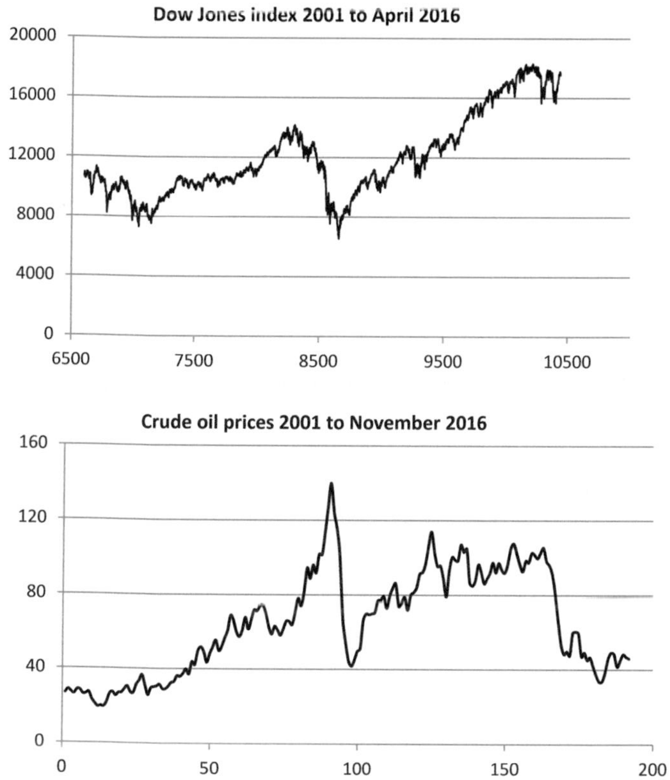

Fig. 12.7 Comparing the Dow Jones index with crude oil prices

References

1. Coyle RG (1996) System dynamics modelling. Chapman and Hall
2. HM Treasury (2013) Investing in Britain's future (www.gov.uk)
3. HMSO (1963) The reshaping of british railways parts 1 and 2. R Beeching BRB (http://www.railwaysarchive.co.uk)
4. Johnston K, Causley J (2013) Take that Dr Beeching: (www.starconference.org.uk)

Chapter 13
Applying Queueing Theory to the Design of a Traffic Light Controller

James Hardy

This case study explains some of the esoteric features of traffic light control and discusses existing systems before introducing a novel control system being developed as a research problem.

Traffic light systems are used to share a small area of contested roadway with some predefined declaration of safety, efficiency or fairness. Each of these terms is open for discussion and can be defined for the context to which it is applied. For example, safety is generally considered to be the easiest to define and commonly implies a lack of collisions between vehicles and/or pedestrians. However, collision effects are by degree; safety is frequently considered to imply that the rate of death or serious injury is reduced in the event of an accidental collision and not that the accident was avoided. Efficiency can refer to actual time, useful application of time, fuel use or generation of undesirable chemical compounds, e.g. CO and CO_2. Fairness is open for even wider interpretation. Fairness can be implied as all travellers experience the same delay at a physical point, experience the same average delay over a period, or that delay should be attributed in a pro rata fashion according to relative queue length or, if the number of lanes is considered, queue capacity. These interpretations are not intended to be exhaustive; they do give a flavour of some of the diverse features of traffic control.

Traffic control systems progress through the light sequence in response to one of two trigger inputs, time and detection. The inputs can be taken in isolation, in sequence or as a function. Simple examples are as follows:

- Fully timed control. The sequence changes based purely on a clock period, the period can be fixed for all times of the day, and all days of the week or can be varied to suit periodicity of traffic flow.

J. Hardy (✉)
University of Derby, Derby, UK
e-mail: j.hardy@derby.ac.uk

S. Berry et al. (eds.), *Guide to Computational Modelling for Decision Processes*, Simulation Foundations, Methods and Applications,
DOI 10.1007/978-3-319-55417-4_13

- Fully detected system. There are several versions of fully detected systems, the most common version being the pedestrian activated pedestrian crossing. Timing is still required, the initial change is delayed, the yellow warning is shown for a period, and the system returns to the initial state after a fixed time period.
- Vehicle actuated. These systems can be highly complex, ranging from isolated junction to city wide fully coordinated. Detection can be made using a wide range of methods including pressure plate, magnetic disturbance, radar, laser ranging and even sonic methods.

Triggering can be in response to an event such as the time that a vehicle has been waiting, detected length of queue. Triggering can also be predictive based on calculated speed of approach, which can also provide traffic calming. Vehicle-actuated (VA) systems rarely rely on detection alone; minimum and maximum timing is applied. This ensures that a minimum number of vehicles move in any cycle and that there is still a periodic change in the event of overloaded or erroneous detection (faulty detector, parked vehicle and accident).

One of the most significant concerns for any traffic control system is the possibility and probability of a gridlock situation. Gridlock occurs when there is a circular effect of traffic flow being recursively blocked by a flow which is ultimately blocked by the initial block. Gridlock generally occurs over a group of several junctions and is not restricted to a small area. The most common root cause for gridlock is a traffic queue from one junction extending back to block another upstream junction. Three prime methods for avoiding gridlock are (1) reliance on moral driving standards to require courteous use of shared junction space, (2) application of "yellow box" junctions with legal enforcement and financial penalties and (3) prevention of queue build-up to a length that causes upstream blocking.

Traffic flow in a system can be considered to have infinite input capacity, especially at the boundary edges of a congested region. Traffic flow in a free-flowing system is highly stochastic; any number of vehicles can arrive at any time with any spacing.

The complexity of traffic flow means that simulation is the most commonly applied analysis tool, but this does not provide a formal proof under all circumstances or even under bounded conditions. A formal theoretical understanding can be derived by result comparison which is only possible when controlled experimentation and measurement are undertaken.

Queueing theory is an obvious tool which could be used to analyse vehicle traffic queues. However, the application of queueing theory becomes complex as the junction under consideration becomes more realistic.

13.1 Unidirectional Queueing

This is the simplest queue formation, with timing and control in only a single direction e.g. East–West. If the junction is considered to be isolated and the entry and exit roads are identical in terms of physical size and number of lanes, then QT can be applied with the following attributes:

Normal distribution arrival, deterministic service distribution, single server, infinite input buffer size, infinite population and fifo service. Using Kendall notation, this is given as $M/D/1/\infty/\infty$/fifo or, more commonly simply $M/D/1$ (Fig. 13.1).

The distribution of service times is considered to be deterministic as it is event driven.

In practical terms, the junction described would be quite rare limited to, for example, a pedestrian crossing in a one-way street. A more common system would be for a pedestrian crossing in a two-way street (Fig. 13.2).

The controller service interval is the same for both queues which are therefore interdependent. In real systems, it is unlikely that traffic will arrive at the same rate from each direction; there is normally a periodic overall inbound or outbound flow bias. Moreover, urban traffic light systems are generally not isolated; boundary control points between urban and inter-urban locations are very rarely considered isolated; the urban system will have a finite population and finite buffer which are not able to satisfy the much higher flow rate of the multilane, high speed inter-urban roadway. In this case, the inter-urban to urban direction will be $M/D/1/\infty/\infty$/fifo with a series $D/D/1/K/\infty$/fifo and the urban to inter-urban route is potentially $M/D/1/K/P$/fifo.

While it is possible to consider this as two individual systems, vehicular traffic queues and congestion are temporal events, both directions must be resolved for a given instance, and multiple instances must be resolved to determine a trend.

13.2 Bidirectional Queueing

The previous scenario might exist at a pedestrian crossing, where vehicle queues exist in only two directions. If the system under consideration is a vehicular crossroads as shown below, the solution is more complex (Fig. 13.3).

In this scenario:

East to West flow is subject to two QT networks in series,
West to East is a single QT network with finite input and population,

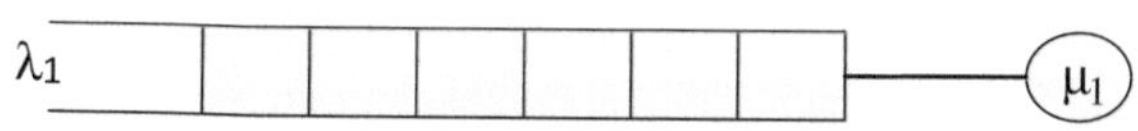

Fig. 13.1 Queue server

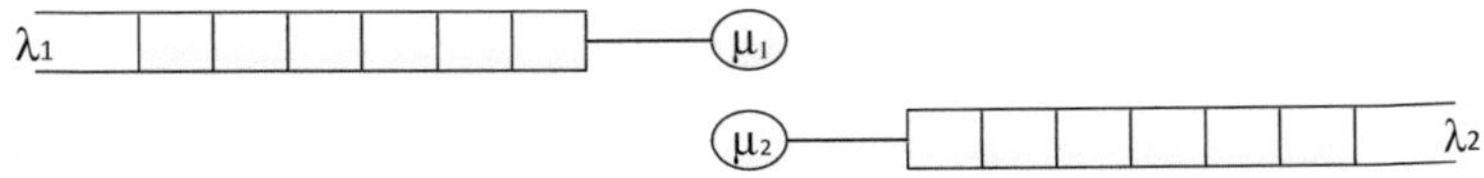

Fig. 13.2 Queue servers at a junction, no direction changes

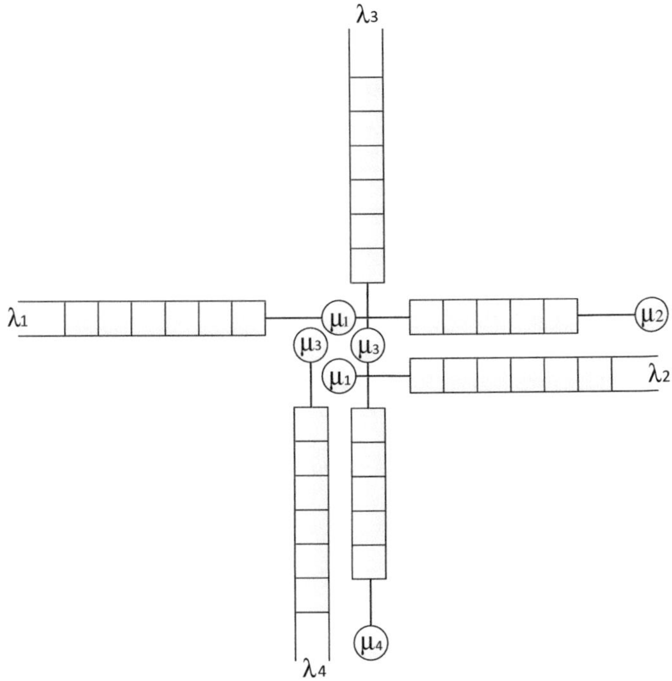

Fig. 13.3 Queueing at a non-isolated junction

North to South flow is subject to two QT networks in series,
South to North is a single QT network with finite input and population.

There is a single controller (μ_1) for both EW and WE and a single controller (μ_3) for both NS and SN. The two controllers μ_1 and μ_3 are temporally exclusive. The controllers μ_2 and μ_4 are isolated, independent and unique.

13.2.1 Complex Junction Queuing

Practical road junctions are further complicated by left and right turns which can be either exclusive or in flow. Each turn can be considered as a separate controller; the input queue for each direction may or may not be unique. Further levels of

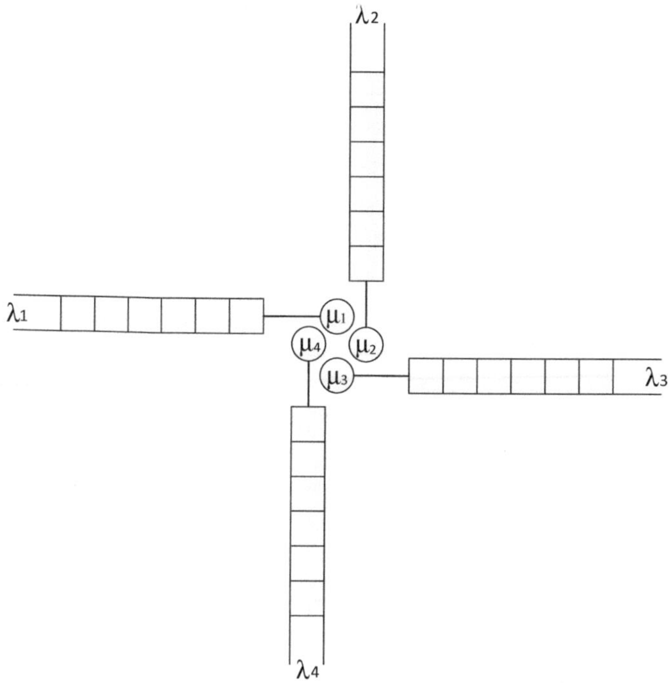

Fig. 13.4 Extended queueing model

complication are provided by junctions with sequential rather than alternate flow permission. These can be junctions with odd numbers of entrances and exits, with unbalanced numbers of entrances and exits (where one-way and two-way traffic meet) or where traffic priority necessitates sequential "round robin" flow permission (Fig. 13.4).

13.2.2 Queueing with Detection

The final layer of complication is provided by detection systems. Very few totally timed traffic light systems exist; detection-based systems are mandated for all new installations. Detection-based systems change in response to vehicle flow, queue length, vehicle waiting time or time period (either when there is no traffic flow or when queueing extends beyond the detection point) (Fig. 13.5).

Systems can, and do, change between detection and time-based triggering dependant on the current state of traffic flow.

Normal vehicular traffic flow presents numerous opportunities to develop a set of queuing theory solutions. The major obstacles are that items in the input queue

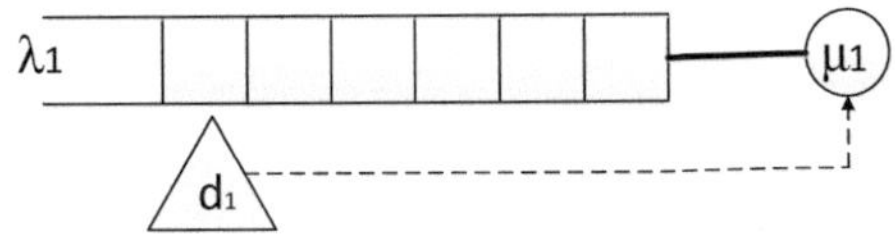

Fig. 13.5 Queue with queue detection

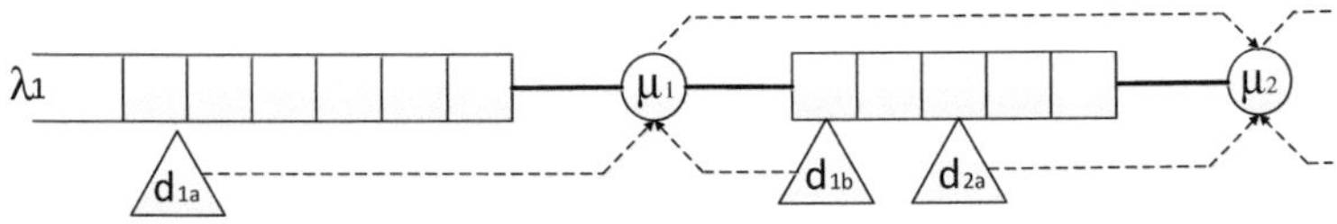

Fig. 13.6 Linking queue detection systems

cannot be dropped, the input flow cannot be stopped or controlled, vehicles can enter and exit input buffers without warning and that multiple queues interact with each other.

13.2.3 Queueing in a Novel Control System

Existing traffic control systems may be reactive or predictive but always rely in information gained from input queue detection. Notwithstanding the complications of real junctions described earlier, our novel controller accepts that there may be a blocked exit due to a downstream junction and avoids allocating flow time to the direction that is not feasible. If the junction is already flowing and the exit becomes blocked, the existing flow permission is ceased. In either case, flow permission moves to the next sequential state where it is anticipated that traffic flow is feasible. In the event that no exit is clear, all lights are held in a transient red state and permission is automatically passed to pedestrian and cycle access. It is assumed, but not mathematically proven, that this action would result in traffic build-up rather than reduction. In order to reduce the traffic build-up, a further control signal is required. When the subject controller takes action based on the information received from the detector located in the junction exit, a signal is sent to the controller which is downstream from the exit detector. The signal is a high priority request for the downstream junction to change to a signal phase which clears traffic arriving from the subject junction (Fig. 13.6).

Due to the fact that this system considers flow feasibility prior to making a signal decision, we refer to this as available forward road capacity (AFRC). If all junctions in an area are equipped with the AFRC system, it is assumed that traffic flowing into uncongested boundary regions will be given higher flow priority and will therefore reduce queue lengths and general congestion.

13.3 Experiments and Discussion

Simulations were carried out on a grid structure comparing control by:

- Timed traffic light changes (TTLC)
- Vehicle-actuated light changes (VALC), and
- Available forward road capacity (AFRC)

Thus traffic light control is either by fixed timing, or stop line detect (VA style) or stop line and exit detect AFRC.

For the AFRC controller, the status of each junction is read, tested and modified in numeric order. The subsequent actions of this controller aim to try to remove blockages and enable the flow of vehicles through the system.

Thus, the action of the AFRC controller at a particular set of traffic lights where the exit is blocked is to turn this traffic light to red and force the next light to turn to green (regardless of current state) to attempt to clear this blockage.

The simulations were four periods and the resultant queueing statistics from each are shown in Fig. 13.7, TTLC blue, VALC red and AFRC black

13.4 Conclusions

These results showing that initially (demand 0–100) not only is there no (real) difference between the performance of VALC and AFRC but also that a system based around the basic TTLC performs well. But as demand continues to increase (demand between 100 and 160), a system based on AFRC outperforms a system based on VALC. Although for demands up to 150, the results do indicate that the simpler TTLC system works quite well and may provide a very good solution; however, for higher demands, it performs very poorly.

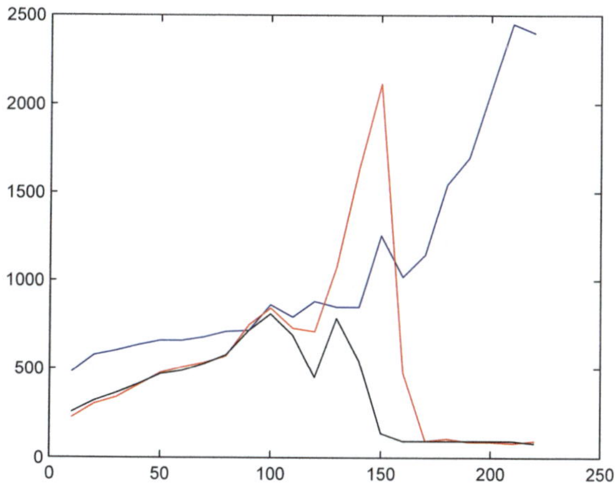

Fig. 13.7 Comparing TTLC, VALC and AFRC control systems

Thus, the choice of the most appropriate traffic control will be dependent upon the expected demand and the cost of the system, however, in summary at:

Low demand	(0–100)	Any system
Medium demand	(100–150)	AFRC or VALC
High demand	(150–160)	AFRC, (note that here TTLC is better than VALC)
Very high demand	(over 160)	VALC or AFRC but not TTLC

Chapter 14
Cellular Automata and Agents in Simulations

Kim Smith, Richard Hill, Stuart Berry and Richard Conniss

14.1 Simulating to Evaluate Message Passing Rules

The average mobile phone and tablet that is now available has more computing power than even a year ago. In particular, the inclusion of Wi-fi capability enables such devices to be used as part of a Wi-fi-based information sharing network without needing to modify the existing mobile telephone infrastructure. In addition to this, the ability to create decentralized wireless ad hoc networks means that there is no reliance upon any form of network infrastructure, such as switches, routers, access points and servers.

Ad hoc networks are wireless networks where the nodes communicate directly with each other, when in range. As devices join together to form the ad hoc network, messages can be sent, received and relayed between devices.

Each device, or node, within the network can act as the originator of the message, the destination of the message, or as a router that passes the message onto other nodes on a path between the source and destination. As such, a message can be routed from node to node until it reaches the intended recipient node.

If it is not possible to clearly identify a path between a sending node and an intended recipient node, then alternative opportunistic network (ON) topology is required. An ON facilitates the propagation of messages amongst nodes that are within a physical proximity that allows successful transmission and reception of messages. Whenever two nodes come within range, they pass messages to each other, due to the potentially constant mobility of nodes, and messages can be propagated to nodes that were initially out of range at the point of message creation.

K. Smith · R. Hill · S. Berry (✉) · R. Conniss
College of Engineering and Technology, University of Derby,
Kedleston Road, Derby DE22 1GB, UK
e-mail: s.berry@derby.ac.uk

S. Berry et al. (eds.), *Guide to Computational Modelling for Decision Processes*,
Simulation Foundations, Methods and Applications,
DOI 10.1007/978-3-319-55417-4_14

ONs utilize a *store-carry-and-forward* paradigm where each node holds a copy of the message, whilst also propagating to other nodes as they come within range. The objective of an ON is to move messages as quickly as possible whilst minimizing any load upon the network itself.

14.1.1 Background

A consistent theme of these case studies is to make use of a real-world scenario to both explain the challenges presented to mobile networked devices, as well as explore the potential of such environments for the emergence of new business opportunities (Fig. 14.1).

Consider the scenario of a town centre or a retail mall where shoppers and potential customers congregate. It is likely that as each user enters a location, their mobile device can attempt to join an ad hoc network. Through this network, messages in the form of adverts for services and applications are propagated amongst the connected devices.

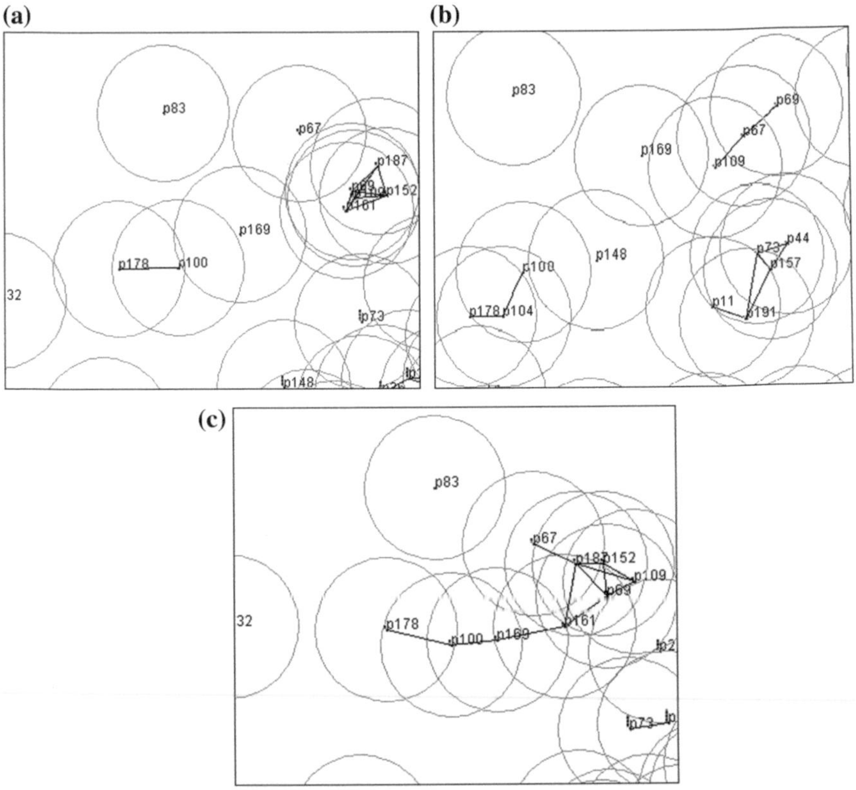

Fig. 14.1 **a** Network at t = 1. **b** Network at t = 2. **c** Network at t = 3

Software applications resident on the mobile devices proactively filters any advertisements that present themselves, retaining only those that the user is interested in. Service providers such as shop keepers/restaurateurs have the ability to create and publish to the ad hoc network adverts for new services.

These adverts propagate through the ad hoc network to each device currently connected, and are selectively displayed or discarded as per the user's preferences, whilst also relaying messages to other devices that fall within transmission and reception range.

14.1.2 Case Studies

The first case study is concerned with simulating message passing protocols in an opportunistic network.

In second case studies, message passing protocols have been simulated, using cellular automata that represent the message carriers, first on a grid structure with all automata/agents moving at the same speed, see Fig. 14.5 for an illustration of such a system, then secondly where the automata/agents can move in any direction at different speeds, see Fig. 14.7.

Case Study 1: Matching Services with Users in Opportunistic Network Environments

Consider the scenario of a town centre or a retail mall where shoppers and potential customers congregate. It is likely that as each user enters a location, their mobile device can attempt to join an ad hoc network. Through this network, messages in the form of adverts for services and applications are propagated amongst the connected devices.

Software applications resident on the mobile devices proactively filters any advertisements that present themselves, retaining only those that the user is interested in. Service providers such as shop keepers/restaurateurs have the ability to create and publish to the ad hoc network adverts for new services.

These adverts propagate through the ad hoc network to each device currently connected, and are selectively displayed or discarded as per the user's preferences, whilst also relaying messages to other devices that fall within transmission and reception range.

As a user leaves that location, messages stored on their device are carried until another location with an ad hoc network is reached. These locations are not connected to each other, and there is no central infrastructure except for the ad hoc wireless system. The propagation of these adverts between locations is achieved through the mobility of users.

The very nature of the intermittently connected network as described, with devices "carrying" adverts between locations, means that an opportunistic ad hoc network would be a natural solution.

In the context of this, there is a need to be both effective and efficient in matching adverts (provision) to the user's device. There are a number of reasons for this. First, users that are bombarded by unwanted messages are likely to turn their device off.

Second, unlike a wired network where the devices are connected to mains power, a mobile device relies on its own scarce resource (battery). All wireless transmission requires energy, and excessive, irrelevant message propagation will increase the power consumption of mobile devices.

Finally, excessive message propagation results in increased traffic for the network.

The overall objective of this work is to facilitate the provision of providing the relevant adverts to interested parties without crippling an individual device or the network as a whole.

Consequently, if a device receives multiple copies of the same message, it will significantly reduce the capacity of the battery. Additionally, the amount of storage on the device to store adverts is another constraint so as not to interfere with the normal operation of the device.

With regard to the network, it too has a limit to its capacity, referred to as bandwidth; this indicates the maximum number and the size of the messages that can be passed across the network. As the number of messages reaches this limit, the network becomes congested and further messages are unable to be sent, resulting in overload.

In order to establish the extent by which messages can be efficiently provisioned in an ON environment, a number of characteristics need to be monitored and evaluated such as

- Quantity of adverts arriving at the interested parties;
- Volume of network traffic generated;
- Time taken for interested parties at various points to receive the advert;
- Proportion of interested parties that were reached;
- Amount of energy consumed by a device/the whole network.

Experiment Design

The base scenario that has been used so far is that of mobile users within a retail shopping mall, which possess devices with wireless network capability.

Within the geography of the shopping mall, users move randomly within the area. Also, users enter and leave the area (and therefore the network) at random.

The mobility of users will be represented via the random waypoint model [1, 2]. The random waypoint model attempts to capture the movement of humans, and each node is given random coordinates in the simulation area (waypoint). Each node moves at a constant velocity directly towards the given waypoint.

At this point, the node pauses and a new waypoint is defined, together with a new random velocity. Simulations were executed using epidemic and other non-context-aware routing protocols.

Whilst this refers to the base scenario, there are also more specialized scenarios that can be envisaged.

For instance, when user leaves a location, they may be carrying advertisements for services that will be recognized by subsequent connections to ad hoc networks in other locations.

In this way, an originator in one location, having identified that a significant amount of custom comes from another location, could target that location. For example, a chain of retail outlets could propagate a voucher that is redeemable in any one of the bricks and mortar stores.

The propagation of these adverts between locations is achieved through the mobility of users; it is the mobility of users, which makes the connections in an ad hoc fashion.

The simulations assume that users move based on the shortest path map-based movement model [3]. The shortest path map-based movement model is one of a number of map-based movement models, where the movement of the node is constrained to a path as defined in a set of map data.

In the map-based model, nodes are able to move randomly along any path, whereas in the case of the shortest path map-based model, nodes follow the shortest route to a point on the map.

This point on the map is chosen either as a random point on the map or from a list of points of interest. In this case, there will only be a single point of interest provided. This opens up the possibility of defining a range of simulations based on a number of context-aware routing protocols.

In this research, the matching of services to user's needs is to be done in an effective and efficient manner, this including not only the transmission protocol but also the matchmaking algorithm itself.

In systems where an external broker is used to carry out the matchmaking, there is a significant overhead due to the time required to communicate to a third party, as well as the extra network load imposed.

However, given the very nature of an ad hoc network without any infrastructure, a community broker node would not be viable due to the extra processing and therefore power dissipated in the process. In such a case, it would be logical for all the processing needs to be confined within the individual device. This arrangement has the additional benefit of reducing concerns regarding user privacy concerns as the user's profile remains within the device and is not communicated to a third party.

Network Layer

In the TCP/IP model, the protocol sits at the network layer. Messages are passed up to the application layer via the transport layer for processing. The proposed matchmaking service might require messages to be passed from the protocol layer up to the application layer. It can be assumed that widespread adoption of the service amongst users would result in myriad user's preferences being instantiated. The end result would be that a significant number of these messages would ultimately be discarded, resulting in wasted processing of raw messages that did not conform to the preferences of the user.

A more efficient method might be to carry out a first pass of the matchmaking at the protocol layer. This would allow messages that the user is broadly interested in to be filtered, and any message types that have not been seen before, to be processed and passed up to the system application layer for user interaction.

Further processing could then be carried out at the application layer, such as adding a newly discovered preference into the profile.

How much processing takes place at each level depends upon the speed of the main processor and the time between received messages. Too many processing cycles would result in dropped messages, and too little processing would reduce the accuracy of the matchmaking.

In order to keep the matchmaking algorithm in close proximity to the protocol layer, a protocol wrapper has been adopted to prevent any direct modifications to the underlying network protocol. Figure 4.1 illustrates this.

In order to simulate this process, the ONE simulator was selected since it was designed specifically for opportunistic or delay-tolerant networks. The ONE simulator is designed to emulate various routing protocols and movement models [4].

The upper layers of the communications stack are not implemented other than a link that can be used to pass messages up to the next layer. The way the network layer is implemented in the ONE simulator, is similar to that required in that the network layer is already in two parts.

Matching Algorithm

The matchmaker algorithm needs to be relatively simple in order to minimize an additional processing overhead, and will need a clear profile to work with. Even in this, there needs to be a level of sophistication built into the matchmaking algorithm; otherwise, the system will not gain acceptance with the users. For example, the system will need to work on multiple levels, i.e. a general descriptor and then a modifier. An example would be a keyword such as restaurant where the subsequent second-level keyword would define what type, i.e. French, Chinese, Mexican, etc. In addition, when an advert for a restaurant that is not in the list of existing restaurants is received, user input will be required. So, for instance, the device receives an advert for a Korean restaurant which is not in the profile, it will be passed to the user so that the user can respond, who would accept or reject the advert, and the profile would be updated automatically. This would provide a learning system that could build up a user's profile on the fly, that is rather than the user been required to initially input a profile, the device would learn the user's preferences based on their response to adverts

In a simple way, if the system received an advert for a restaurant of type Chinese, which is already in the user's profile, it would therefore be matched and the advert would be passed onto the user.

Alternatively, if the user's profile shows a dislike for French and an advert for a French restaurant arrives, the profile would indicate that this is not a preferred option and the advert would not be passed to the user. Then, there would be the special case where there is currently nothing in the profile, which would be passed to the user. The user's response would then cause it to be added to the profile as either a like or dislike.

An additional concept would be where there were two adverts for restaurants which are both shown as likes on the user's profile. The matchmaker algorithm (see Chap. 16 for the detailed algorithm) could possibly then make a decision based on the distance to the two restaurants or in a simple case display both adverts.

The message is passed as a list of keywords together with a number that signifies the number of keywords in the message (levels).

The "user profile" currently consists of a number of sets of keywords, one for each level to be matched. Also, there are two complete groups of sets: one for "likes" and the other for "hates", and example set is shown below. For example, a three-level information database is shown in Fig. 14.2.

Whilst for a given customer enquiry, this information structure can be refined to include "likes" and "dislikes" to represent the preferences.

As an example, consider the preference lists:

Likes :	[Restaurant;	Level 1
	Asian, African, Pacific;	Level 2
	Chinese]	Level 3
Hates	[European, American	Level 2
	Japanese, French]	Level 3

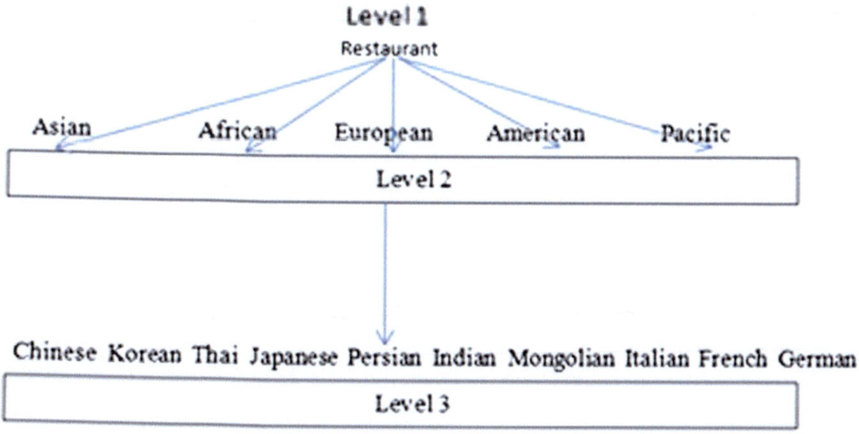

Fig. 14.2 Sample three level database

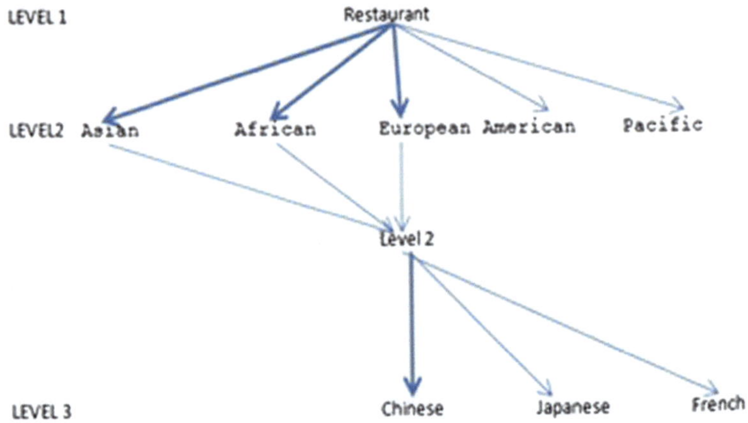

Fig. 14.3 Adding information to the database

Incorporating this additional information gives the revised structure (Fig. 14.3).

The algorithm (see Chap. 20) first enters a whilst loop that will step through all the levels unless the current level that is been tested is either in the hates group or does not exist in either group. If all levels in the message match those in the likes group, the **pass_level** will equal the levels and the message will be displayed. If any level matches an entry in the hates group, **decn** will be set to 2 and the message will be ignored. Whilst if at any level there is no match with either group, the message will be displayed for user decision.

Sample messages together with the result from the search are shown below:

message 1	[Restaurant; American; Brazilian]	would fail at level 2,
message 2	[Restaurant; African; Moroccan]	would be referred at level 3.

Simulation Results

The matchmaking algorithm has been simulated in MATLAB. To be able to evaluate the effectiveness of the search, in relation to the number of levels to be considered, three simulations were carried out.

In each simulation, a node attempted to contact a neighbouring node (mobile device) and the message compared with this neighbouring device's user profile ("likes" and "hates" lists). This determined the level (1, 2 or 3) and message status at which the search was completed (success, refer or reject), and further, these results were used to investigate the cost/benefit of the depth of search used (level of information to be considered).

The three messages simulated were as follows:

Simulation Run 1 message

Restaurant `likes' High probability
Asian
Indian

Simulation Run 2 message

DIY `likes' Medium probability
Chain store
Carpets

Simulation Run 3 message

Food `likes' Very High probability
Farmers market
Indian

In the case study, each mobile device could have up to five members in each of their "likes" and "hates" sets. Tables 5.1, 5.2, 5.3, 5.4, 5.5 and 5.6 define the probabilities that each keyword at each level of the proposed message will be liked or disliked by a neighbouring device. For each of the keywords in the message, a weighting was applied to indicate the user's preference in their device profile.

The next tables show the effects (proportion of correct decisions) had the search been halted at each level and the implied work carried out.

The results from the first simulation P_likes(H, M, VH) show that halting the matching algorithm after two levels (only) results in a 2% maximum error, where an error occurs when a message has been passed incorrectly (Fig. 14.4).

An interesting result highlighted by these simulations is that the depth of search required, so that most decisions will be correct, seems to depend upon the proportion of devices expected to like the level 1 category. Table 14.1e shows that the

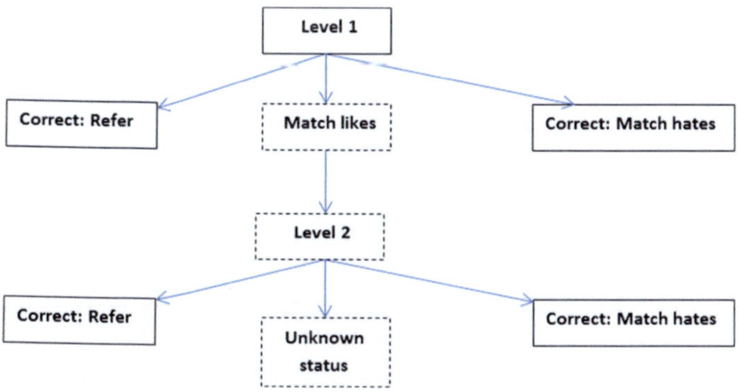

Fig. 14.4 Illustrating the operation of the first two decision levels

Table 14.1 Simulation run

Level	Likes probabilities		Hates probabilities	
(a) *Simulation run 1*				
Simulation 1				
1	0.3	High	0.1	Low
2	0.1	Medium	0.1	Low
3	0.45	Very high	0.05	Very low
(b) *Simulation run 2*				
Simulation 2				
1	0.1	Medium	0.1	Medium
2	**0.05**	**Low**	0.3	High
3	0.3	High	0.2	High
(c) *Simulation run 3*				
Simulation 3				
1	0.55	Very high	0.2	Medium
2	0.25	High	0.3	High
3	**0.05**	**Low**	0.05	Very low

Level	Simulation 1	Simulation 2	Simulation 3
(d) *Summary of results from all simulation runs*			
1	0.71	0.90	0.46
2	0.98	0.99	0.87
3	1.00	1.00	1.00

Simulation	*P* (level 1 liked)	Percentage of correct decisions halting at	
		Level 1	Level 2
(e) *Summary of results from all simulation runs*			
2	0.1	90	99
1	0.3	71	98
3	0.55	46	87

greater the probability of a message being liked at level 1, the greater the depth of search required to reach the correct decision (to pass or not to pass the message).

Thus, additional knowledge concerning the “liked” status of the options (probability) within the query could indicate an appropriate depth of search required to give (mostly) correct decisions, to pass or not pass the message.

Additionally, the simulation indicated how the “cost” measured in terms of the number of tests carried out is related to the liked status of the top-level option within the query. The number of option comparisons for each level of search when N devices are interrogated is shown in Table 14.2, showing that a message where the top-level category is very popular will cause more work than a message where the top-level category is unpopular, and a theoretical consideration of an m-level search is given in Chap. 20.

Table 14.2 Summary of results from all simulation runs

Case	P(L1 liked)	Level 1	Level 2	Level 3
2	0.1	N	1.10 N	1.11 N
1	0.3	N	1.29 N	1.31 N
3	0.55	N	1.54 N	1.68 N

Measurement Framework

The protocol benchmark used for comparison of the performance of opportunistic network protocols is the epidemic protocol. The measurement framework for the protocol has already been established in prior work [5, 6]. However, a similar benchmark does not exist for the process of service matchmaking, and it is important for the validation of experimental findings to establish pertinent metrics.

An initial set of characteristics to monitor is likely to comprise the following:

- Quantity of messages that have been successfully matched with a user's preference;
- Time taken to achieve a match;
- Volume of traffic imposed upon the network;
- Latency: the time taken to match a service advertisement to a user;
- Quantity of processor cycles consumed by a mobile device within the network. This will help infer the quantity of energy consumed by the process.

Once these characteristics have been identified and verified, other challenges can be evaluated.

In particular, this will permit investigation into the effects of differing user and device behaviour. A user may arbitrarily decide to simplify the matchmaking process by deliberately using broad terms to express their preferences. Such users may be satisfied with a general set of advertisements appearing upon their devices.

Conversely, a user may develop very specific preferences that require several levels of validation before a successful match is achieved. This will necessitate more processing cycles, increased network traffic and, as a result, more energy from the device power source.

The application developer may also wish to either delegate the matchmaking service to the protocol wrapper, or alternatively use the functionality provided and extend it by augmenting new functionality.

Such an approach could imbue a device with a greater degree of sophistication and autonomy in terms of managing the service advertisement management process. It is feasible that this could also be a dynamic behaviour that responds to environmental and system effects such as a low battery, where power needs to be conserved.

One further aspect is to investigate the effect of user mobility patterns upon the message propagation and matchmaking process, since a user's lack of a preference for a service advertisement should not prevent messages spreading in an ON environment.

The preliminary results for the matchmaking algorithm are promising, and the next step is to implement the algorithm within the protocol wrapper architecture in the ONE simulator in order to start optimizing the process.

Conclusions

Opportunistic networking has considerable potential for new business models, particularly since it avoids the need for traditional infrastructure, yet there is a proliferation of users with mobile, Wi-fi-enabled devices.

Whilst such devices have typically been smart phones, the wider acceptance and resulting dependence upon "smart" technology mean that emerging devices, such as smart watches and such like, serve as enablers for greater connectedness between users and business organizations.

The realities of ON can already be realized through ad hoc network connections and a variety of network protocols. However, the potential of more sophisticated uses of this technology also presents challenges for the research community, such as the need to be able to efficiently manage and provision network services in an ON environment.

This research makes use of prior work to establish a measurement framework for ON, and augments this by proposing a means by which the propagation of messages can be managed both by users and, more transparently, by the mobile devices themselves, through the use of a matchmaking algorithm and protocol wrapper.

The protocol wrapper enables matchmaking to be added to the ON protocol of choice, without requiring modification of the underlying protocol. This architecture permits a more autonomous approach to managing the matching of service advertisements to the preferences of users, whilst also providing access to application developers who may choose to extend the functionality even further at the application layer.

Preliminary simulation of the matchmaking algorithm indicates that in excess of 61% of service advertisements can be discarded at the first level, which suggests that there may be a measurable reduction in power consumption as a result of reduced processing on mobile devices. As mobile devices reduce in size and become more pervasive and embedded, this is of particular interest to the research and business communities.

Case Studies 2a: Effectiveness of Message Transmission/Travel

In this first simulation, each automaton moves along the grid, Fig. 14.5, representing movements of people in a shopping centre; at each junction, it can choose its next direction of travel (same direction most likely, reversed direction least likely and options to turn left or right). In the second simulation, the automata/agents move in a randomly chosen direction at a randomly chosen velocity. When "meeting" another automaton, the carrier will pass the message if the receiving automaton is both able to receive messages and selected to receive messages.

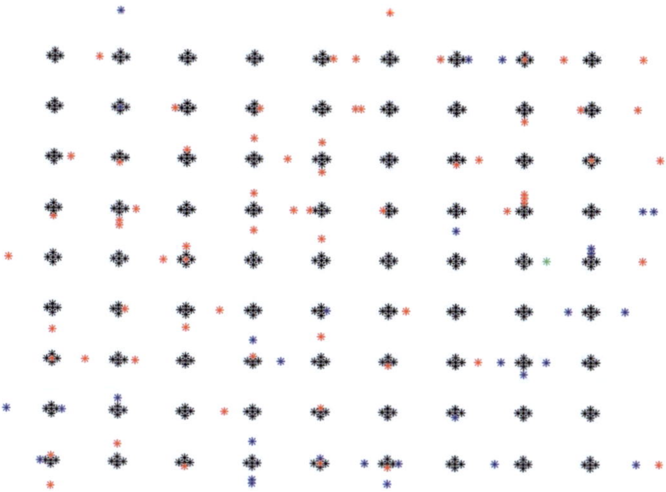

Fig. 14.5 Movement of customers

Table 14.3 Simulation results comparing protocols and efficiency

Device density (%)	Area density (%)	Directional		Non-directional (random)		Non-directional	
		Time	Messages	Time	Messages	Time	Messages
2	25.1	153	187	131	279	114	290
1.5	18.8	150	125	140	227	128	249
0.7	8.8	269	125	225	228	200	245
0.2	2.5	578	112	528	235	397	210
0.1	1.25	807	100	719	258	631	263
0.04	0.5	1502	163	1134	212	888	230

In these simulations, a single message enters the grid with a target destination, and when two devices are "close", messages are passed according to one of the rules:

- To any other device with sufficient capacity and passing close enough to the message carrier
- Randomly to a fixed proportion of devices passing close enough to the message carrier
- Any other device travelling in the correct direction
- Randomly to another device travelling in the correct direction.

For the first model, simulations were carried out with the density of message carriers (proportion of available spaces occupied) of 1.5, 0.7, 0.2, 0.1, 0.04%.

Each configuration was simulated 20 times, and the average results, time for the message to reach the destination and number of messages passed are shown in Table 14.1 (Table 14.3).

The results (see Table 14.1) from all cases where cost was measured in terms of the number of messages passed showed that

$$\text{Time(directional)} > \text{Time(random non directional)} > \text{Time(non directional)}$$

$$\text{Cost(directional)} < \text{Cost(random non directional)} < \text{Cost(non directional)}$$

Time models (p percentage loading), high correlations

Directional (D)	$T_D = 13.5p^{-0.60}$
Non Directional-random(ND(rd))	$T_{ND(rd)} = 13.6p^{-0.57}$
Non Directional (ND)	$T_{ND} = 13.5p^{-0.54}$

For a high percentage loading of 10%, $p = 0.1$, the message passing time estimates for the three models are as follows:

$$T_D = 54$$
$$T_{ND(rd)} = 51$$
$$T_{ND} = 47$$

Notice that as p tends to 1, the times to pass messages converge, and at a loading of (about) 44.5%, the time differences, directional to non-directional, are less that 5% (Fig. 14.6).

Thus, it would seem that

Low density	"non-direction" passes the message much quicker (non-direction 40% reduction in time, but 41% more messages passed)
High density	"non-direction" passes the message quicker (but times close although non-direction does pass more messages).

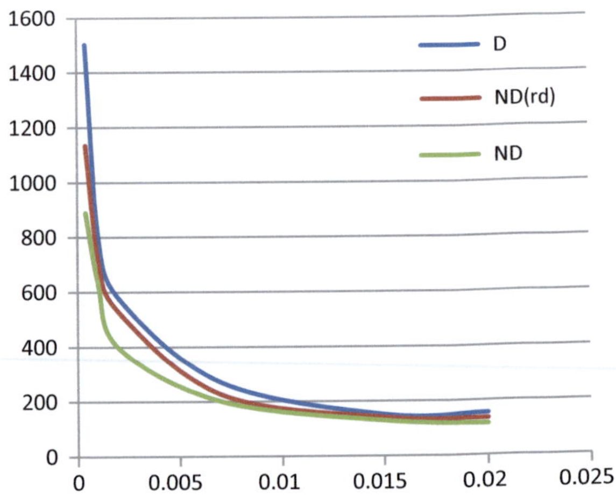

Fig. 14.6 Comparing the results from the message passing protocols

Recommendations

Low density:	"non-direction", real difference in time to pass message
High density:	"direction", no real difference in times to pass messages and fewer messages passed.

Message passing models, low correlation, messages passed (M) loading (p)

Directional passing	D	$M = 3.6p + 98$
Non directional Random passing	ND(rd)	$M = 2.9p + 224$
Non directional	ND	$M = 2.3p + 230$

Case Studies 2b: Effectiveness of Message Transmission/Travel

The third case study is based on message passing within an "open area/town square", where people can move in all directions, see Fig. 14.7. The objective of the case study is to measure the time required to be able to receive the message at a given destination.

The simulations were carried out with initial density of message carriers (proportion of available spaces occupied) of 7.5, 5.0, 2.5, 2.0, 1.5, 1.0, 0.5%.

Each configuration was simulated 10 times, and the average results, time for the message to reach the destination and number of messages passed are shown in Table 14.4.

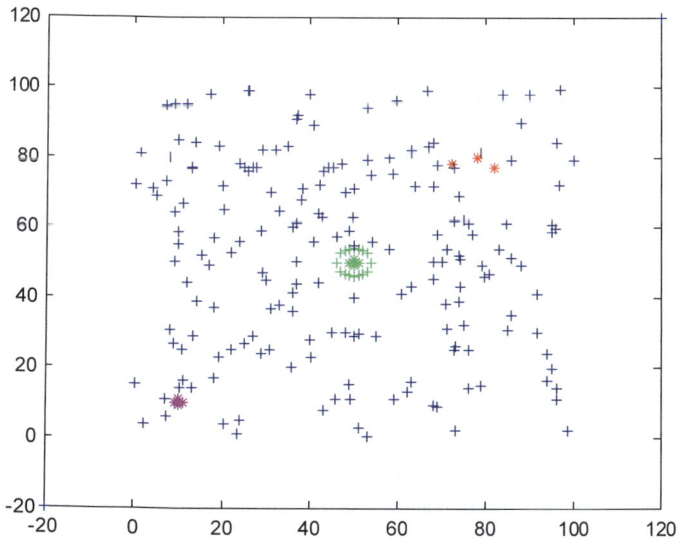

Fig. 14.7 Target point for message at (50,50). Note: *blue colored plus* denotes a customer and *red colored asterisks* a customer carrying a message

Table 14.4 Time to destination and number of messages passed, protocol pass message on all occasions

Density (p) (%)	Time to target (T)	Number of messages passed (P)
0.5	131	16
1	69	21
1.5	57	23
2	54	41
2.5	53	45
4.0	44	116
5.0	42	146
7.5	22	158

Table 14.5 Time to destination and number of messages passed, protocol pass message on 50% of occasions

Density (p) (%)	Time to target (T)	Number of messages passed (P)
0.5	172	14
1	104	26
1.5	59	22
2	55	38
2.5	55	40
5.0	45	120
7.5	28	132

From these results, the following approximate models can be deduced:

$$\begin{aligned} T &= 80p^{-0.52} & r^2 &= 0.91 & &\text{all values of } p \\ T &= 66 - 5.3p & r^2 &= 0.98 & &p > 1\% \\ P &= 24p - 1.5 & r^2 &= 0.92 & & \end{aligned}$$

Repeating the simulation when are the messages passed on only 50% of occasions gave the results (Table 14.5).

From these results, the following approximate models can be deduced:

$$\begin{aligned} T &= 97p^{-0.57} & r^2 &= 0.92 & &\text{all } p \\ T &= 65.5 - 4.5p & r^2 &= 0.99 & &p > 1\% \\ P &= 21p - 0.5 & r^2 &= 0.96 & & \end{aligned}$$

The models derived from the two message passing protocols are compared in Fig. 14.8, showing that as the percentage loading increases, the time differential becomes less significant, but Fig. 14.8 shows that as the loading increases, the percentage increase in messages passed tends towards (approximately) 15%.

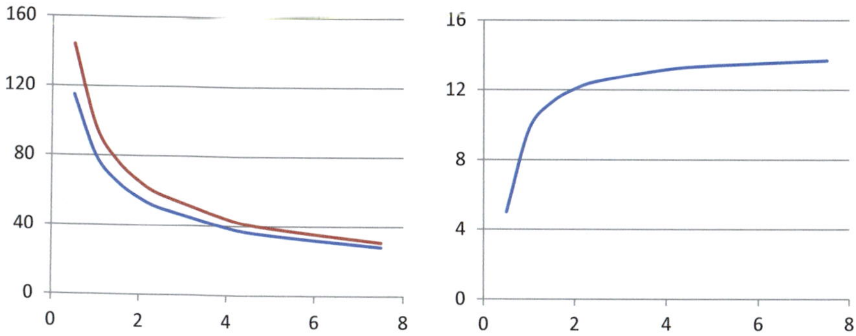

Fig. 14.8 Comparing message passing protocols

14.2 Fire Evacuation Modelling

The aim is, for a given environment, to investigate the effect of the following:

- Number of people (cellular automata) present
- Number of exits
- Position of exits
- Size of room/exits
- Rate of spread of a fire.

Here, although time to exit the room is important, the expected number of casualties is of greater importance.

The optimal design is now influenced by the need of the customers to avoid the spreading fire; so it is expected that this extra constraint could act to change the previous conclusions.

This case study is developed in the three stages:

- movement of people exiting a room
- spread of a fire in a room
- full model where people are attempting to exit a fire in a room.

14.2.1 Modelling Movements in a Room

Within a closed environment, bar or shopping centre, customers will tend to move towards certain points, usually displays or service positions.

In a simulation of this scenario, customers are placed at random and move towards a service point, nearest or required, but with an allowed randomness; so a customer could move towards a more distant point. Figure 14.9a shows an early stage in the cellular automata simulation when customers are attempting to access a service point, and Fig. 14.9b shows the (less random) paths chosen in the agent-based simulation where each agent retains a memory of its most recently chosen direction.

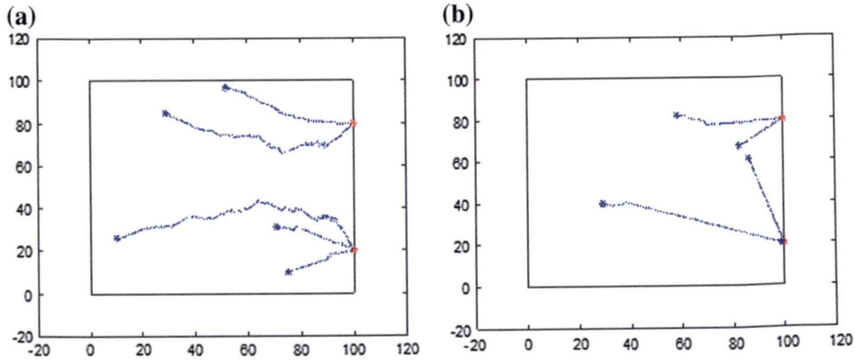

Fig. 14.9 **a** Random choice of exit. **b** Leaving by used entrance

Room Evacuation Modelling

Building evacuation drills is a common part of everyday life in the twenty-first century. Workplaces, universities, hospitals and other institutions by law must have preplanned procedures to enable the occupants of their premises to escape safely in case of fire. New building designs must abide by legislation to meet minimum safety requirements, and the fire service constantly campaigns about the dangers of fire and the need to be prepared. This case study shows how cellular automata may be used to construct models to evaluate both building design and fire escape procedures. The previous simulation is amended so that the destinations are "doors" on the boundary of the room and having reached the exit the customers/automata leave the simulation.

Two models can be developed, the first where the automata head towards the nearest exit, but with some randomly choosing to head towards the further exit and a second model where most of the agents choose to head towards the "entry" door and a few moving towards the other (emergency) exit. Figure 14.10 shows plots of the automata paths to the exit from both models; in each model, there are 100 automata.

The objective, of the simulations, is to investigate the effect of the location of the emergency exit and the capacity (width) of the "entry" door on the time to evacuate the area (because the entry door aims to control people coming into this area and it will tend to have a controlled capacity and width.

In all the simulations, the room had dimensions 100×100 with

the "entry" located at the point (100,100) and
emergency exits located at points (100,0), (100,50), (100,80), (100,90).

Each agent chooses their exit point on the basis of the following:

Distance, the closest is more favoured but choice is random and
The influence of their point of entry into the system.

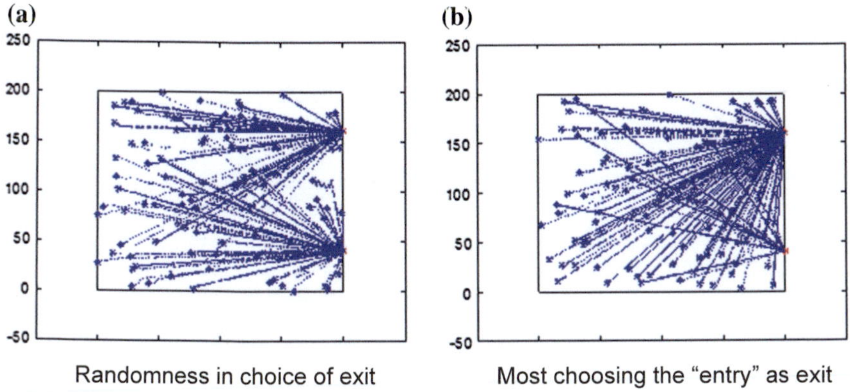

Fig. 14.10 Illustrating the effect of the different exit choice strategies

The results from these simulations (low, high and very high demand) are summarized in Table 14.6a–c. These results indicate that although the location of the emergency exit, at high levels of demand, has some effect on the time required to fully evacuate the area, the capacity of the "entry" point, during evacuation, has a much greater effect on the time to clear the area.

These results lead to the summaries in Table 14.6d–f indicating that the capacity of the "entry" point is more important than the location of the emergency exit with respect to the time required to evacuate a building. In cases of high occupancy, the width of the entry/exit point becomes very significant, savings of over 50% in total evacuation time.

However they do indicate that when there is a tendency for customers to aim to leave using their original entry point then an emergency exit should be located close to this entry point (as any entry point will have a limited capacity) to enable a speedier evacuation of the building.

14.2.2 *Model 1: Modelling the Spread of a Fire*

The location of the fire is generated randomly and the probability that it will spread into an adjacent cell is a function of the states (on fire or not) of the 8 surrounding cells.

Figure 14.11a shows a fire starting at a single source, and Fig. 14.11b shows the fire developing.

Similarly, Fig. 14.11c shows the fire starting at two points, and Fig. 14.11d shows how this fire develops.

The next stage adds barriers into the room, walls and doorways, for example, to model their effect on the spread of fire in an enclosed area.

Table 14.6 Results from low, high and very high demand simulations

(a) Low demand				
Emergency exit	Demand = 200	Time for last to leave "entry" width		
	Leaving by time	1	2	3
(100,0)	"entry" point	180	133	125
	Emergency exit	114	100	95
	Last to leave	180	133	125
(100,50)	"entry" point	171	131	127
	Emergency exit	97	95	95
	Last to leave	171	131	127
(100,80)	"entry" point	168	130	125
	Emergency exit	98	104	105
	Last to leave	168	126	125
(100,90)	"entry" point	168	135	125
	Emergency exit	93	100	100
	Last to leave	168	135	125
(b) High demand				
Emergency exit	Demand 500	Time for last to leave "entry" width		
	Leaving by time	1	2	3
(100,0)	"entry" point	439	222	155
	Emergency exit	112	123	119
	Last to leave	439	222	155
(100,50)	"entry" point	427	209	144
	Emergency exit	95	100	103
	Last to leave	427	209	144
(100,80)	"entry" point	423	213	142
	Emergency exit	105	90	115
	Last to leave	423	213	142
(100,90)	"entry" point	420	209	143
	Emergency exit	107	110	105
	Last to leave	420	209	143
(c) Very high demand				
Emergency exit	Demand 750	Time for last to leave "entry" width		
	Leaving by time	1	2	3
(100,0)	"entry" point	670	338	225
	Emergency exit	120	120	124
	Last to leave	670	338	225
(100,50)	"entry" point	639	329	216
	Emergency exit	102	100	98
	Last to leave	639	329	216

(continued)

Table 14.6 (continued)

(c) Very high demand				
Emergency exit	Demand 750	Time for last to leave "entry" width		
(100,80)	"entry" point	629	311	208
	Emergency exit	120	120	115
	Last to leave	629	311	208
(100,90)	"entry" point	585	292	204
	Emergency exit	115	120	116
	Last to leave	585	292	204

(d) Location of emergency exit

Demand	(100,0)	(100,90)	Percentage saving
200	180	168	7
500	439	420	4
750	670	585	13

(e) Width of main exit: Emergency exit at (100,0)

Demand	1	2	3	Percentage saving
200	180	133	125	31
500	439	222	155	65
750	670	338	225	66

(f) Width of exit: Emergency exit at (100,90)

Demand	1	2	3	Percentage saving
200	168	135	125	31
500	420	209	143	66
750	585	292	204	65

This shows how the fire is restrained by the wall and spreads through the doorway (Fig. 14.11e, f).

The final Fig. 14.11g, is an example of a more complex enclosure, where the fire is again retarded by the walls and spreads through the open doorways.

14.2.3 Modelling the Effect of Fire and the Movement of People

These models investigated the effect of the location and size (capacity) of the exits on the survival of the people.

Model 1: A fixed room without any obstacles was created, and people (cellular automata) were inserted at random locations. The number of occupants varied between 1 and 80 (room loading between 1 and 80%); in each case, the occupants would move towards the exit provided that their route was not blocked by the fire, causing a detour, or by other people, causing queues at an exit. The starting position of the fire was generated randomly, and the simulations until the room had been evacuated.

The results from these simulations are shown in Fig. 14.12.

These results can be modelled by the logistic curve

$$C = \frac{82}{(1 + 7.2e^{-0.08P})}$$

where C is the percentage number of casualties and P is the population.

Model 2: Investigating the Effect of Exit Width

Simulations were carried out, where the width of the exit varied, again with between 1 and 80 people present. The results are summarized in Fig. 14.13.

These simulations have shown that for normal room loading (office or residential), there is an advantage to having wider (2 width) exit points, normally, the door and a specific (extra) fire exit, and for very high loadings, class/lecture rooms, music concerts and sporting events, there seems to be a need for even wider exits (3 or more width).

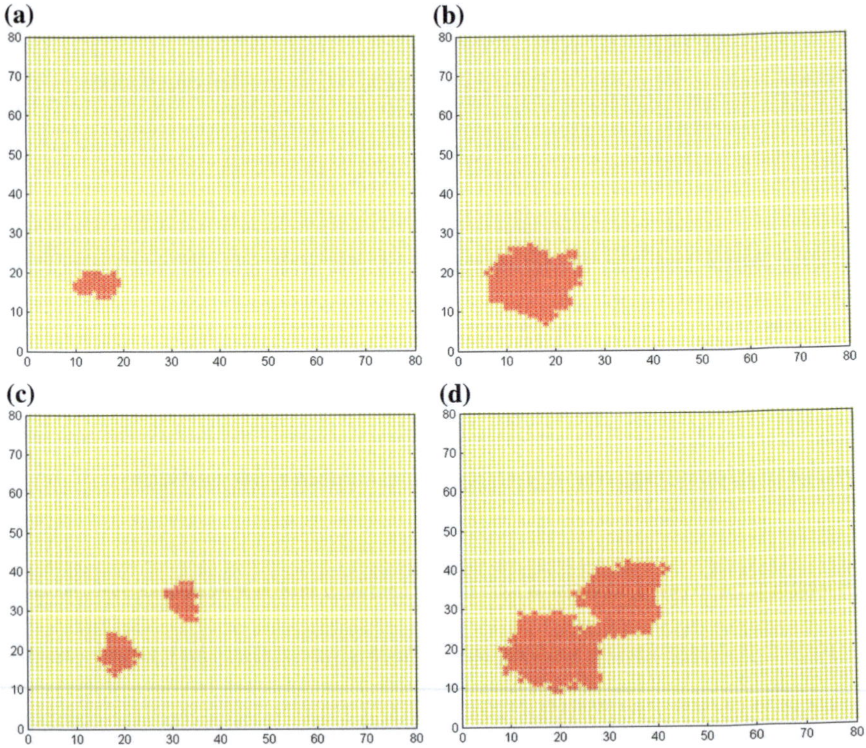

Fig. 14.11 Simulating the spread of a fire

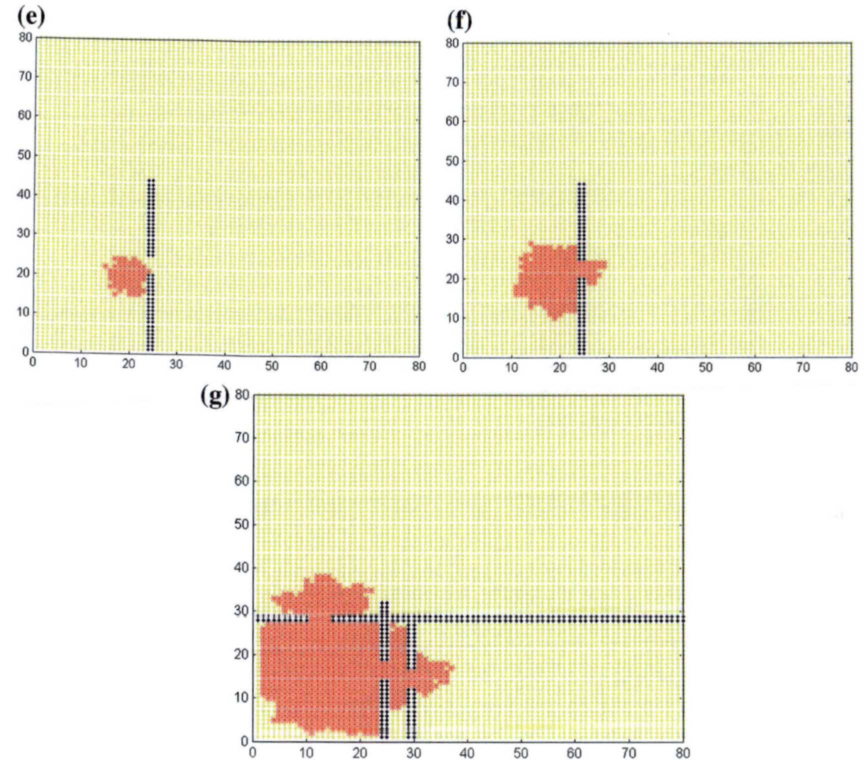

Fig. 14.11 (continued)

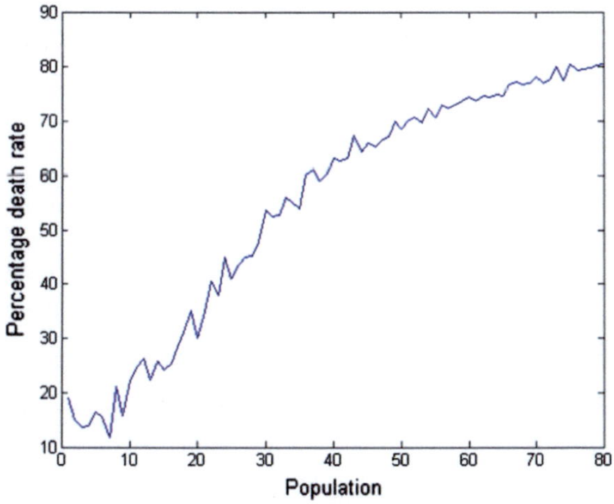

Fig. 14.12 The effect of population size on the number of casualties

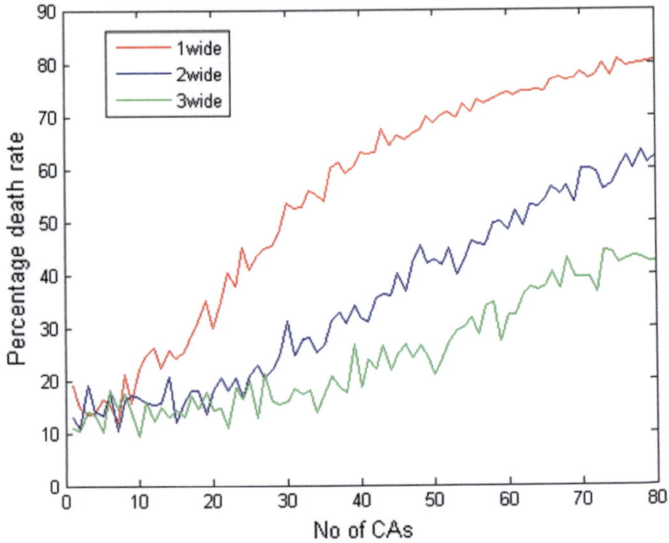

Fig. 14.13 The effect of exit width on the number of casualties

Model 3: Investigating the Effect of the Exit Location

Three locations for a single exit were considered. The results showed that in the case of a randomly located fire, the location of the exit does not influence the number of casualties Fig. 14.14. Hence, it would seem that when designing a facility with a single exit, when a fire could occur anywhere, the precise location of the exit is not an important design feature.

Summary: The simulations demonstrated that in the event of a fire occurring in an enclosed area, the location of the exit is much less significant than the capacity of this exit. In practice, this implies that the entry point (most used as an exit in emergencies), which aims to control/restrict entry, must be able to increase its (exit) capacity in an emergency.

Note: Choice of Emergency Exit

Each agent carries a memory of entry points/exit points and uses this information in their choice of emergency exit.

For example, for a room with several possible exits with attractiveness (knowledge-based or user-based) $m_1, m_2, .., m_n$, the probability of heading towards exit i is given by

$$P(\text{exit}_i) = \frac{S_i}{\sum S_j} \quad \text{where} \quad S_i = m_i/d_i$$

This can be extended through the addition of a crowd effect, where the agents tend to follow the crowd when the probability is given by

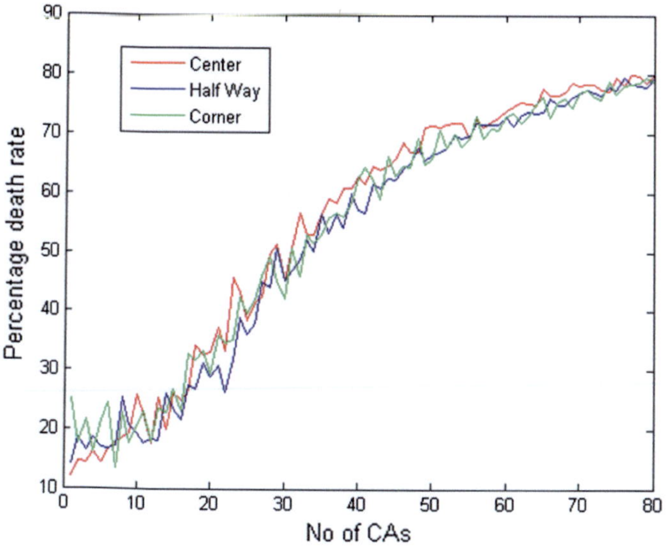

Fig. 14.14 The effect of exit location on number of casualties

$$P_t(\text{exit}_i) = \frac{n(d_i)S_i}{\sum n_t(d_j)S_j}$$

where $n_t(d_i)$ relates to the attractiveness of exit i at time t.

References

1. Camp T, Boleng J, Davies VA (2002) Survey of mobility models for ad hoc network research. Wireless communications & mobile computing (WCMC): Special issue on mobile ad hoc networking: research, trends and applications, vol 2, pp 483–502
2. Johnson DB, Maltz DA (1996) Dynamic source routing in ad hoc wireless networks. In: Mobile computing, Kluwer Academic Publishers, pp 153–181
3. Choffnes DR, Bustamante FE (2005) An integrated mobility and traffic model for vehicular wireless networks. In: Proceedings of the 2nd ACM international workshop on vehicular ad-hoc networks, pp 69–78
4. Yu S, Al-Jadir L, Spaccapietra S (2005) Matching user's semantics with data semantics in location-based services, 1st workshop on Semantics in Mobile Environments (SME'05)
5. Smith A, Berry S (2012) Evaluation of a framework for measuring efficiency in opportunistic ad-hoc networks, emerging intelligent data and web technologies. In: 3rd international conference
6. Smith A, Hill R (2011) Towards framework for the evaluation of efficient provisioning in opportunistic ad-hoc networks. In: Proceedings of the 2011 international conference on P2P, IEEE Computer Society

Chapter 15
Three Big Data Case Studies

Marcello Trovati and Andy Baker

15.1 Case Study 1: Criminology

The utilisation of Big Data within the criminology field has allowed a revaluation of the traditional assessment and investigation of available data sources, pushing criminological research towards new frontiers [1, 2]. However, the enormous amount of data, which is now available from numerous sources focusing on criminology , has created both challenges and opportunities in the discovery of innovative approaches to prevent, detect and predict crime.

In particular, an important aspect of current research is the automation of the decision-making process applied to criminology, see Fig. 15.1, and to enable an automated assessment of the available data, to facilitate the knowledge discovery process.

15.1.1 Text Analysis of Datasets

Text fragments were analysed via the Stanford Parser [3] and Python NLTK [2], and the probabilistic relationships among the different concepts were extracted via text patterns. The text patterns considered included, first of all, the following quadruples:

$$(\mathrm{NP1}, \mathrm{MOD}, \mathrm{PROB}, \mathrm{keyword}, \mathrm{NP2})$$

M. Trovati (✉)
Department of Computer Science, Edge Hill University,
Ormskirk L39 4QP, UK
e-mail: M.Trovati@derby.ac.uk; trovatim@edgehill.ac.uk

A. Baker
College of Engineering and Technology, University of Derby,
Kedleston Road, Derby DE22 1GB, UK

S. Berry et al. (eds.), *Guide to Computational Modelling for Decision Processes*,
Simulation Foundations, Methods and Applications,
DOI 10.1007/978-3-319-55417-4_15

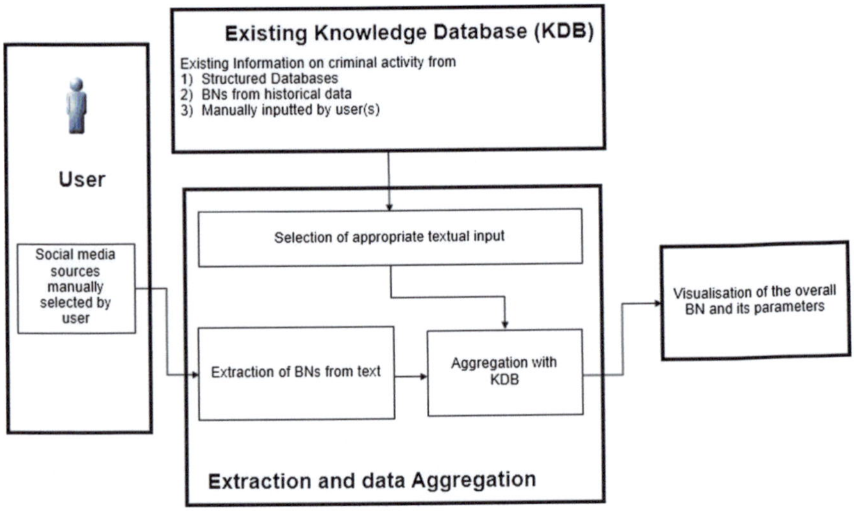

Fig. 15.1 General architecture of Bayesian networks for crime detection as discussed in Trovati [1]

where

- NP1 and NP2 are the noun phrases, where a noun is the head word.
- keyword refers to probabilistic terms collected and identified in an ontology [1].
- MOD is the keyword modality, which can be either positive or negative, depending on whether it confirms or negates the existence of a specific probabilistic relationship.
- PROB includes keywords associated with probability, such as may, might, could and likely.

Similarly, we also define the following pattern:

$$(\text{MOD}, \text{VerbKeyword}, \text{NP1}, \text{NP2}).$$

In such case, its components are the same as in the previous pattern, with the only difference in the characterisation of VerbKeyword, which includes a verb followed by link, cause and correlation.

An example of the first pattern is the following sentence:

A wealthy background is not likely to influence mental well-being.

This suggests that a wealthy background and mental well-being are probably not linked by a relation. In particular, the modality, which refers to the keyword "not", captures such lack of relationship. Note the presence of a PROB keyword, namely "likely". For more details and full theoretical justification, please refer to Trovati and Bessis [1].

15.1.2 Node Extraction

In the manual definition of DNs, the identification of the appropriate nodes typically depends on a variety of parameters, including personal choice. There could be, in fact, some concepts, such as "*day*", which could be considered per se, even though when an attribute, for example, "*cloudy*" is associated with them. Alternatively, "*cloudy day*" might be considered as a single entity. Clearly, this task is very complex to fully address when extracting nodes automatically from text, due to the dependency on the user's preferences, as well as on the deep semantic understanding, which would be required.

In this context, each node refers to a concept consisting of one or more terms extracted from text, directly identifiable from structured datasets, or from textual sources to extract collections of nouns, as well as keywords defined by the user.

Similar to Trovati [1], semantically equivalent noun collocations, such as "unlawful act", would be considered equivalent to, for example, "criminal act". In the implementation of the system, WordNet was used to identify semantically equivalent objects. In particular, if ⟨adj 1, noun 1⟩ and ⟨adj 2, noun 2⟩ such that adj 1 is semantically similar to adj and so are noun 1 and noun 2, then these would be regarded as the same entity and combined into a single node v.

This node would subsequently be incorporated into the corresponding stemming root unless the user wishes to use a different root. For example, words such as "*cats*" and "*pussy cat*" would have the same root "*cat*".

15.1.3 Extraction of Dependency Networks

Once the relevant nodes are identified, suitable network(s) are created depending on the different relations between them defined by the text analysis results.

In particular, the text patterns discussed above provide the links between the concepts in NP1 and NP2. In the approach used in this context, the nodes contained in NP1 are pairwise connected with the nodes from NP2, aiming to maximise recall rather than precision.

Once the network or fragments of networks have been defined, their properties can be further investigated, which will provide an insight into the probabilistic relations, based on the threshold $pt = 0.5$ [1].

15.1.4 Description of the Dataset

Three unstructured datasets were considered for validation purposes. The first contained research questionnaires containing information concerning change management process during times of austerity within an English police service.

A total of 103 interviews were transcribed for subsequent text analysis. The other two datasets were the Reuters News Corpus and the Brown Corpus [3]. The former contains news stories for text mining, information retrieval, and machine learning research and evaluation, with over 800,000 manually categorised newswire stories. They contain texts from 500 sources, categorised by the relevant genre, such as news, editorial and reviews. In particular, the extraction of relevant information obtained from these two datasets would provide further relevant insights. In fact, while the above interviews contain specific information relevant to a specific criminology aspect, the Reuter and Brown Corpora can provide further intelligence and knowledge based on its size and contextual width.

15.1.5 Implementation

The main implementation was carried out in Python 2.7 with the NLTK and NetworkX libraries. NLTK provides an interface to more than 50 corpora and lexical resources such as WordNet, which integrates text processing libraries to classify and analyse textual data as well as semantic reasoning capabilities [1].

The NetworkX library is designed to define, analyse and investigate the proper [4] ties of complex networks [4]. This was used to define the networks extracted from the textual sources, and subsequently analyse them via the numerous methods available in this library. In fact, it contains several algorithms to fully assess the topology of networks.

15.1.6 Evaluation

All the interviews were merged into a text file, which was subsequently analysed as based on the following keywords:

- Leadership
- Collaboration
- Austerity
- Financial and staff cuts
- Partnership
- Rehabilitation
- Punishment
- Risk management
- Organisational change
- Service user
- Community penalties
- Public image

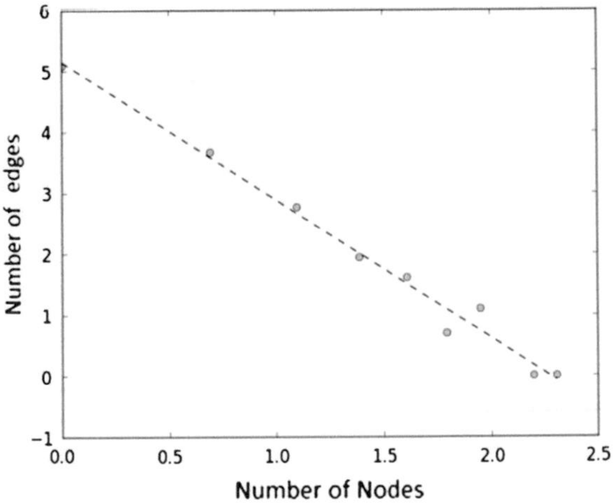

Fig. 15.2 Scale free structure of the network extracted

- Drugs
- Violence
- Privatisation
- Structural change
- Offender
- Politics
- Police
- Prison
- Sex offenders and
- Mentally disordered offenders.

To maximise the recall of relevant information, we also considered their lexical variations based on WordNet.

The analysis defined a network with over 15,000 nodes and over 48,000 edges, exhibiting a scale-free structure as depicted in Fig. 15.2.

Using the method described in Trovati and Bessis [1, 5], the parameter k was assessed to be approximately 2.9. Subsequently, all the datasets were analysed to identify suitable nodes, based on the set of keywords described above, and corresponding synonymy properties.

15.1.7 Dependency Network Extraction

Once the nodes and their mutual connections were identified, a group of criminologists manually assessed the different entries and evaluated approximately 80 concepts, which were subsequently compared with the automated extraction of the corresponding influences. This generated a recall of 71% and a precision of approximately 62%.

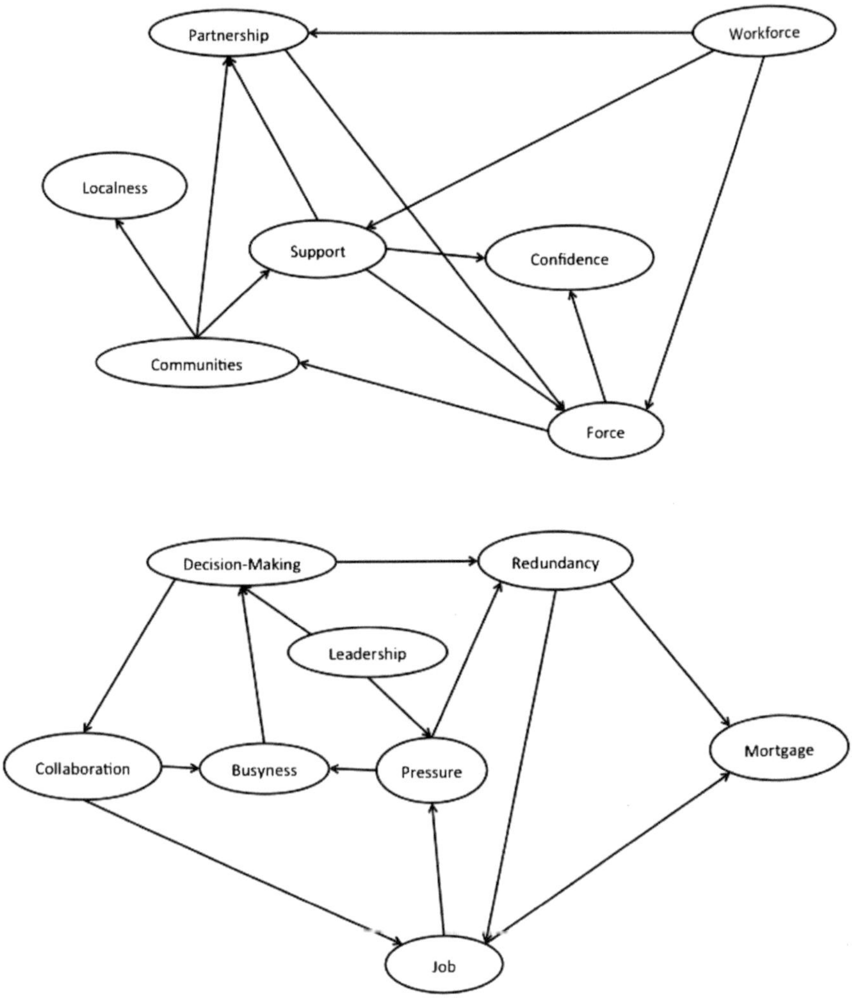

Fig. 15.3 Two DNs extracted from the datasets

Once the probability extraction algorithm has been carried out, only relations with an associated probability above the probability threshold were kept. Figure 15.3 depicts some DNs extracted.

The same group of experts also evaluated the above dependency networks, who agreed on their topological structure. However, it was noted that an edge between localness and support should be present.

15.2 Case Study 2: Computational Objectivity in the PHQ-9 Depression Assessment

The Patient Health Questionnaire (PHQ-9) is the most common depression assessment tool, which aims to identify the severity and type of depression an individual may be suffering from. It is based on the Diagnostic and Statistical Manual of Mental Disorders (DSM-IV) criteria, and it consists of nine questions (each utilised by general practitioners and mental health professionals in the diagnosis of depression [5]. Each question is scored with values ranging between 0 ("not at all") and 3 ("every day") with a general score of 27, as shown in Table 15.1.

The validity of the PHQ-9 must be constantly monitored. A measure of internal consistency is the Cronbach's alpha, which highlights the extent to which all items in a questionnaire measure the same concept. More specifically, this further assesses the efficacy of PHQ-9 measure depression [5]. It is computed as

$$\alpha = \frac{K}{K-1}\left(1 - \frac{\sum_{i=1}^{K} \sigma_Y^2}{\sigma_X^2}\right)$$

where

K is the number of items in the test,
σ_X^2 is the variance of observed total questionnaire scores and
σ_Y^2 is the variance of component i for the current sample of people.

Cronbach's alpha is within the interval $[0, 1]$, whose categorisation is as follows:

- $\alpha \geq 0.9$ excellent
- $0.7 \leq \alpha < 0.9$ good
- $0.6 \leq \alpha < 0.7$ acceptable
- $0.5 \leq \alpha < 0.6$ poor
- $\alpha < 0.5$ unacceptable

The Cronbach's alpha for the PHQ-9 has been assessed as 0.85, suggesting it is a good resource for detecting depression. However, due to the importance of depression, it is crucial to ensure it can be upgraded to the "excellent" category, that is when $\alpha \geq 0.9$.

Table 15.1 Depression severity score

Outlined depression severity PHQ-9 score	
1–4	None
5–9	Mild
10–14	Moderate
15–19	Moderately severe
20–27	Severe

Table 15.2 Relative importances

Intensity of importance	Definition	Explanation
1	Equal importance	Two elements contribute equally to objective
3	Moderate importance	Experience and judgement slightly favour one element over the other
5	Strong importance	Experience and judgement strongly favour one element over the other
7	Very strong importance	Experience and judgment very strongly favour one element over the other
9	Extreme importance	The evidence favouring one element over another is of the highest possible order of affirmation

Analytical hierarchy process (AHP) aims to provide a set of tools to facilitate the decision-making process with a wide set of applications to a variety of multidisciplinary contexts [5]. The main idea behind AHP is the evaluation of the importance of specific contributors, which are likely to have an influence on an objective, by carrying out a pairwise comparison between each and every contributing element. This process is based on a rating scale, as depicted in Table 15.2, to determine the importance of each element when compared with other ones.

Since the PHQ-9 consists of nine questions, each of them can be regarded as contributors, so that the AHP can be applied aiming to achieve the relevant diagnosis. To achieve objectivity, selected experts in the field would need to consider each pairwise combination of questions so that they could decide and rate which of the questions they consider more important. Once all the questionnaires have been completed, an overall mean average can be computed, allowing the construction of a final analysis pairwise comparison table.

AHP is an efficient method, which can be interactively modified at any stage of development. Furthermore, it allows a thorough investigation of the problem, so that any discrepancies can be addressed effectively. Furthermore, the AHP questionnaire could be distributed to a large number of experts representing particular statistical cross sections of a population.

15.2.1 A Text Mining Approach

In order to improve the above method, text mining techniques were implemented to automate the use of the AHP described above. Each question in the PHQ-9 questionnaire contains specific keywords, including the following:

- Pleasure
- Interest
- Hopeless

- Feeling down
- Depressed
- Asleep
- Sleep
- Tired
- Energy
- Appetite
- Overeating
- Failure
- Feeling bad
- Concentrating
- Moving slowly
- Fidgety
- Restless
- Death and
- Hurting

Such collections of keywords can provide an indication of the type and frequency of their use. In fact, the more often they occur, the more likely they are to be relevant in depression detection via the PHQ-9. PubMed contains over 24 million citations and abstracts for biomedical literature, life science journals, and online books [5].

Over 260,000 abstracts from PubMed were identified by considering the keywords "PHQ-9" and "depression". Subsequently, the combination of the keywords corresponding to each of the nine questions was extracted from the above abstract, and each of the items was then fed into the AHP method to assess the importance of the questions.

The results were evaluated by a number of health professionals, who had agreed on the suggested ordering. Furthermore, the weight of each question was determined by renormalising the number of occurrences of the corresponding keywords, which proved more difficult to agree upon by the health professionals, with average of 40%. This can be easily explained by the fact that this step would require a much deeper understanding of the issues regarding depression assessment compared to their ranking.

15.2.2 Conclusion

In this chapter, two study cases have been described, which highlight the importance of Big Data science and its applications. The exponential amount of data which is being continuously created poses new challenges, which, if harnessed, could advance the state-of-the-art technology in a variety of fields, as well as providing new research direction to further develop theoretical approaches to data acquisition, analysis and visualisation.

15.3 Case Study 3: Admissions Project

The objective was the construction of a predictive model to estimate the likelihood that a given student would progress from an initial enquiry or application to enrolling onto their chosen course of study.

The initial dataset comprised of 15 variables with categorisation of the values as shown.

LOC (Location): six-level factor with values

DERBY, NEARBY BIG CITY, NEARBY, OTHER UK, SCOTWALESNI, NOTUK

AGE: six-level factor with values

18, 19, 20–24, 25–29, 30–34 and 35+

SEX: two-level factor with values

Male, Female

P2 (POLAR 2): five-level factor with values
P3 (POLAR 3): five-level factor with values

One, Two, Three, Four and Five

ETH (Ethnicity): seven-level factor with values

Asian, Black, Mixed, Missing, Not applicable, Other, White

DIS (Disability): eleven-level factor with values

- A No disability,
- B A social/communication impairment such as Asperger's syndrome/other autistic spectrum disorder,
- C Blind or have a serious visual impairment uncorrected by glasses,
- D You are deaf or have a serious hearing impairment,
- E A long-standing illness or health condition such as cancer, HIV, diabetes, chronic heart disease or epilepsy,
- F A mental health condition, such as depression, schizophrenia or anxiety disorder,
- G A specific learning difficulty such as dyslexia, dyspraxia or AD(H)D,
- H Physical impairment or mobility issues, such as difficulty using your arms or using a wheelchair or crutches,
- I A disability, impairment or medical condition that is not listed above,
- J Two or more impairments and/or disabling medical conditions,
- M Missing

SOCIO (Socioeconomic background): eight-level factor with values

Higher managerial and professional occupations,
Intermediate occupations,
Lower managerial and professional occupations,
Lower supervisory and technical occupations,
Not applicable,
Not classified / missing,
Routine occupations,
Semi-routine occupations,
Small employers and own account workers.

DOM (Domicile country): three-level factor with values

EU (excluding UK), Not EU, UK

TYPE (Type of education institution): eight-level factor with values

Academy, Further education, Grammar school, Independent school, Missing, Other, Sixth form college, State

DEF (Deferred status): two-level factor with values

Yes, No

LIVE (Where is student living?): three-level factor with values

Home, Other, University

CAMP (Campus attending): three-level factor with values

B, C, Main

PLC (Placed status): two-level factor with values

Placed, Unplaced

ROUTE (Acceptance route). seven-level factor with values

Adjustment, Direct clearing, Extra, Firm choice, Insurance choice,
Main scheme clearing, Unplaced

Thirteen parameters were chosen Fig. 15.4 representing the relative importance each variable towards the target value "Placed".

This information was used to construct a predictive model.

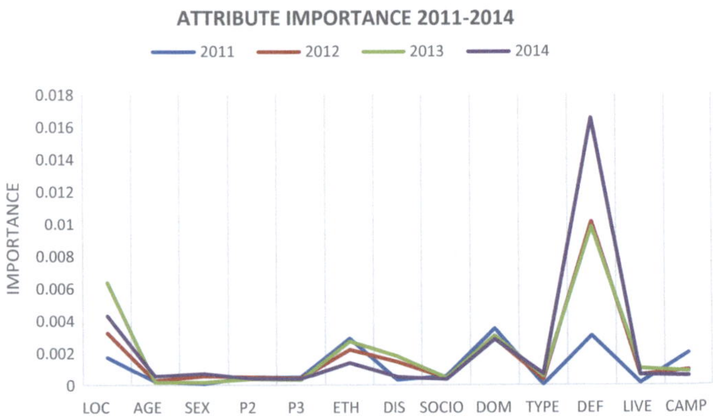

Fig. 15.4 Relative importance of the chosen factors

Table 15.3 Comparing the results from the model with enrolment data

	Number of applicants	Number enrolling	Projected enrolments	Error (%)
2012	16,827	3002	2673	+10.97
2013	17,874	3172	3156	+0.50
2014	19,901	3615	3642	−0.75

Using the available data for the years 2012, 2013 and 2014 as an initial set of test data, the predictive model showed a 0.753% error, wrongly forecasting the final status of the applicants.

As an initial test, of the modelling process, a random sample of 2000 was chosen from the 2014 applications dataset and used as input data into the model constructed from the (full) 2013 data, the results using this test data were that for 97% of the cases the model accurately forecast the students decision.

Applying 2012–2014 versus the previous year's data, using location and postcode only, the following overall projections were available (Table 15.3).

These results, using historic data, indicated that this approach is able to provide information regarding expected (new student) enrolment using information from students UCAS data.

Figure 15.5 provides a visual representation of the admissions data, each arrow representing a causal relationship between variables.

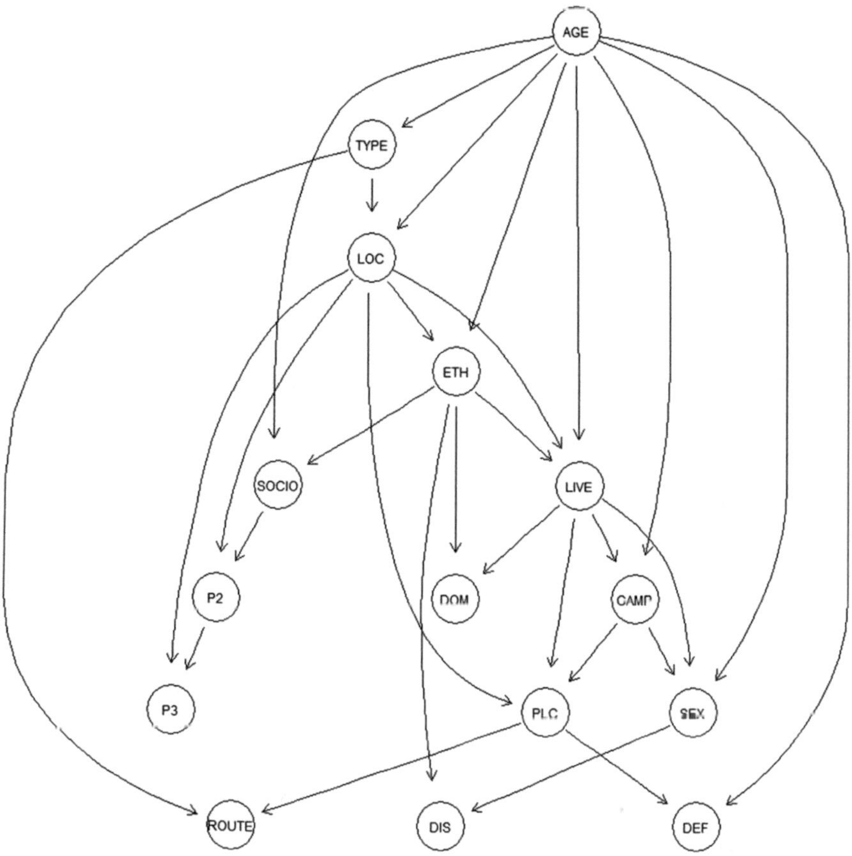

Fig. 15.5 Causal relationships between parameters

References

1. Trovati M, Bessis N (2015) An influence assessment method based on co-occurrence for topologically reduced big data sets. *Soft Computing*
2. Bird S, Loper E, Klein E (2009) Natural language processing with python O'reilly media Inc
3. Manning CD, Schutze H (1999) Foundations of statistical natural language processing. MIT Press, Cambridge
4. Hagberg AA, Schult DA, Swart PJ (2008) Exploring network structure, dynamics, and function using networkX. Proceedings of the 7th Python in Science Conference (SciPy 2008)
5. Johnson A, Holmes P, Craske L, Trovati M, Bessis N, Larcombe P (2015) A computational objectivity in depression assessment for unstructured large datasets. Proceedings of IBDS-2015

Part III
Appendices

Chapter 16
Appendix A: Queueing Theory

Stuart Berry

16.1 Introductory Model

The usual introductory models are derived in the situation where both arrivals and service times are random and there are n-like servers, the M/M/n model. The results from this model provide an indication of the suitability of the required service in an application.

16.2 Analysis of a M/M/* Service System: Derivation of Formula

Arrivals and service times follow an exponential distribution at average rates of k per time unit (for arrivals $k = \alpha$ and for service $k = \beta$); thus

$$f(t) = k\mathrm{e}^{-kt}$$

and the probability of an event/arrival/service completed in the (small) interval:

$$a = t \quad \text{to} \quad b = t + \delta t$$

is given by

S. Berry (✉)
College of Engineering and Technology, University of Derby,
Kedleston Road, Derby DE22 1GB, UK
e-mail: s.berry@derby.ac.uk

S. Berry et al. (eds.), *Guide to Computational Modelling for Decision Processes*,
Simulation Foundations, Methods and Applications,
DOI 10.1007/978-3-319-55417-4_16

$$\int_a^b ke^{-kt}\mathrm{d}t = e^{-ka} - e^{-kb}$$

From which, it follows that

$$e^{-ka} - e^{-kb} \cong (1 - kt + \cdots) - (1 - k(t + \delta t) + \cdots) \cong k\delta t$$

Therefore, it follows that in a small interval of time

$$\begin{aligned} P(\text{arrival}) &= \alpha\delta t \\ P(\text{service ends}) &= \beta\delta t \end{aligned}$$

To complete the analysis, consider the situation where there are n customers in the system (queueing or being served) and consider how there can (still) be n customers a time δt later (Fig. 16.1).

Thus, the state where there are n customers in the system at time $(t + \delta t)$ could have been achieved by:

(i) $(n - 1)$ customers at time t, one arrival, none leaving;
(ii) (n) customers at time t, no arrival, none leaving; and
(iii) $(n + 1)$ customers at time t, no arrival, one leaving.

Giving the probability of n customers in the system at time $t + \delta t$ as

$$P_n(t + \delta t) = \lambda\delta t(1 - \mu\delta t)P_{n-1}(t) + (1 - \mu\delta t)(1 - \lambda\delta t)P_n(t) + \mu\delta t(1 - \lambda\delta t)P_{n+1}(t)$$

Expanding and neglecting terms $O(\delta t^2)$ gives

$$P_n(t + \delta t) = (\lambda\delta t)P_{n-1}(t) + (1 - \mu\delta t - \lambda\delta t)P_n(t) + (\mu\delta t)P_{n+1}(t)$$

or
$$P_n(t + \delta t) - P_n(t) = \lambda\delta t P_{n-1}(t) + (-\mu\delta t - \lambda\delta t)P_n(t) + \mu\delta t P_{n+1}(t)$$

Dividing by δt and letting $\delta t \to 0$ lead to a set of simultaneous differential equations.

Steady-state results can now be derived from these results:

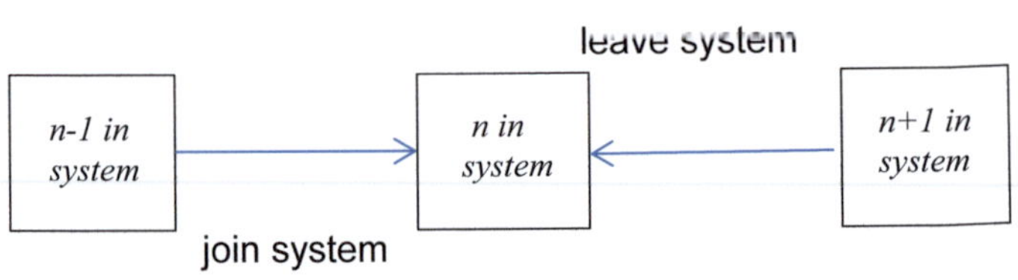

Fig. 16.1 Defining queues

$$\text{for } n > 0: \frac{dP_n(t)}{dt} = \lambda P_{n-1}(t) + (-\mu - \lambda)P_n(t) + \mu P_{n+1}(t) = 0$$

$$\text{and for } n = 0: \frac{dP_0(t)}{dt} = -\lambda P_0(t) + \mu P_1(t) = 0$$

Giving the resultant probabilities

$$P_1 = \rho P_0,\ P_2 = \rho P_1 \ or\ P_2 = \rho^2 P_0$$

$$\text{Note as } \sum_{i=0}^{\infty} P_i = 1 \quad \text{then} \quad P_0 \sum \rho^i = 1 \quad \text{thus} \quad P_0 = 1 - \rho$$

Using these results, it follows that the average number in system is given by

$$\sum_{i=1}^{\infty} i P_i = \sum_{i=1}^{\infty} i \rho^i P_0 = \left(\rho + 2\rho^2 + 3\rho^3 + \cdots\right) P_0 = \rho\left(1 + 2\rho + 3\rho^2 + \cdots\right)(1 - \rho)$$

$$= \rho\left(\frac{1}{(1-\rho)^2}\right)(1-\rho) = \frac{\rho}{(1-\rho)}$$

Average number in queue from

$$\sum_{i=1}^{\infty} (i-1)P_i = \sum_{i=1}^{\infty} i \rho^i P_0 - \sum_{i=1}^{\infty} P_i$$

$$= \frac{\rho}{(1-\rho)} - \rho = \frac{\rho^2}{(1-\rho)}$$

Average time in queue from

$$= \text{Average time for system (on arrival) to clear}$$
$$= \text{Average number in system} \times \text{Average service time}$$
$$= \frac{\rho}{(1-\rho)} \frac{1}{\mu}$$

Average time in system

$$= \text{Average time in queue} + \text{Average time to be served}$$
$$= \frac{\rho}{(1-\rho)} \frac{1}{\mu} + \frac{1}{\mu}$$
$$= \frac{1}{(1-\rho)} \frac{1}{\mu}$$

The general formula for the average number in the queue is given by:

$$\frac{\lambda^2\sigma^2+\rho^2}{2(1-\rho)}$$

For $M/M/1$ this reduces to $\frac{\rho^2}{(1-\rho)}$

$M/D/1$ this reduces to $\frac{\rho^2}{2(1-\rho)}$

16.3 Additional Results

For a situation where there is limited waiting space (many applications) if there is capacity for only k customers, here

$$P_n=\rho^n P_0 \quad \text{for } n\leq k \quad \text{otherwise } P_n=0$$

Thus $\sum_{i=0}^{k} P_i=\sum_{i=0}^{k}\rho^i P_0=1$

$$\sum_{i=0}^{k}\rho^i=(1-\rho^{k+1})/(1-\rho)$$

Then $P_0=(1-\rho)/\left(1-\rho^{k+1}\right)$

Also it follows that $P(\text{customer turns away})=P(\text{system full})=P(k)=\rho^k P_0$

Notice that

(1) when the service rate is dependent upon the state of the system, for example the case when there is more than one server, it follows that

$$P_n(t+\delta t)=(\lambda\delta t)P_{n-1}(t)+(1-\mu_n\delta t-\lambda\delta t)P_n(t)+\left(\mu_{n+1}\delta t\right)P_{n+1}(t)$$

$$P_i=\frac{\prod\lambda}{\prod\mu_j}P_0$$

(2) if arrivals also depend upon the state of the system

$$P_n(t+\delta t)=(\lambda_{n\ 1}\delta t)P_{n\ 1}(t)+(1-\mu_n\delta t-\lambda_n\delta t)P_n(t)+\left(\mu_{n+1}\delta t\right)P_{n+1}(t)$$

$$P_i=\frac{\prod\lambda_j}{\prod\mu_j}P_0$$

For example, if

$$\mu_j = k\lambda_j \quad \text{then} \quad \rho = \tfrac{1}{k}$$
$$\text{and} \quad P_i = \left(\tfrac{1}{k}\right)^i P_0 \quad \text{with} \quad P_0 = 1 - \tfrac{1}{k}$$

$$\text{Or if} \quad \mu_j = k\lambda_{j-1} \quad j = 2, 3, \ldots \text{and} \quad \mu_1 = \mu \quad \text{then} \quad P_i = \frac{\prod_{j=2}^{i-1} \lambda_j}{\prod_{j=2}^{i-1} \mu_j} P_0$$

$$\text{Or when } \mu_j = \mu j = 1 \ldots k \quad \text{and} \quad \mu_j = 2\mu j = (k+1) \ldots$$

$$P_i = \rho^i P_0 \quad i = 1 \ldots k \quad \text{and}$$
$$P_i = \frac{\rho^i}{2^{k-i}} P_0 \quad i = (k+1) \ldots$$

Example When there are 4 customers in the system, the service capacity is doubled and this leads to the probabilities:

$$P_1 = \rho P_0, P_2 = \rho^2 P_0, P_3 = \rho^3 P_0, P_4 = \frac{\rho^4}{2} P_0, P_5 = \frac{\rho^5}{2^2} P_0, \ldots$$

Leading to an expression to determine P_0

$$P_0\left(1 + \rho + \rho^2 + \rho^3 + \frac{\rho^4}{2} + \frac{\rho^5}{2^2} + \cdots\right) = 1, \text{ or}$$
$$P_0\left(\left(1 + \rho + \rho^2\right) + \rho^3\left(1 + \frac{\rho}{2} + \frac{\rho^2}{2^2} + \cdots\right)\right) = 1$$

Chapter 17
Appendix B: Function Optimisation Techniques Genetic Algorithms and Tabu Searches

Val Lowndes and Mirko Paskota

17.1 Introducing Genetic Algorithms

This technique was developed by John Holland introducing to attempt to overcome the problems faced by more classical techniques, in problems where, for example, there are multiple objectives.

Example:

To determine the maximum (integer) value of the function $f(x) = 15x - x^2$, $0 \leq x \leq 16$.

The method proceeds by generating a set (population) of (random) trial solutions, represented as binary numbers.

Stage 1 (Table 17.1)

Stage 2

From these test values (genetic algorithm strings), select a new set of strings to be processed.

Method 1: **Tournament selection** randomly choose 8 pairs of strings from the list above and carry forward the best string from each pair; for example, Table 17.2 shows the selection of the first 4 new strings.

Stage 3

Generate new strings. For each pair, generate a "crossover" point (Table 17.3).

Repeating stages 1 and 2 until the process converges.

Stage 4

Each element in these new strings is considered for **random mutation**, typically if selected its value is reversed.

V. Lowndes (✉)
University of Derby, Kedleston Road, DE22 1GB Derby, UK
e-mail: V.P.Lowndes@derby.ac.uk

M. Paskota
College of Engineering and Technology, University of Derby,
Kedleston Road, DE22 1GB Derby, UK

S. Berry et al. (eds.), *Guide to Computational Modelling for Decision Processes*,
Simulation Foundations, Methods and Applications,
DOI 10.1007/978-3-319-55417-4_17

Table 17.1 Genetic algorithm initial population

String	x		$f(x)$
1	12	01100	36
2	3	00011	36
3	11	01011	44
4	15	01111	0
5	11	01011	44
6	7	00111	56
7	14	01110	14
8	8	01000	56

Table 17.2 Tournament selection for new strings

			Selected string
1	01100	36	01100
4	01111	0	
5	01011	44	01011
7	01110	14	
5	01011	44	00111
6	00111	56	
6	00111	56	00111
4	01111	0	

Table 17.3 Crossover to generate new strings

	Selected strings	Cut point	New strings		Function value
1	**011**00	3	01111	15	0
5	010**11**		01000	8	56
6	**0011**1	4	00111	7	56
6	0011**1**		00111	7	56
5	**01**011	2	01100	12	36
1	01**100**		01011	11	44
3	**01011**	5	01011	11	44
4	01111		01111	15	0

The bold values illustrating the construction of the new strings

Table 17.4 Illustrating the mutation operation

String	0	1	1	0	1
Random number	0.0845	0.0007	0.5004	0.0001	0.9332
New string	0	0	1	1	1

Table 17.5 Selection by roulette wheel

String	x		$f(x)$	Cumulative	Percentage	Random numbers	String
1	12	1100	36	36	0.126	0.284	2
2	3	11	36	72	0.252	0.560	5
3	11	1011	44	116	0.406	0.414	5
4	15	1111	0	116	0.406	0.716	6
5	11	1011	44	160	0.559	0.414	5
6	7	111	56	216	0.755	0.129	2
7	14	110	14	230	0.804	0.457	5
8	8	1000	56	286	1.000	0.979	8

Repeating these stages until the process converges.

Typically, if an attached random number is less than 0.001, its value is reversed (Table 17.4).

Alternative Stage 2: **Roulette selection** (Table 17.5).

Then, crossover and mutation are carried out as before.

An approach using genetic algorithms, and tournament selection, was used to determine a "good" solution to the problem:

$$\text{Maximise } (31x - x^2) \qquad \text{Optimal function value} = 240.25$$

using 8 strings: mutation rate 0.001 (mutate 0.1% of binary values at random).

Stopping when the variance in the 8 function values is less than 0.1, convergence.

Both sets of results show how the genetic algorithm acts to move the search towards a good solution. The successive values of the population variance shown in Fig. 17.1 indicate the convergence in the process.

17.2 Tabu Search

Tabu search is a deterministic search procedure, and if the starting point is repeated then the end point will be unchanged.

This process tries to extend the search by prohibiting the return (of the search) to an already visited point until a given number of steps have been completed.

For example,

to minimise	$f(x,y) = (x-1)^2 + (y-8)^2$
from the starting point	$[x,y] = [5,5]$
with a step size	$h = 1$ (large just used for an introduction)

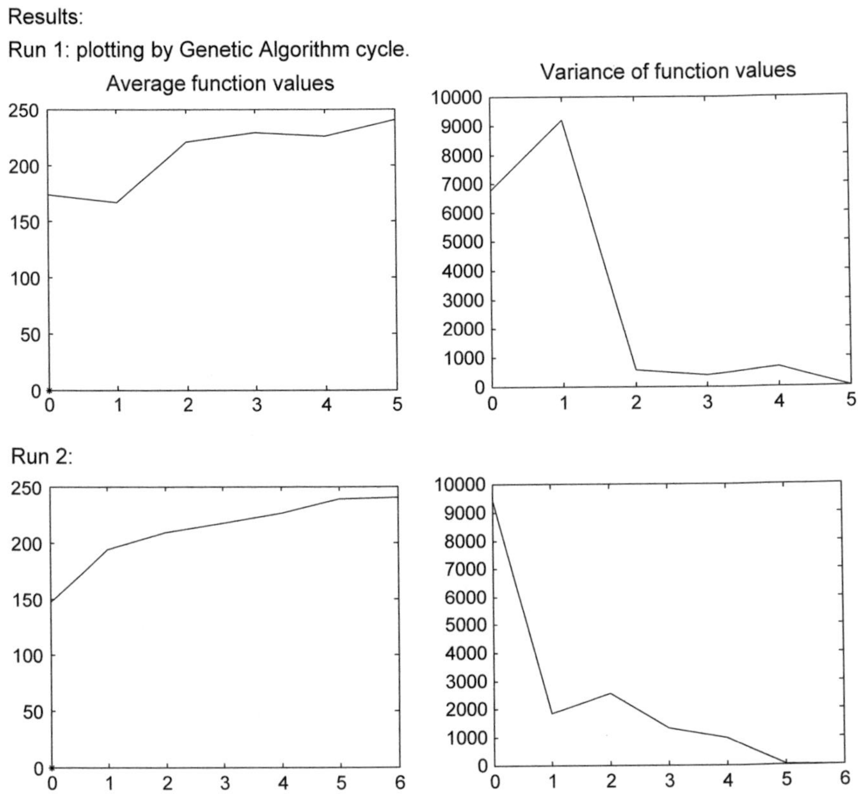

Fig. 17.1 Results from the use of a genetic algorithm

Step 1

Current point	$[5, 5;\ f(5, 5) = 25]$
Tabu list	empty

Defining the neighbourhood points, there are 4 such points and function values:

$$\begin{array}{ccc} & [5,6;\ 20] & \\ [4,5;\ 18] & [5,5;\ 25] & [6,5;\ 34] \\ & [5,4;\ 32] & \end{array}$$

The best neighbouring point is [4, 5; 18].
Move to this point and place [5, 5; 25] on the tabu list.
End of Step 1:

Current point	[4, 5; 18]
Tabu list	[5, 5; 25]

Step 2:

Current point	[4, 5; 18]
Tabu list	[5, 5; 25]

Defining the neighbourhood points, there are 4 such points and function values:

	[4, 6; 13]	
[3, 5; 13]	[4, 5; 18]	[5, 5; TABU]
	[4, 4; 25]	

The best neighbouring point is [4, 6; 13]
Move to this point and place [4, 5; 18] onto the tabu list.
End of Step 2:

Current point	[4, 6; 13]
Tabu list	[5, 5; 25]
	[4, 5; 18]

and so on.

The search pattern, starting from the point [5, 5] passes through the best point at [1, 8], function value 0, optimal solution found, the search path is shown in Fig. 17.2.

Tabu list contains the points

5, 5; 4, 5; 3, 5; 3, 6; 2, 6; 2, 7; 1, 7; 1, 8; 2, 8; 2, 9

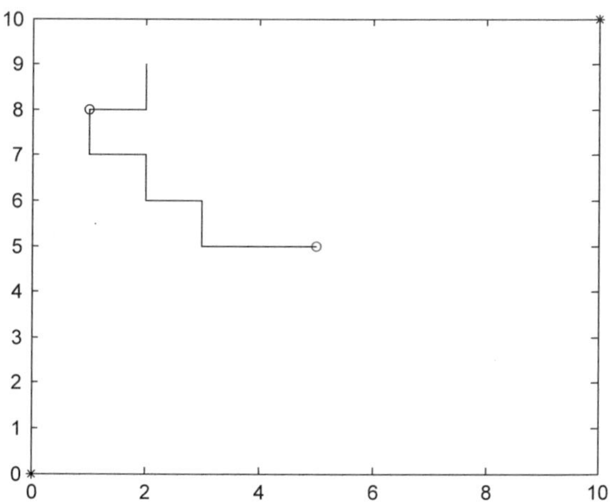

Fig. 17.2 Tabu search path

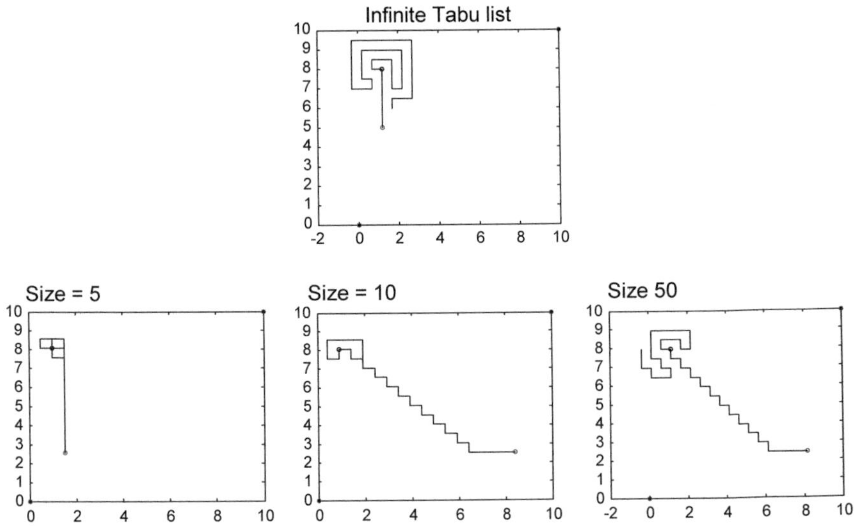

Fig. 17.3 Tabu search paths with tabu list size

Here, the tabu list has infinite length; however, often a finite length is used so that a point could be visited more than once. As a consequence, the size of the tabu list can influence the quality of the determined solution.

The next plots were generated using a step size of 0.5 with the optimal solution located at the point (1, 8).

The plots in Fig. 17.3 show that when the tabu list is very short, the search cannot escape from a good point.

Note that the difficulty in using a tabu search (only) to disclose the optimal solution of a problem is also related to the size of the search space, the location of the starting point and the step size.

Example to illustrate the effect of step size. Here, the optimal solution is at the point (1, 1).

The starting points were generated randomly to 2 decimal places with a step size for the search of 0.1.

The first plot shown in Fig. 17.4 demonstrates that the search has successfully located the optimal point but the search shown in the second plot has not located the optimal solution, thus showing that a successful search is dependent upon the starting point and the step size.

Consequently, the search could produce a good solution rather than the optimal solution.

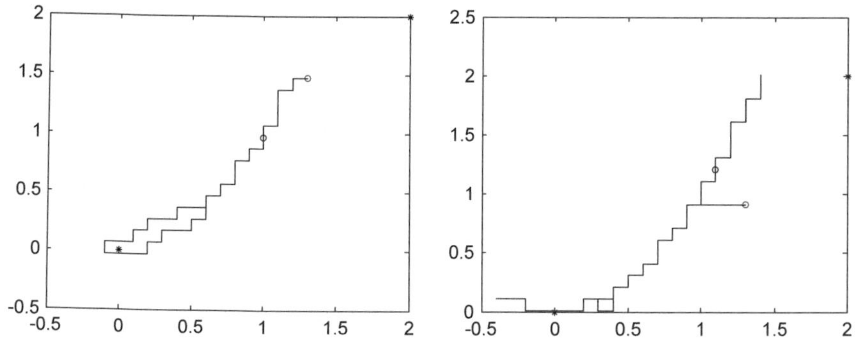

Fig. 17.4 Tabu search showing the effect of step size

Fig. 17.5 Defining 4 point line searches

17.3 Comparing "Line Search" Methods, Derivative and Non-derivative-based Approaches with Heuristic Approaches

The objective is to determine the location of the minimum value for the function $y = f(x)$, only one independent variable, comparing the results from line searches with those obtained using an approach based on genetic algorithms.

Many of these methods are "said" to be applicable, where the function is quasi-convex (or concave).

Non-differential-based Line Search Methods

This approach follows the methodology (Fig. 17.5):

Search space: Initial range a to b, successively reduced, function values known

Two new points: determine the function values at c and d

Update reduces the search space:

$f_3 < f_2$ new search space c to b

$f_2 < f_3$ new search space a to d.

repeating until c and d are sufficiently close.

17.3.1 General Case

The "distance" between c and d reduces at each stage; at the first step, the search space is a to b, then if

$$ab = 1;$$
$$ad = x;\ cb = x;$$
$$ac = bd = 1 - x \quad \text{hence} \quad cd = (2x-1)$$

Then, at the second step, the search space is reduced from a to c.
The four points are now as follows:

$$a, c', d' \text{ and } b' \ (= d)$$

then

$$ab' = x;$$
$$ad' = x^2;$$

Hence, central width $= c'd' = x\,(2x-1)$,
and at the n-th step, central width $= (2x-1)x^{n-1}$.
Example: $x = 0.8$ (Table 17.6)

$$\text{Note} : x = 1 \quad \text{after } n \text{ steps width} = 1, \text{ no search carried out}$$
$$x = 0.5 \quad \text{at initial step width} = 0, \text{ no search carried out}$$

17.3.2 Golden Section Method

This approach uses the interval division approach but reduces the calculations required by an appropriate selection of the internal points so that at each stage there is only one new function value to be calculated:

At stage 0 $\quad ab = 1;\ ad = x;\ cb = x$
At stage 1 $\quad c' = d;\ a' = c;\ b' = b;\quad d$ is now the only new point.

Table 17.6 Illustrating the reduction in the [0.2, 0.8] search space

Step, search points	a	c	d	b	Central width cd	Width cb
0	0	0.2	0.8	1.0	0.6	0.8
1	0	0.16	0.64	0.8	0.48	0.64
2	0	0.128	0.512	0.64	0.384	0.512

Table 17.7 Illustrating the reduction in the Golden Ratio search space

Stage	a	c	d	b	Width cd	Width cb
0	0.000	0.382	0.618	1.000	0.236	0.618
1	0.000	0.236	0.382	0.618	0.146	0.382
2	0.000	0.146	0.236	0.382	0.090	0.236
3	0.000	0.090	0.146	0.236	0.056	0.146
4	0.000	0.056	0.090	0.146	0.034	0.090

Table 17.8 Fibonacci numbers

Term	1	2	3	4	5	6	7	8	9	10	11	12
Number	1	1	2	3	5	8	13	21	34	55	89	144

As a consequence, it follows that

$$ad = x \quad \text{and} \quad ac' = 1 - x^2 \ (c'd = x^2,\ ad = 1)$$
$$x = 1 - x^2 \quad \text{or} \quad x^2 + x - 1 = 0$$
$$\text{giving} \quad x = 0.6180 \quad \text{the Golden Ratio.}$$

The first 4 steps in a search could have the search points (x values) (Table 17.7). The question is as follows:

"Is the reduction in calculations counterbalanced by a greater number of stages to achieve a given convergence?"

17.3.3 Fibonacci Search

This makes use of the Fibonacci sequence, $F_{n+1} = F_n + F_{n-1}$, and the first terms in this sequence are shown in Table 17.8.

Here, the internal points are calculated from the following:

$$\text{point } c = a_k + \frac{T_{n-k-1}}{T_{n-k+1}}(b_k - a_k) \quad \text{and} \quad \text{point } d = a_k + \frac{T_{n-k}}{T_{n-k+1}}(b_k - a_k)$$

Note: adding gives

$$c + d = a + b \quad \text{as} \quad T_{n-k-1} + T_{n-k} = T_{n-k+1}$$

where n defines the starting point, number of iterations. Using $n = 10$, (Table 17.9).

Notice that these values are very similar to those obtained for a search using the golden Ratio.

Table 17.9 Fibonacci search

Stage	a	c	d	b	Width cd	Width cb
1	0	0.382	0.618	1.000	0.236	0.618
2	0	0.236	0.382	0.618	0.146	0.382

Table 17.10 Comparing golden Ratio and Fibonacci search

Search step	Fibonacci numbers	Fibonacci search		Active space	Golden section		Active space
	1						
9	2	0.333	0.667	0.014	0.382	0.618	0.013
8	3	0.400	0.600	0.021	0.382	0.618	0.021
7	5	0.375	0.625	0.035	0.382	0.618	0.034
6	8	0.385	0.615	0.056	0.382	0.618	0.056
5	13	0.381	0.619	0.090	0.382	0.618	0.090
4	21	0.382	0.618	0.146	0.382	0.618	0.146
3	34	0.382	0.618	0.236	0.382	0.618	0.236
2	55	0.382	0.618	0.382	0.382	0.618	0.382
1	89	0.382	0.618	0.618	0.382	0.618	0.618

Table 17.11 Catalan numbers

Term	1	2	3	4	5	6	7	8	9	10	11	12
Number	1	1	2	5	14	42	132	429	1430	4862	16,796	58,786

The full search space table shows that differences, in the size of the active space, only occur at the final stages of the search (Table 17.10).

17.3.4 Catalan Search

This makes use of the Catalan sequence, $C_1, C_2, C_3, C_4, \ldots, C_{n+1}, \ldots$, where neighbouring terms of the sequence are related according to the recursion $C_{n+1} = \left(\frac{2(n+1)}{n+2}\right) C_n$ (see, for instance, Larcombe [1]. The first few terms are explicitly shown in Table 17.11.

Here, the internal points are calculated from the following:

$$\text{point } c = a_k + \frac{C_{n-k}}{C_{n-k+1}}(b_k - a_k) \quad \text{and} \quad \text{point } d = a_k + \left(1 - \frac{C_{n-k}}{C_{n-k+1}}\right)(b_k - a_k)$$

Table 17.12 Catalan search

Search step	Catalan numbers	Catalan search		Active space
9	2	0.400	0.600	0.030
8	5	0.357	0.643	0.051
7	14	0.333	0.667	0.079
6	42	0.318	0.682	0.118
5	132	0.308	0.692	0.174
4	429	0.300	0.700	0.251
3	1430	0.294	0.706	0.358
2	4862	0.289	0.711	0.508
1	16,796	0.286	0.714	0.714
	58,786			

Giving the search regions, again starting with $n = 10$ (Table 17.12).

This approach has the (slight) advantage that it rejects less of the current search space; however, convergence (to a required accuracy) will be slower.

Comparing the effectiveness and efficiency of Golden Ratio Search and Genetic Algorithms applied to 2 variable problems.

Both approaches were applied to determine the minimum point for the (unimodal) function

$$f(x,y) = (x - 1.3467)^2 + (y - 7.5643)^2$$

with a search space 0–10.

The results tending to show that, in this example, the interval division approach is both more effective and more efficient.

Golden ratio	0.006 s $x = 1.3436; y = 7.5656$
Genetic Algorithm	40 strings, 20 cycles,
	0.04 s $x = 1.308; y = 7.57$
	0.04 s $x = 1.354; y = 7.290$
	0.04 s $x = 1.300; y = 7.610$
	40 strings, 50 cycles
	0.06 s $x = 1.345; y = 7.560$
	0.06 s $x = 1.300; y = 7.603$

Both approaches were also applied to determine the minimum point for the function, chosen as an extreme case:

$$f(x,y) = \left((x - 1.3467)^2 + (y - 7.5643)^2 + k\right)\left((x - 7.9)^2 + (y - 0.1)^2\right)$$

Note : $k = 0$ a function with two minima and $f(x, y) = 0$ at both points
$k > 0$ the minimum is at the point $x = 7.9$, $y = 0.1$.

Initially, set $k = 0.1$.

$$\text{Here the Optimal Solution} \quad x = 7.9,\ y = 0.1, \quad f(x, y) = 0$$

Applying the golden Ratio (area search) gives the non-Optimal Solution.

$$0.007\ \text{s} \quad x = 1.3427;\ y = 7.5656$$

While the genetic algorithm can obtain the optimal solution, the quality of the solution is dependent upon the number of strings and the maximum number of cycles.

The results were as follows:

40 strings, 20 cycles

0.04 s $x = 8.514;\ y = 0.2816 \quad f(x, y) = 42.9262$

100 strings, 20 cycles

0.07 s $x = 7.7830;\ y = 0.2322 \quad f(x, y) = 2.9697$

0.07 s $x = 1.2979;\ y = 7.490 \quad f(x, y) = 10.590$

0.08 s $x = 1.2800;\ y = 7.513 \quad f(x, y) = 10.577$

0.07 s $x = 7.7921;\ y = 0.1090 \quad f(x, y) = 1.1398$

100 strings, maximum 50 cycles or convergence

0.19 s $x = 8.0988;\ y = 0.01216 \quad f(x, y) = 4.0422$

0.17 s $x = 8.0101;\ y = 0.0950 \quad f(x, y) = 1.2182$

0.08 s $x = 7.9111;\ y = 0.0990 \quad f(x, y) = 0.0123$

0.07 s $x = 7.5263;\ y = 0.1401 \quad f(x, y) = 13.1945$

Here, there is no (or little) change in the time for the interval division search, but the golden search converges to a suboptimal point; note: at each stage, the search space is reduced by 62%, hence possibly excluding the global optimal solution (as here).

Whereas the approach using genetic algorithms is able to determine the location of optimal solution but with greater computation time, a tabu search from the genetic algorithms converged value can then obtain the optimal solution (Fig. 17.6).

Then, by setting $k = 1$, the Optimal Solution is $x = 7.9$, $y = 0.1$, $f(x, y) = 0$.

Note: the golden section search will give the non-optimal solution as before.

The results from an investigation into the use of a genetic algorithm approach were as follows:

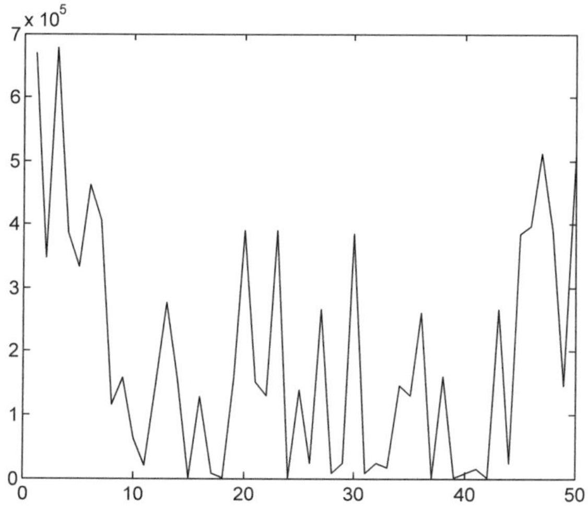

Fig. 17.6 Plot of best function value at each genetic algorithm cycle

40 strings, 20 cycles

0.04 s $x = 7.418;\ y = 0.9368 \quad f(x, y) = 76.2418$

0.07 s $x = 7.6434;\ y = 0.2206 \quad f(x, y) = 7.6434$

0.10 s $x = 7.9077;\ y = 0.0866 \quad f(x, y) = 0.0239$

40 strings, 50 cycles

0.07 s $x = 1.5241;\ y = 7.4695 \quad f(x, y) = 98.804$

0.10 s $x = 7.8623;\ y = 1.2093 \quad f(x, y) = 103.20$

0.08 s $x = 8.5313;\ y = 0.0735 \quad f(x, y) = 43.376$

100 strings, maximum 50 cycles or convergence

0.20 s $x = 8.0242;\ y = 0.1020 \quad f(x, y) = 1.5626$

0.20 s $x = 7.9225;\ y = 0.1054 \quad f(x, y) = 0.0535$

0.19 s $x = 8.0122;\ y = 0.2201 \quad f(x, y) = 2.6842$

0.18 s $x = 7.8483;\ y = 0.3934 \quad f(x, y) = 8.4046$

40 strings, 100 cycles

0.08 s $x = 7.9121;\ y = 0.0795 \quad f(x, y) = 0.0567$

0.08 s $x = 7.8979;\ y = 0.22; \quad f(x, y) = 1.4096$

0.09 s; $x = 7.8861;\ y = 0.1352; \quad f(x, y) = 0.1417$

0.09 s; $x = 1.4399;\ y = 6.9997; \quad f(x, y) = 118.594$

100 strings, maximum 50 cycles or convergence

0.17 s $x = 7.9009;\ y = 0.0766 \quad f(x, y) = 0.0548$

0.20 s $x = 7.8707;\ y = 0.0997 \quad f(x, y) = 0.0852$

0.18 s $x = 8.6911;\ y = 0.4405 \quad f(x, y) = 78.398$

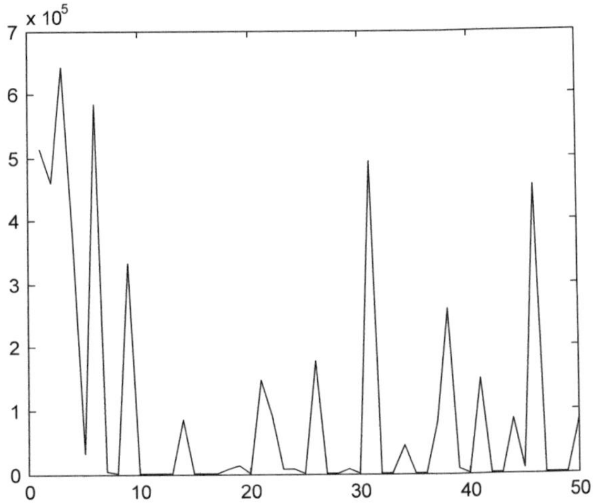

Fig. 17.7 Best cost at each cycle

Notice that for the (nicer) function

$$f(x,y) = \left((x-1.3467)^2 + (y-7.5643)^2 + 10\right)\left((x-4.9)^2 + (y-6.1)^2\right)$$

Applying a genetic algorithm with 40 strings for a maximum of 50 cycles gave the following (Fig. 17.7):

$$\begin{array}{ll} 0.06\text{ s} & x = 4.89,\ y = 6.11,\ f(x,y) = 0.0045 \\ 0.08\text{ s} & x = 4.94,\ y = 6.10,\ f(x,y) = 0.0290 \\ 0.09\text{ s} & x = 4.88,\ y = 6.22,\ f(x,y) = 0.2174 \end{array}$$

Applying a tabu search methodologies to the function:

$$f(x,y) = \left((x-1.3467)^2 + (y-7.5643)^2 + 0.1\right)\left((x-4.9)^2 + (y-6.1)^2\right)$$

Solution at $x = 4.9;\ y = 6.1$

Starting from a random point, typical search paths are shown in Figs. 17.8a–d. In each case, the starting point is shown by *, the end point by * and the best point found by *. Notice, search a does not find the best point, searches b-d find the best point.

The searches were carried out for a fixed number of steps and the resulting search paths are shown in Figs. 17.8a–d. Searches shown in Figs. 17.8b–d disclosed the optimal point so in general the search would need to consist of many steps to be reasonably sure that the best solution has been found.

The function values, and best known value, at each step in a typical search are shown in Fig. 17.9.

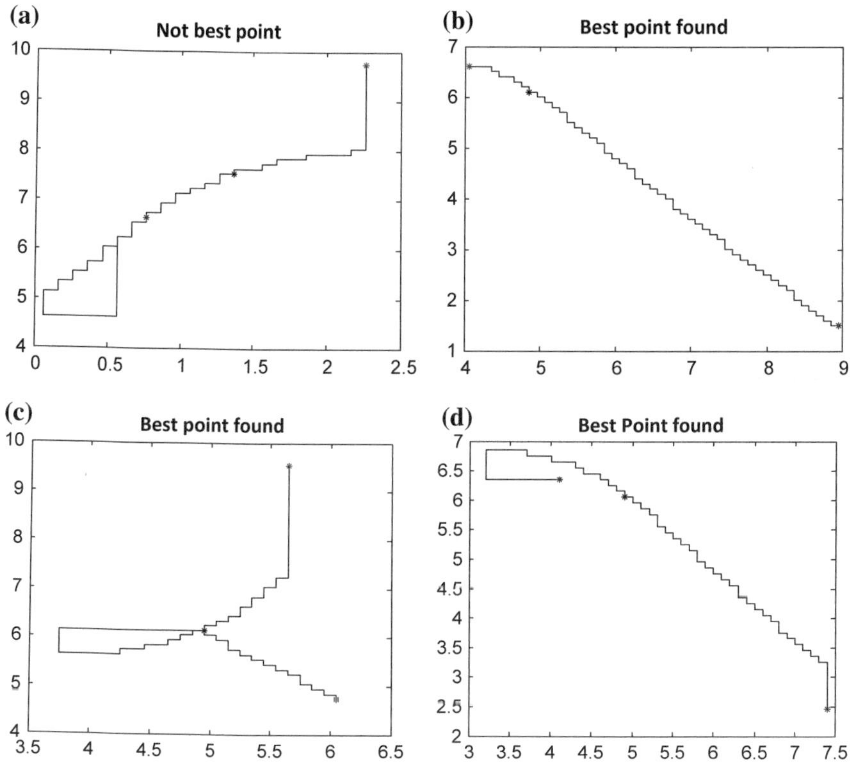

Fig. 17.8 **a** Not best point, **b–d** best point found

Fig. 17.9 Best cost and best cost at each cycle

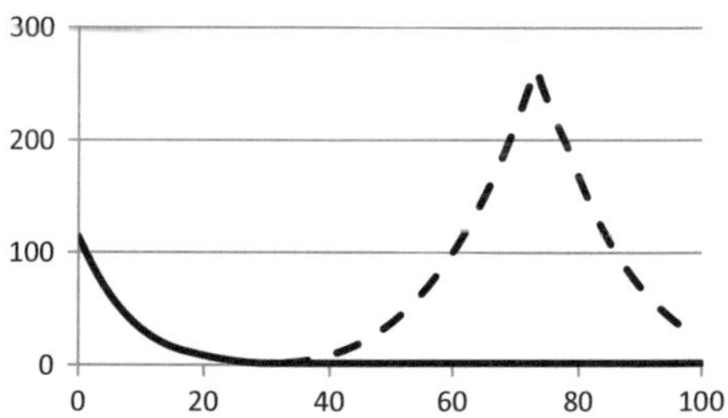

Repeating the searches for the "simpler" function

$$f(x,y) = \left((x-1.3467)^2 + (y-7.5643)^2 + 10\right)\left((x-4.9)^2 + (y-6.1)^2\right)$$

Showed that here the search is better able to determine the location of the optimal solution (Fig. 17.10).

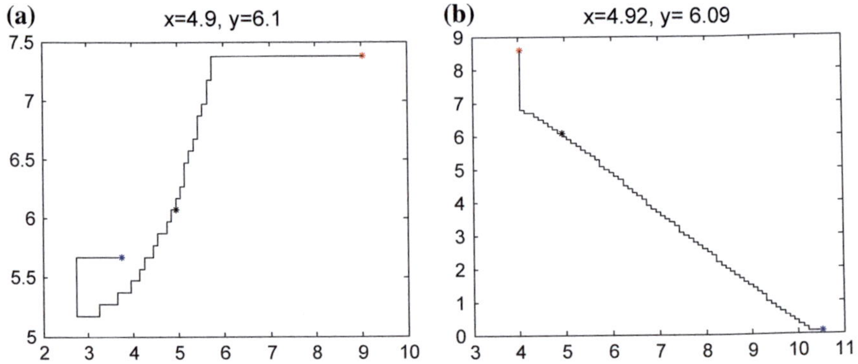

Fig. 17.10 Showing that both searches **a** and **b** Paths taken by tabu search

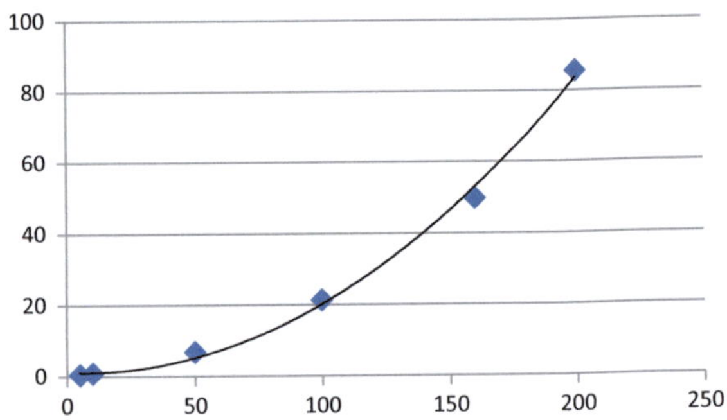

Fig. 17.11 Solution time and search length

$$x = 4.9,\ y = 6.1 \quad x = 4.92,\ y = 6.09$$

Here, if 4 decimal place accuracy is required, the search space (x and y between 0 and 10) has 10^{10} points, and if the starting point is (0, 0) and the solution is at (10, 10), the path to the solution will consist of $2\left(10^5\right)$ steps; consequently, a search would have more than this number of steps.

Solution times: see Fig. 17.11 plotting number of points (N) searched (in thousands) against time for search (T) gave the model:

$$T = 0.0022N^2 - 0.0313N + 1.994$$

$$\begin{array}{lll} \text{Note : when}\, N = 1{,}000{,}000 & \text{then} & T = 2170.7\ \text{s (over 36 min)} \\ N = 5{,}000{,}000 & \text{then} & T\ \text{is over 15 h!!} \end{array}$$

Hence, an exhaustive or a full tabu search is not viable.

Combining Tabu Search with Genetic Algorithms

This combination aims to employ the advantages of both approaches to develop a more efficient and effective search procedure.

Using as a starting point for the tabu search, the best point generated by the genetic algorithm, here starting from the point:

$$x = 4.88,\ y = 6.22,\ f(x, y) = 0.2174$$

The tabu search gave the location of the optimal solution $x = 4.9,\ y = 6.1$ with the path to this solution shown in Fig. 17.12. The tabu search required 0.11 s and the genetic algorithm 0.20 s giving a total solution time of 0.31 s.

Calculus-based Searches

If the function, $f(x)$, is quasi-convex, it will have the form shown in Fig. 17.13a with the derivative having the form shown in Fig. 17.13b.

Fig. 17.12 Tabu search from genetic algorithm best point

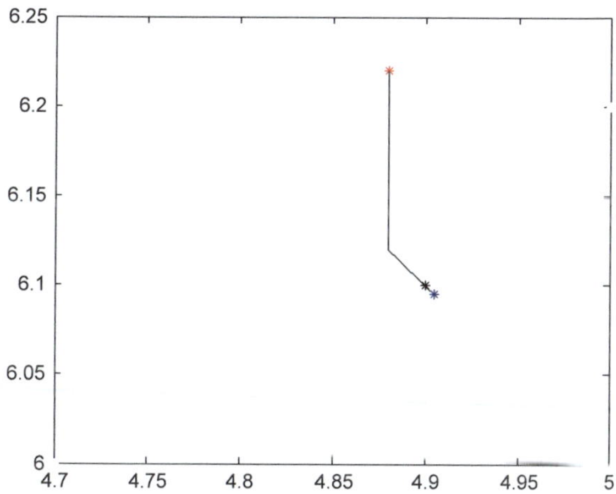

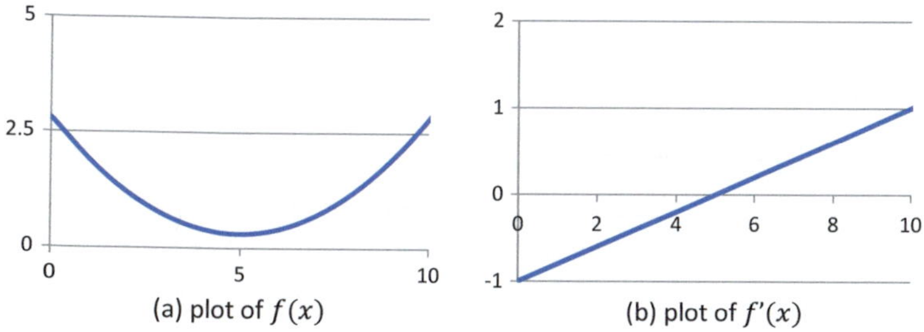

Fig. 17.13 **a** Plot of $f(x)$, **b** Plot of $f'(x)$

Table 17.13 The bisection method an illustrative example

a	Mid-point	Mid-point slope	b	New range
0	3	−4	6	*m to b*
3	4.5	0.5	6	*a to m*
3	3.75	−0.0625	4.5	
3.75	4.125	0.00781	4.5	
3.75	3.9375	−0.001	4.125	

Interval division-based searches: these can be applied to the derivative of the function, see previous section.

Bisection and gradient method

At the mid-point of the range, $m = (a + b) / 2$, find the value of $f'(m)$,

$$\begin{array}{lll} f'(\text{midpoint}) < 0 & \text{new range} & \text{midpoint to } b \\ f'(\text{midpoint}) > 0 & \text{new range} & a \text{ to midpoint} \end{array}$$

Example:

$$f(x) = (x{-}4)^4 \quad \text{hence } f'(x) = 4(x{-}4)^3 \quad \text{search range, } 0-6$$

(Table 17.13)

Continue until converged to the acceptable tolerance; at each step, the search space is halved, and thus, after n steps, the search space has become

$$(b - a)\, 0.5^n$$

Therefore, reduction to a, the space 10^{-y} requires $n = (y + \log(b - a)) / 0.3010$ steps

$$\text{For example}: \quad \begin{array}{llll} (b-a) = 1 & \text{and} & y = 4 \text{ gives} & n = 14 \text{ steps} \\ (b-a) = 100 & \text{and} & y = 4 \text{ gives} & n = 20 \text{ steps} \end{array}$$

Newton's Iterative Formula

Starting with the Taylor series

$$f : \Re \to \Re : f(x) = f(x^k) + (x - x^k) f'(x^k) + \frac{(x - x^k)^2}{2!} f''(x^k) + O\left[(x - x^k)^3\right]$$

Table 17.14 Newton's method an illustrative example

Step n	x_n	x_{n+1}	$f'(x)$	$f(x)$
0	8.000	6.667	75.88030	52.99319
1	6.667	5.778	22.48305	12.39372
12	4.031	4.021	0.00004	2.40000
13	4.021	4.014	0.00001	2.40000

an approximate solution to the equation $f(x) = 0$ can be determined using the iterative scheme

$$x^{n+1} = x^n - f(x_n)/f'(x_n)$$

Then, as the optimum point for the function $f(x)$ occurs when $g(x) = f'(x) = 0$, the required iterative scheme is

$$x^{n+1} = x^n - g(x_n)\,/\,g'(x_n)$$
$$\text{or}\quad x^{n+1} = x^n - f'(x_n)\,/\,f''(x_n)$$

Example: to determine the optimal value of the function

$$f(x) = (x-4)^4 + 2.4$$
$$g(x) = f'(x) = 4\,(x-4)^3 \quad \text{and}$$
$$g'(x) = f''(x) = 12\,(x-4)^2 \quad \text{with}$$
$$x_0 = 8.$$

To give the iterative scheme $x_{n+1} = x_n - 4(x_n-4)^3/(12(x_n-4)^2)$.
Starting with $x_0 = 8$, this gives the solution (Table 17.14):

If this process can be considered to have converged, the optimal value is at the point $x = 4.014$.

Multi-variate Unconstrained Optimisation

Definition of Unconstrained Optimisation:

$$\text{Minimise}\{f(x) \quad \text{such that} \quad x \in \Re^n\}$$

Here, there exists

an initial point	x^0	
the solution at step k	x^k	
the direction at step k	d^k	to be determined
the step size at step k	α^k	to be determined

the iterative scheme is described by

$$\boldsymbol{x}^{k+1} = \boldsymbol{x}^k + \boldsymbol{\alpha}^k \boldsymbol{d}^k$$

Assumptions: f must be sufficiently smooth and the solution set, and

$$S(x^0) = \{x \in \Re^n \text{ such that } f(x) \leq f(x^0)\} \text{ is bounded,}$$

General algorithm for iterative methods for unconstrained optimisation consisting of 4 steps (Table 17.15):

The direction $d^k = -f'(x^k)$, if f is one dimensional,

slope negative $(f'(x^k))$ then increase x^k, x^k is less than the optimal point

slope positive $(f'(x^k))$ then decrease x^k, x^k is more than the optimal point

Example, one variable: $f = x^2$, $\frac{df}{dx} = 2x$ and let $x^0 = 4$ (Table 17.16)

Table 17.15 Defining the steps in the search procedure

Step		
1	Choose an initial point x^k	
2	Find $\nabla f(x^k)$	If $\nabla f(x^k) = 0$ **STOP***
3	Find $d^k = -\nabla f(x^k)$ the next direction. Chosen so that f decreases	
4	Determine the step size α^k by solving Minimise$\{f(x^k + \alpha^k d^k) \mid \alpha \geq 0\}$	
5	Let $x^{k+1} = x^k + \alpha^k d^k$	

Table 17.16 Illustrative example of the search

Step	Calculation	
1	$x^0 = 4$	
2	$\frac{df}{dx} = 2x \Rightarrow \frac{df}{dx}(x = x^0) = 8$	
3	$d^0 = -8$	w direction
	$x^1 = x^0 + \alpha^0 d^0 = x^0 - 8\alpha^0$	$x^1 = 4 - 8\alpha$
4	$\{f(x^0 + \alpha^0 d^0)\} = (4 - 8\alpha)^2$	
	$\frac{df}{d\alpha} = -16(4 - 8\alpha) = 0 \Rightarrow \alpha = 0.5$	New step size
5	$x^1 = x^0 + \alpha^0 d^0 = 4 + 0.5(-8) = 0$	New point $x^1 = 0$
2	$\frac{df}{dx} = 2x \Rightarrow \frac{df}{dx}(x = x^1) = 0$, STOP	Optimal point found

Fig. 17.14 Path to solution

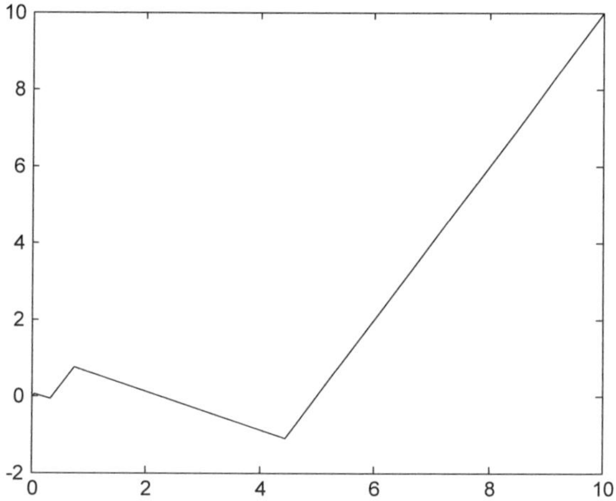

Example, two variables

Here

$$x = \begin{bmatrix} x_1 \\ x_2 \end{bmatrix}; \quad f(x_1, x_2), \ \nabla f = \begin{bmatrix} \frac{\partial f}{\partial x_1} \\ \frac{\partial f}{\partial x_2} \end{bmatrix} \text{ and } d^k = -\nabla f\left(x^k\right)$$

when $\nabla f\left(x^k\right)$ is used in finding d^k, the group of methods used are called "gradient methods".

The path towards the solution, $f(x, y) = x^2 + 2y^2$, is shown in Fig. 17.14

The optimal solution is $x = \begin{bmatrix} 0 \\ 0 \end{bmatrix}$, and the successive directions were

$$d^0 = \begin{bmatrix} -2 \\ -4 \end{bmatrix}, d^1 = \begin{bmatrix} -8/9 \\ 4/9 \end{bmatrix}, d^2 = \begin{bmatrix} -2 \\ -4 \end{bmatrix}, d^3 = \begin{bmatrix} 8/9 \\ 4/9 \end{bmatrix}$$

These methods (gradient methods) are described by the iterative scheme

$$x^{k+1} = x^k + \alpha^k d^k \quad \text{where} \quad d^k = \nabla f\left(x^k\right)$$

However, a parabolic approximation of the curve at the point x^k is likely to be better than a linear approximation.

Cauchy's method used a linear approximation, and Newton's method uses a parabolic approximation.

Newton's Method

This is derived from the Taylor series

$$f:\Re^n\to\Re: f(x)=f(x^k)+\nabla f(x^k)(x-x^k)+H(x^k)\frac{(x-x^k)^2}{2!}+o\left[(x-x^k)^3\right]$$

where the Hessian matrix is given by $H=\begin{bmatrix}\frac{\partial^2 f}{\partial x_1\partial x_1} & \cdots & \frac{\partial^2 f}{\partial x_1\partial x_n}\\ \vdots & \vdots & \vdots\\ \frac{\partial^2 f}{\partial x_n\partial x_1} & \cdots & \frac{\partial^2 f}{\partial x_n\partial x_n}\end{bmatrix}$

Differentiating the Taylor series leads to the iterative scheme

$$x^{k+1}=x^k-H^{-1}(x^k)\nabla f(x^k)$$

Example Minimise$\{x^2+2y^2\}$ with $x^0=\begin{bmatrix}x\\ y\end{bmatrix}=\begin{bmatrix}3\\ 3\end{bmatrix}$

Here, $\nabla f=\begin{bmatrix}2x\\ 4y\end{bmatrix}$ and $H=\begin{bmatrix}2 & 0\\ 0 & 4\end{bmatrix}$; thus, $H^{-1}=\frac{1}{8}\begin{bmatrix}4 & 0\\ 0 & 2\end{bmatrix}=\begin{bmatrix}0.5 & 0\\ 0 & 0.25\end{bmatrix}$

giving Newton's iterative scheme

$$x^1=\begin{bmatrix}3\\ 3\end{bmatrix}-\begin{bmatrix}0.5 & 0\\ 0 & 0.25\end{bmatrix}\begin{bmatrix}6\\ 12\end{bmatrix}=\begin{bmatrix}3\\ 3\end{bmatrix}-\begin{bmatrix}3\\ 3\end{bmatrix}=\begin{bmatrix}0\\ 0\end{bmatrix}$$

Properties of the method:

Positive	Rapid convergence
Negative	The Hessian matrix is sometimes hard to find and (even worse) sometimes cannot be defined

Reference

1. Larcombe PJ, Wilson PDC (1998) On the trail of the catalan sequence. Bull IMA 34(4): 114–117

Chapter 18
Appendix C: What to Simulate to Evaluate Production Planning and Control Methods in Small Manufacturing Firm's

Val Lowndes and Stuart Berry

A set of case studies confirmed that there exists a group of small (micro-sized) manufacturing firms (SMFs) that operate similar production systems and have similar requirements from a production planning and control system (PP&C). The features common to all the case study firms are the small number of production stages together with the fact that all jobs follow the same route through the factory.

The case studies confirmed that:

- The small manufacturing firms are likely to be organised as flow shops with multiple servers at each stage.
- Typically, there will be three stages in the production process {prepare, make and finish}.
- In a larger firm, the number of processors at each stage will be chosen so that the "daily" production capacities at each stage are approximately the same (balanced workshop) whilst in the smaller firm one stage will tend to dominate the manufacturing process (have the least capacity per time unit).
- In the smaller firms, the manager has an active (crucial) role in the planning and control function whilst a large firm will tend to have implemented a formal system.
- Often in a small manufacturing firm, there exists a non-formal but strict planning and control system, managed by the firms owner/manager.
- A small manufacturing firm has few non-production resources, specifically few non-production workers.

V. Lowndes
University of Derby, Kedleston Road, DE22 1GB Derby, UK
e-mail: V.P.Lowndes@derby.ac.uk

S. Berry (✉)
College of Engineering and Technology, University of Derby,
Kedleston Road, DE22 1GB Derby, UK
e-mail: s.berry@derby.ac.uk

S. Berry et al. (eds.), *Guide to Computational Modelling for Decision Processes*,
Simulation Foundations, Methods and Applications,
DOI 10.1007/978-3-319-55417-4_18

The case studies have also indicated the importance to the firm of being able to quote accurate delivery times at that time when an order has been received. Therefore, the simulations need to be able to evaluate the suitability of the alternative approaches to the estimation of delivery lead times for use in a small manufacturing firm.

The case studies have indicated two approaches in use in such firms:

- fixed lead times regardless of the work in hand (attainable or not)
  - normally a fixed delivery time, larger than the manufacturing time
  - with all firms quoting the same time, or
- calculating and quoting (to the customer) realistic lead times.

In the first approach, the firm needs to know a realistic quotable value, and in the second, the firm needs to be able to estimate the job lead time on arrival.

Thus indicating the scope for the simulation exercises, testing possible delivery times in the first case and evaluating methods to enable the forecasting of production times in the second case.

To be able to simulate these firms, the simulation model needs to be able to satisfy the requirements of each firm:

- A few stages in the process, often three.
- The workshop is organised as a flow shop.
- Possibly multiple servers at each stage, with preferred servers at some stages.
- Seasonal demand patterns.
- Random arrival.
- Deterministic job times or very low variance on the job times.
- Orders both single items and multiple items.
- Several product types, typically three.

The results from the simulation models constructed around these small manufacturing firms confirmed that these firms can operate effectively without a large resource base; in particular, it established that quoting a fixed delivery time, larger than the manufacturing time, can produce an efficient and effective system planning and control system.

This policy leads the firms to use the minimal information set model given in Fig. 18.1, through the removal of "unnecessary" connections; here, there is no direct contact between sales and production, no longer needed the quoted time is such that delivery is virtually certain in this time period.

The second simulation demonstrated that the firms can provide these reliable estimates using either the "order book" or a planning board system.

Comparing the use of the order book/work in progress implies an "active" regression-based system, (time $\propto$ number in system, for example) with the use of a planning board "passive" system which selects the first viable completion time from the planning board. This comparison indicated that the passive "white board" system is the most appropriate system being more accurate, requiring less information and less management input.

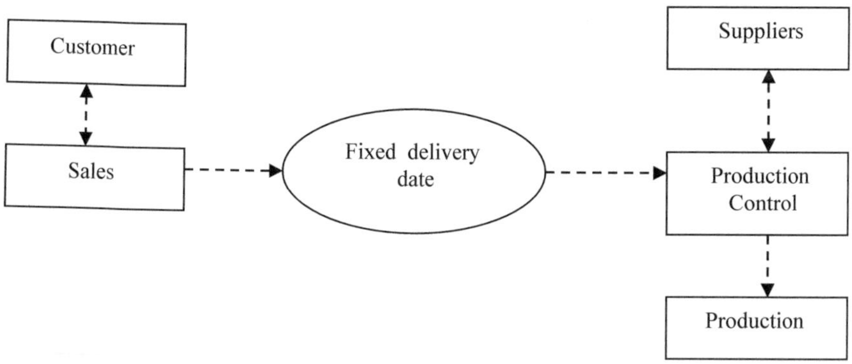

Fig. 18.1 Information flow model in small manufacturing firms

So how do the firm get it right?

The investigation has shown that crucially in smaller firms the manager has an active (critical) role in the planning and control function whilst a large firm will tend to have implemented a formal system.
In fact managers in small firms have often derived and implemented a "sophisticated" control system from their "hands on" experiences. It must be noted, however, that this strength for a small firm can become a weakness if the firm grows in size and the manager is no longer able to carry out the role of controller.

The information from the case study has indicated that the development of effective and efficient planning and control systems has resulted from:

- A detailed knowledge of the firms operations,

  dominant stages and
  the manager actively control input, starting (only) what can be finished without delays

- A knowledge of the behaviour of their competitors

  Delivery dates

- A knowledge of the requirements of their customers

  What they will accept with respect to delivery, guaranteed dates.

These lead to the information flows shown in Fig. 18.1 with jobs entering the system in a FIFO order, based on the date of the order being placed, thus minimising the planning with the manager allocating overtime working, on the basis of the number of jobs waiting, to ensure that all the jobs will be ready on time for delivery to the customers.

Chapter 19
Appendix D: Defining Boolean and Fuzzy Logic Operators

Val Lowndes

19.1 Definition Boolean Logic

If an element x

	is contained in set A then	$\mu_A(x) = 1 \quad \text{if} \quad x \in A$
	is not contained in set A then	$\mu_A(x) = 0$
As		$\mu_{A \cup B}(x) = 1 \quad \text{if} \quad x \in A \text{ or } x \in B$
and		$\mu_{A \cap B}(x) = 1 \quad \text{if} \quad x \in A \text{ and } x \in B$
Then it follows that		$A \cup B \to \mu_{A \cup B}(x) = \max[\mu_A(x), \mu_B(x)]$
and		$A \cap B \to \mu_{A \cap B}(x) = \min[\mu_A(x), \mu_B(x)]$

19.2 Definition Fuzzy Logic

If an element x

	has some membership in set A then	$\mu_A(x) = k_A \quad 0 < k_A \leq 1$
	has no membership in set A then	$\mu_A(x) = 0$
define		$A \cup B \to \mu_{A \cup B}(x) = \max[\mu_A(x), \mu_B(x)]$
and		$A \cap B \to \mu_{A \cap B}(x) = \min[\mu_A(x), \mu_B(x)]$

thus $\quad 0 \leq \mu_{A \cup B}(x) \leq 1 \quad \text{and} \quad 0 \leq \mu_{A \cap B}(x) \leq 1$

V. Lowndes (✉)
University of Derby, Kedleston Road, DE22 1GB Derby, UK
e-mail: V.P.Lowndes@derby.ac.uk

S. Berry et al. (eds.), *Guide to Computational Modelling for Decision Processes*,
Simulation Foundations, Methods and Applications,
DOI 10.1007/978-3-319-55417-4_19

Chapter 20
Appendix E: Assessing the Reinstated Waverly Line

Stuart Berry and John Stubbs

20.1 Transport Modelling

Travel within London. Data from travel-in-london-report-7 within this report travellers have been categorised by:

$$\begin{aligned} \text{RAIL} &= \{\text{railway, underground, Dockland Light Railway}\} \\ \text{ROAD} &= \{\text{bus, tram, taxi, car, car passenger, motor cycle}\} \\ \text{WALK} &= \{\text{those walking the full journey to their destination}\} \end{aligned}$$

Although the cycling provision has been given a higher priority by transport planners than the provision for walkers, the available data (see Fig. 20.1) show that the greater need is the provision appropriate areas for walkers.

S. Berry (✉) · J. Stubbs
College of Engineering and Technology, University of Derby,
Kedleston Road, DE22 1GB Derby, UK
e-mail: s.berry@derby.ac.uk

S. Berry et al. (eds.), *Guide to Computational Modelling for Decision Processes*,
Simulation Foundations, Methods and Applications,
DOI 10.1007/978-3-319-55417-4_20

Fig. 20.1 Forecast annual return trips per year

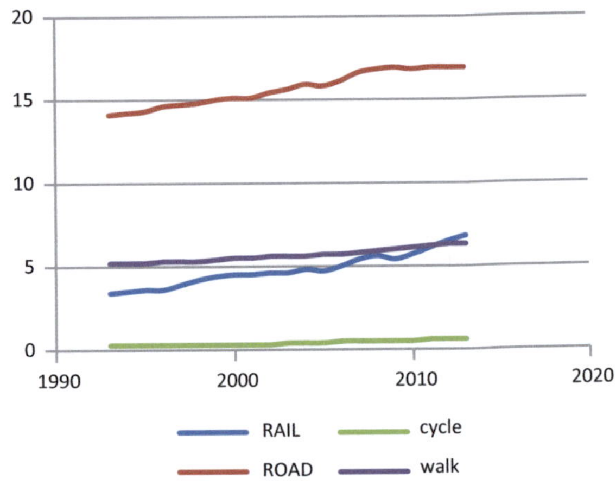

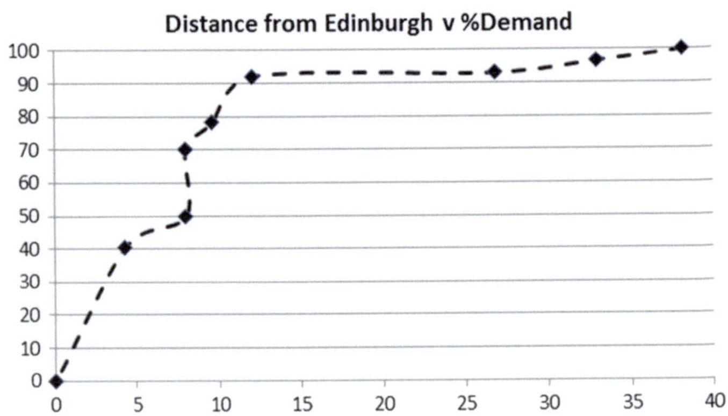

Fig. 20.2 Cumulative percentage demand and distance from Edinburgh

20.2 Assessing the reinstated Waverley Line

The halts Shawfair to Tweedbank have been reopened resulting from the line extension from Newcraighall to Tweedbank. The (forecast) demand data seems to indicate that this reopening should have stopped at Gorebridge (over 90% of potential customers carried) giving a new track length of 7.8 miles and a total track

Table 20.1 Forecast annual return trips in the opening year

	Cumulative % demands from Edinburgh			Cumulative % new halt demands from Shawfair		Distance from Edinburgh	
Tweedbank	21621	647136	100.0	21621	100.0	38.2	New stations
Galashiels	23431	625515	96.6	23431	94.4	33	
Stow	5843	602084	93.0	5843	88.3	26.8	
Gorebridge	90019	596241	**92.1**	90019	86.8	**12.1**	
Newtongrange	52918	506222	78.2	52918	63.5	9.6	
Eskbank	130525	453304	70.0	130525	49.8	8	
Shawfair	61860	322779	49.8	61860	16.0	8	
Newcraighall	986	260919	40.3			4.3	Existing stations
Waverley	220533	259933	40.1				
Haymarket	35329	39400	6.0				
Edinburgh Park	4071	4071	0.6				
Total	647136			386217			

distance 12.1 miles, rather than a new track length of 33.9 miles and total distance of 38.2 miles to Tweedbank. The shorter railway could then have been viable, an economic proposition (Fig. 20.2 and Table 20.1).

Chapter 21
Appendix F: Matching Services with Users in Opportunistic Network Environments

Stuart Berry

21.1 Search Algorithm

The first iteration of the algorithm is

$$
\begin{array}{lll}
\text{decn} & \leftarrow 0 & \text{\% decision} \\
\text{test\_level} & \leftarrow 1 & \text{\% current level tested} \\
\text{pass\_test} & \leftarrow 0 & \text{\% current level passed}
\end{array}
$$

```
while (test_level ≤ levels) AND (decn=0)
        If((ISEMPTY({new_message_type(test_level)}∩ {likes(test_level)})=TRUE)
                If((ISEMPTY({new_message_type( test_level)}∩{hates( test_level)})=FALSE)
                        decn ←2
                %dont want to display this message
                else
                        decn ←3
                %category not seen, do we want to
                add it to likes or hates
                end
        else
                pass_test ← pass_test +1
```

S. Berry (✉)
College of Engineering and Technology, University of Derby, Kedleston Road, DE22 1GB Derby, UK
e-mail: s.berry@derby.ac.uk

S. Berry et al. (eds.), *Guide to Computational Modelling for Decision Processes*, Simulation Foundations, Methods and Applications,
DOI 10.1007/978-3-319-55417-4_21

```
        end
        test_level ← test_level +1
end

if pass_level = levels
        "display message"
End

if decn = 2
        "message ignored"
end

If decn = 3
        If message STATUS is good
           %"display message", add to likes?.
           %new info at level 'pass_level + 1'
        else
                "message ignored"
        end
end
"STATUS is good", depends on parameters to be derived.
```

21.2 Investigating Effective and Efficient Depths of Search

An investigation of the basic case provides a mathematical basis for truncating the depth of search dependent upon the form of the query.

21.2.1 Formal Definition of the Search Conditions

The basic search considers only those problems defined formally by:

L_{ij} = "True" option i at level j is liked, and
H_{ij} = "True" option i at level j is not liked

It then follows that the following describe the matching procedures

$$[H_{ik} = \text{“True”}] \rightarrow [H_{i,j,k+1} = \text{“True”}]$$

If an option not liked at level k implies that lower level options are not relevant in this search, and

$$[L_{ik} = \text{“True”}] \rightarrow [L_{i,k-1} = \text{“True”}]$$

hence to be required to interrogate an option at level k option, the higher level options in the query will have been liked.

finally if

$$[\mathrm{L}_{ik}] \text{ and}\neg\left[\mathrm{L}_{i,k+1} \vee \mathrm{H}_{i,k+1}\right] = \text{“True”}$$

the query will be referred at level $k + 1$.

21.3 Analysis

If p_{ij} is the probability that option i is liked at level j, and q_{ij} is the probability that option i is not liked at level j (note, $r_{ij} = 1 - p_{ij} - q_{ij}$ is the probability of referral at level j given that the query has been liked at higher levels).

One level search

If a one level search (only) is implemented it follows that

$$P(\text{correct decision}) = (1-p_{1j}) + \prod_{k=1}^{n} p_{ik} \prod_{k=1}^{n} p_{ik}$$

and if p_{1j} is small ($p_{1j} = \varepsilon$)

$$P(\text{correct decision}) = 1 - \varepsilon\left(1 - \prod_{k=2}^{n} p_{ik}\right),$$

giving

$$(1-\varepsilon) < P(\text{correct decision}) < 1, \text{close to } 1.$$

But if p_{1j} is large (p_{1j} = P) it follows that

$$P(\text{correct decision}) = 1 - P\left(1 - \prod_{k=2}^{n} p_{ik}\right)$$

$$(1-P) < P(\text{correct decision}) < 1$$

Thus, it follows that if p_{1j} is small, a one level search (only) would be appropriate and efficient.

Two level search

The second level will only be reached if the query has “passed” at the first level; hence, it follows that if the search was halted at this level, then:

$$P(\text{correct decision}) = (1 - p_{2j}p_{1j}) + \left(1 - \prod_{k=2}^{n} p_{ik}\right)$$

and if p_{2j} is small ($p_{2j} = \varepsilon$)

$$P(\text{correct decision}) = 1 - \varepsilon p_{1j}\left(1 - \prod_{k=2}^{n} p_{ik}\right),$$

giving

$$(1 - \varepsilon) < P(\text{correct decision}) < 1.$$

Thus, it follows that if p_{1j} is not small and p_{2j} is small, a two level search (only) would be appropriate and efficient.

***m* level search**

Assuming that the probabilities p_{ij} are not small, for $i = 1$ to $m - 1$, it follows that

$$P(\text{correct decision}) = \left(1 - \left(1 - \prod_{k=2}^{n} p_{ik}\right) + \prod_{k=1}^{n} p_{ik}\right)$$

and if p_{im} is small, then it would be efficient and effective to halt the search at this level.

Thus, a simple rule is to stop at the level when *P*(likes) is very small, an option within the query is unpopular hence reducing the cost of the search process.

References

1. Aarts E, Lenstra JK (ed) (2003) Local Search in combinatorial optimization. Princeton University Press, USA
2. Ausiello G et al (1999) Complexity and approximation combinatorial optimisation problems. Springer, Berlin
3. Back T (1996) Evolutionary algorithms in theory and practice. Oxford University Press, USA
4. Bagdasar O, Popovici N (2015) Local maximum points of explicitly quasiconvex functions. Optim Lett 9(4):769–777. doi:10.1007/s11590-014-0781-3
5. Bandemer H, Gottwald S (1996) Fuzzy sets, fuzzy logic, fuzzy methods. Wiley, USA
6. Bazaraa MS, Sherali HD, Shetti CM (2006) Non linear programming. Wiley, USA
7. Beasley JE (ed) (1996) Advances in linear and integer programming. Oxford University Press, USA
8. Berry S, Lowndes V (2003) Deriving a memetic algorithm to solve heat flow problems—University of Derby Technical Report.
9. Berry S, Parkes C (2016) Green transport planning paradoxes. Mathematics Today 52(5)
10. Blazewicz J et al (2001) Scheduling computer and manufacturing processes. Springer, Berlin
11. Braess D, Nagurney A, Wakolbinger (2005) On a paradox of transport planning (a translation of the 1968 article). Transp Sci 39(4):446-450
12. Buzacott JA, Shanthikumar JG (1993) Stochastic models of manufacturing systems. Prentice Hall, USA
13. Catoni S, Pallottino S (1991) Traffic equilibrium paradoxes. Transp Sci 25(3):240–244
14. Chambers L (1995a) Evolutionary algorithms in practical handbook of genetic algorithms. CRC Press, USA
15. Chambers L,(1995) Practical handbook of genetic algorithms vol 1. CRC Press, USA
16. Chambers L (1995) Practical handbook of genetic algorithms vol 2. CRC Press, USA
17. Chambers L (1999) Practical handbook of genetic algorithms vol 3. CRC Press, USA
18. Chen H (2008). Homeland security data mining using social network analysis. In: IEEE international conference on intelligence and security informatics. Springer, Berlin
19. Cheybani S, Kertesz J, Shreckenberg M (1998) Correlation functions in the Nagel-Schreckenberg model. J Phys A 31:9787–9799
20. Cipriani TA, Leachman RC (1993) Optimization in industry. Wiley, USA
21. Cipriani TA, Leachman RC (1994) Optimization in industry, vol 2. Wiley, USA
22. Conti M, Giordano S, May M, Passarella A (2010) From opportunistic networks to opportunistic computing. Commun Mag IEEE 48:126–139
23. Corne D, Dorigo M, Glover F (1999) New ideas in optimization. McGraw Hill, USA
24. Coyle RG (1996) System dynamics modelling: ISBN 9780412617102. Transp Res Ser B, 18(2) (1984):101–110.

S. Berry et al. (eds.), *Guide to Computational Modelling for Decision Processes*,
Simulation Foundations, Methods and Applications,
DOI 10.1007/978-3-319-55417-4

25. Danks D, Griffiths TL, Tenenbaum JB (2002) Dynamical causal learning. NIPSMIT Press, pp 67–74.
26. Deb K (2001), Multi objective optimization using evolutionary algorithms. Wiley, USA
27. Dhamija R, Tygar JD (2005) The battle against phishing: dynamic security skins. In: Proceedings of the 2005 symposium on usable privacy and security, ACM.
28. Dhamija R, Tygar JD (2006) Why phishing works. In: Proceedings of the SIGCHI conference on human factors in computing systems, pp 581–590
29. Ding C, Song S, Zhang Y (2008) Paradoxes of traffic flow and economics of congestion pricing. In: UNR joint economics working paper series, working paper no 08-007
30. Dolan A, Aldous J (1999) Networks and algorithms. Wiley, USA
31. Dorn J, Froeschl KA (1993) Scheduling of production processes. Ellis Horwood, UK
32. Englemore RS, Terry A (1979) Structure and function of the CRYSALIS system. In: Proceedings of IJCAI-79, pp 250–256
33. Engelmore RS, Morgan AJ, Nii HP (1988) Hearsay-II. In: R Engelmore, T Morgan (eds) Blackboard systems. Addison-Wesley, USA. pp 25–29
34. Esser J, Schreckenberg M (1997) Microscopic simulation of urban traffic based on cellular automata. Int J Modern Phys 18(5):1025–1036
35. Eugster PT, Garbinato B, Holzer A (2005) Location-based publish/subscribe. In Proceedings of the fourth IEEE international symposium on network computing and applications, IEEE Computer Society, 279–282
36. Farrar K (1979) Soundfield microphone. Parts 1 & 2—wireless World
37. Forrester J (1961) Industrial Dynamics, MIT Press
38. Gardner B, Martin K (1994) HRTF measurements of a KEMAR dummy-head microphone. http://sound.media.mit.edu/KEMAR.html
39. Gee ES, Smith CH (1993) Selecting allowances for jobshop performance. Int J Prod Res 31 (8):1839–1852
40. Gelenter D (1983) Generative communication in Linda. ACM Trans Program Lang Syst 7 (1):80–112
41. Gerzon MA (1992) General methatheory of auditory localisation—92nd AES convention, Vienna. Preprint 3306
42. Ghazi A, Laskey K, Wang X, Barbará D, Shackelford T, Wright E, Fitzgerald J (2006) Detecting threatening behavior using Bayesian networks. C4I Papers
43. Guan T, Zaluska E, De Roure D (2008)A semantic service matching middleware for mobile devices discovering grid services. In: Proceedings of the 3rd international conference on advances in grid and pervasive computing. Springer, Berlin, pp 422–433
44. Hadley G (1970) Systems, non linear and dynamic programming. Addison Wesley, USA
45. Hadley G (1971) Linear programming. Addison Wesley, USA
46. Hamed K (2009) Near-term travel speed prediction utilizing Hilbert–Huang transform. Comput-Aided Civil Infrastruct Eng 24:551–576
47. Holland JH (1994) Adaption in natural and artificial systems. MIT Press, USA
48. Holmes RB (1975) Geometic functional analysis and its applications. Springer, Berlin
49. Hui P, Crowcroft J, Yoneki E (2007)BUBBLE rap: social-based forwarding in delay tolerant networks. In: Proceedings of the 2nd ACM international workshop on mobility in the evolving internet architecture (MobiArch), pp 241–250
50. Jakobsson M (2005) Modeling and preventing phishing attacks. In: Financial cryptography and data security. Springer, Berlin, pp 89–89
51. John R, Birkenhead R (eds) (2001) Developments in soft computing. Physica-Verlag
52. Johnson LA, Montgomery DC (1974) Operational research in production planning, scheduling and inventory control. Wiley, USA
53. Kalrath J, Wilson JM (1997) Business optimisation. MacMillan, UK
54. Kirchener A, Schadschneider A (2002) Simulation of evacuation processes using a bionics-inspired cellular automaton model for pedestrian dynamics. Physica A: Stat Mech Appl 312(1–2):260–276

55. Kreher DL, Stinson DR (1999) Combinatorial algorithms. CRC Press, USA
56. Keränen J, Ott T, Kärkkäinen T (2009) The ONE simulator for DTN protocol evaluation. In: Proceedings of 2nd international conference on simulation tools and techniques (SIMUTools'09), Rome, Italy, ICST New York, NY USA. ISBN 978-963-9799-45-5, pp 55:1–55:10
57. Kuokka D, Harada L (1995) Matchmaking for information agents, Readings in Agents, Morgan Kaufmann, pp 672–678.
58. Lau RY, Yunqing X, Yunming Y (2014) A probabilistic generative model for mining cybercriminal networks from online social media. Comput Intell Mag 9(1):31–43
59. Lévy P (1994) Collective intelligence: mankind's emerging world in cyberspace. Basic Books
60. Li HX, Yen VC (1995) Fuzzy sets and fuzzy decision making. CRC, USA
61. Lindgren A, Doria A, Scheln O (2003) Probabilistic routing in intermittently connected networks. In: Proceedings of the fourth ACM international symposium on mobile ad hoc networking and computing (MobiHoc 2003), pp 19–20
62. Luc DT (1989) Theory of vector optimisation. Springer, Berlin
63. Luc DT, Schaible S (1997) Efficiency and generalised concavity. J Optim Theory 147–153
64. Man KF, Tang KS, Kwong S (1999) Genetic algorithms. Springer, Berlin
65. McKenna JA (2004) The internet and social life. Ann Rev Psychol 55:573–590
66. Mendel J (2000) Uncertain rule-based fuzzy logic systems. Prentice Hall, USA
67. Morecroft JDW (2015) Strategic modelling and business dynamics: ISBN: 978-1-118-84468-7
68. Morton TE, Pentico DW (1993) Heuristic scheduling systems. Wiley, USA
69. Nagel K, Schreckenberg M (1992) A cellular automaton model for freeway traffic. Phys France 2221–2229
70. Neuman J (1951) The general theory of automata. In: Cerebal mechanisms in behavior: the Hixon symposium. Wiley, USA, pp 1–41
71. Osman IH, Kelly JP (1997) Meta heuristics theory and application, Kluwer
72. Panwalkar SS, Iskander W (1977) A survey of scheduling rules. Oper Res 25:45–61
73. Parker RG (1995) Deterministic scheduling theory. Chapman Hall, USA
74. Papadimitriou CH, Steglitz K (1998) Combinatorial optimisation, Dover
75. Pardalos PM, Resende MGC (ed) (2002) Handbook of applied optimization, Oxford
76. Pinedo M (1995) Scheduling, theory applications and systems. Prentice Hall, USA
77. Poelmans JE (2010) Formal concept analysis in knowledge discovery: a survey. In: International conference on conceptual structures (ICCS), pp 139–153. Springer, Berlin
78. Popovici N (2005) Pareto reducible multicriteria optimisation problems. Optimization 54 (2005):253–263
79. Schoder D, Gloor P, Metaxas P (2013) Social media and collective intelligence—ongoing and future research streams. KI—Künstliche Intelligenz 27(1):9–15
80. Schwartz MJ (2011) Epsilon fell to spear-phishing attack. Retrieved 17 June 2014 from http://www.darkreading.com/attacks-and-breaches/epsilon-fell-to-spear-phishing-attack/d/d-id/1097119?
81. Shashidhar J, Chen N (2011) A phishing model and its applications to evaluating phishing attacks. International cyber resilience conference
82. Smith A, Berry S, Hill R (2015) Efficient matching of services with users in opportunistic network environments. Int J Adapt Innovative Syst, 99–117.
83. Smith A, Berry S (2012) Evaluation of a framework for measuring efficiency in opportunistic ad-hoc networks. In: Third international conference on emerging intelligent data and web technologies (EIDWT), 2012, pp 61–65
84. Smith A, Hill R (2011) Towards a framework for the evaluation of efficient provisioning in opportunistic ad hoc networks. In: Proceedings of the 2011 international conference on P2P, parallel, grid, cloud and internet computing, IEEE Computer Society, pp 32–36

85. Smith A, Hill R (2011) Measuring efficiency in opportunistic ad hoc networks. Int J Interconnected Netw 12(3):32–36
86. Stahlschmidt S, Tausendteufel H (2013) Bayesian networks for sex-related homicides: structure learning and prediction. J Appl Stat 40(6):1155–1171
87. Stoer J, Witzgall C (1970) Convexity and optimisation in finite dimensions. Springer, Berlin
88. Sycara K, Widoff S, Klusch M, Lu JL (2002) LARKS: dynamic matchmaking among heterogeneous software agents in cyberspace, in cyberspace. Auton Agents Multi-Agent Syst 173–203
89. Unbehauen (ed) Control systems, robotics and automation: fuzzy and intelligent control systems, vol 17. Oxford: EOLSS, pp 303–316
90. Turing AM (1950) Computing Machinery and Intelligence. Mind 49:433–460
91. T1: Canal development and decline (www.ukcanals.net)
92. T2: Passenger Rail Usage, Office of Rail and Road, (www.orr.gov.uk)
93. T3: On the Move (www.racfoundation.org)
94. T4: The false paradise of Metroland, (www.spectator.co.uk)
95. T5: Railways Developments (www.victorianweb.org)
96. T6: Travel in London (www.tfl.gov.uk)
97. Vahdat A, Becker D (2000) Epidemic routing for partially connected ad-hoc networks. Technical report CS-200006, Duke University
98. Verma A, Srivastava A (2011) Integrated routing protocol for opportunistic networks. Int J Adv Comput Sci Appl 2(3)
99. Wiggins B, Paterson-Stephens I, Schillebeeckx P, The analysis of multi-channel sound reproduction algorithms using HRTF data—19th AES surround sound conention, Schoss
100. Williams HP (1993) Model solving in mathematical programming. Wiley, USA
101. Williams HP (2009) Logic and integer programming. Springer, Berlin
102. Williams HP (2013) Model building in mathematical programming. Wiley, USA
103. Woolsey RED (1982) The fifth column: production scheduling as it really is. Interfaces 12 (6):115–118
104. Yoneki E, Hui P, Chan S-Y, Crowcroft J (2007) A socio-aware overlay for publish/subscribe communication in delay tolerant networks. In: Proceedings of the 10th ACM symposium on modelling. Analysis and simulation of wireless and mobile systems (MSWiM), pp 225–234.
105. Zadeh LA (1973) Outline of a new approach to the analysis of complex systems & decision processes. IEEE Trans Syst Manage Cybern SMC 3(1)
106. Zalinescu C (2002) Convex analysis in general vector spaces world scientific, River Edge
107. Zalzala AMS, Fleming PJ (1997) Genetic algorithms in engineering systems; IEE Digital Library
108. Zhu S, Weibing Z, Jihui H, Xu C (2016) The effects of overtaking strategy in the Nagel-Schreckenberg model. Eur Phys J B

Index

S. Berry et al. (eds.), *Guide to Computational Modelling for Decision Processes*,
Simulation Foundations, Methods and Applications,
DOI 10.1007/978-3-319-55417-4

MIX
Papier aus verantwortungsvollen Quellen
Paper from responsible sources
FSC® C105338

If you have any concerns about our products,
you can contact us on
ProductSafety@springernature.com

In case Publisher is established outside the EU,
the EU authorized representative is:
Springer Nature Customer Service Center GmbH
Europaplatz 3, 69115 Heidelberg, Germany

Printed by Libri Plureos GmbH
in Hamburg, Germany